Logarithmic scales

	Interval	Correspondance*) interval	ratio
exp	exponential interval	n exp	e^{n}
dex	interval in powers of 10	n dex	10^{n}
dB	decibel: interval in 0.1 powers of 10	n dB	$10^{0.1n}$
mag	interval of magnitudes	n mag	$10^{-0.4n}$

*) Correspondance of logarithmic interval and ratio of intensity variable.
1.000 dex = 10.000 dB = 2.303 exp = - 2.5 mag

Landolt-Börnstein / New Series

Landolt-Börnstein

Numerical Data and Functional Relationships
in Science and Technology

New Series
Editor in Chief: O. Madelung

Units and Fundamental Constants in Physics and Chemistry

Nuclear and Particle Physics (Group I)

Atomic and Molecular Physics (Group II)

Solid State Physics (Group III)

Macroscopic Properties of Matter (Group IV)

Geophysics (Group V)

Astronomy and Astrophysics (Group VI)

Biophysics (Group VII)

Landolt-Börnstein

Numerical Data and Functional Relationships in Science and Technology

New Series / Editor in Chief: O. Madelung

Group VI: Astronomy and Astrophysics

Volume 3

Astronomy and Astrophysics

Extension and Supplement to Volume 2

Subvolume a

Instruments, Methods, Solar System

Editor: H.H. Voigt

Contributors:

J.W. Baars, H. Beer, C.J. Durrant, U. Graser, B. Guinot, M. Hoffmann, U. Hopp, W.-H. Ip, E.K. Jessberger, B. Klecker, D. Lemke, K. Meisenheimer, E. Möbius, H. Palme, J. Rahe, H.-J. Röser, J. Schubart, R. Schwenn, J. Solf, G. Soltau, R. Staubert, R. Stewart, J. Trümper, V. Vanysek, G. Weigelt, R. Wolf

Springer-Verlag

Berlin Heidelberg NewYork

London Paris Tokyo

Hong Kong Barcelona Budapest

ISSN 0942-8011 (Astronomy and Astrophysics)

ISBN 3-540-56079-3 Springer-Verlag Berlin Heidelberg New York
ISBN 0-387-56079-3 Springer-Verlag New York Berlin Heidelberg

Library of Congress Cataloging in Publication Data
Zahlenwerte und Funktionen aus Naturwissenschaften und Technik, Neue Serie
Editor in Chief: O. Madelung
Vol. VI/3a : Edited by H.H. Voigt
At head of title: Landolt-Börnstein. Added t.p.: Numerical data and functional relationships in science and technology.
Tables chiefly in English.
Intended to supersede the Physikalisch-chemische Tabellen by H.H. Landolt and R. Börnstein of which the 6th ed.
began publication in 1950 under title: Zahlenwerte und Funktionen aus Physik, Chemie, Astronomie,
Geophysik und Technik.
Vols. published after v. 1 of group I have imprint: Berlin, New York, Springer-Verlag
Includes bibliographies.
1. Physics--Tables. 2. Chemistry--Tables. 3. Engineering--Tables.
I. Börnstein, R.(Richard), 1852-1913. II. Landolt, H. (Hans), 1831-1910.
Physikalisch-chemische Tabellen. IV. Title: Numerical data and functional relationships in science and
technology.
QC61.23 502`.12 62-53136

Typesetting and Dataconversion: Lewis & Leins, Berlin
Printing: Mercedes-Druck, Berlin
Binding: Lüderitz & Bauer, Berlin

63/3020 - 5 4 3 2 1 0 - Printed on acid-free paper

Editor

H. H. Voigt
Universitätssternwarte, Geismarlandstraße 11, D-37083 Göttingen, Germany

Contributors to subvolume VI/3a

J. W. Baars
Steward Observatory-SMTO, University of Arizona, Tucson, AZ 85721, USA
Radio astronomy instrumentation

H. Beer
Kernforschungszentrum Karlsruhe, Institut für Kernphysik III, Postfach 3640, D-76021 Karlsruhe, Germany
Abundance of elements in the solar system

C. J. Durrant
Department of Applied Mathematics F 07, University of Sydney, Sydney, NSW 2006, Australia
Solar telescopes, Sun

U. Graser
Max-Planck-Institut für Astronomie, Königstuhl, D-69117 Heidelberg, Germany
Photoelectric photometry, CCD

B. Guinot
Bureau International des Poids et Mesures, Pavillon de Breteuil, F-92312 Sèvres, France
Time determination

M. Hoffmann
Observatorium Hoher List, D-54550 Daun, Germany
Asteroids (minor planets)

U. Hopp
Max-Planck-Institut für Astronomie, Königstuhl, D-69117 Heidelberg, Germany
Optical telescopes

W.-H. Ip
Max-Planck-Institut für Aeronomie, Max-Planck-Str. 2, D-37191 Katlenburg-Lindau, Germany
Planetary physical data

E. K. Jessberger
Max-Planck-Institut für Kernphysik, Saupfercheckweg, D-69117 Heidelberg, Germany
Meteors and meteorites

B. Klecker
Max-Planck-Institut für Extraterrestrische Physik, Giessenbachstraße, D-85748 Garching, Germany
Energetic particles in interplanetary space

D. Lemke
Max-Planck-Institut für Astronomie, Königstuhl, D-69117 Heidelberg, Germany
Infrared techniques

K. Meisenheimer
Max-Planck-Institut für Astronomie, Königstuhl, D-69117 Heidelberg, Germany
Photoelectric photometry, CCD

E. Möbius
Institute for the Study of Earth, Oceans and Space (EOS), Science and Engineering building, University of New Hampshire, Durham, NH 03824, USA
Gases of non-solar origin in the solar system

H. Palme
Max-Planck-Institut für Chemie, Abteilung Kosmochemie, Saarstraße 23, D-55122 Mainz, Germany
Abundance of elements in the solar system

J. Rahe
Code SL, NASA Headquarters, Washington, DC 20546, USA
Comets

H.-J. Röser
Max-Planck-Institut für Astronomie, Königstuhl, D-69117 Heidelberg, Germany
Photoelectric photometry, CCD

J. Schubart
Astronomisches Recheninstitut, Mönchhofstr. 12-14, D-69120 Heidelberg, Germany
Astronomical constants
Mechanical data of planets and satellites

R. Schwenn
Max-Planck-Institut für Aeronomie, Max-Planck-Str. 2, D-37191 Katlenburg-Lindau, Germany
Interplanetary plasma and magnetic field (solar wind)

J. Solf
Max-Planck-Institut für Astronomie, Königstuhl, D-69117 Heidelberg, Germany
Spectrometers and spectrographs

G. Soltau
Institut für Angewandte Geodäsie, Karl-Rothe-Straße 10-14, D-04105 Leipzig, Germany
Astronomical latitude and longitude

R. Staubert
Astronomisches Institut der Universität, Waldhäuser Str. 64, D-72076 Tübingen, Germany
X-ray and γ-ray instruments

R. Stewart
Australia Telescope National Facility, CSIRO-Radiophysics Laboratory, PO Box 76, Epping, NSW 2121, Australia
Radioemission of the Sun

J. Trümper
Max-Planck-Institut für Extraterrestrische Physik, Giessenbachstraße, D-85748 Garching, Germany
X-ray and γ-ray instruments

V.Vanysek
Astronomical Institute, Charles University Prague, Švédská 8, 15000 Praha 5, Czech Republic
Comets

G. Weigelt
Max-Planck-Institut für Radioastronomie, Auf dem Hügel 69, D-53121 Bonn, Germany
Optical high-resolution methods

R. Wolf
Max-Planck-Institut für Astronomie, Königstuhl, D-69117 Heidelberg, Germany
Optical telescopes

Preface

Twelve years ago the volume "Astronomy and Astrophysics", LB NS VI/2, subvolume a, has been published in the Landolt Börnstein New Series. Since then many fields of astronomy have evolved enormously, and new knowledge has been gained. Thus it has been agreed upon for some time that a new volume should be published. However, there has been some discussion among the authors whether this volume VI/3 should be an "extension and supplement" to volume VI/2 or a completely new edition. Opinions among the authors were divided about fifty to fifty.

The decision on this issue has been influenced by the intention to publish an astronomy volume in the new series "Data in Science and Technology". This series is containing extracts of Landolt-Börnstein volumes, giving the most important data without detailed explanation. These volumes have a low price so that the scientist can afford it to have them on his desk for daily use, whereas the Landolt-Börnstein volumes normally are available in libraries only. Thus it has been decided to design the Landolt-Börnstein volume VI/3 as an extension and supplement to the previous volume VI/2, and later on to compile the data for the scheduled "Data in Science and Technology" volume from all three Landolt-Börnstein volumes VI/1, VI/2, and VI/3 dedicated to astronomy and astrophysics. Subvolume VI/3a is presented herewith.

The concept "extension and supplement" is now followed more strictly than in volume VI/2, which itself has been extending and supplementing the then preceding volume VI/1. Especially the decimal numbering of chapters and sections is kept unchanged with respect to volume VI/2. The following pages give a synoptic list of contents for subvolumes VI/2a and VI/3a. There are of course fields which are now complete or where important new data are not available, as, for instance, for section 1.4 (Photographic emulsions) or subsection 3.3.5.3.3 (Interplanetary propagation of solar cosmic rays). Such sections are deleted in the present volume without changing the decimal numbering of the sections. Often a reference to the corresponding section or page in volume VI/2a is given in the text. Partly, even for tables in volume VI/2a, only extensions are given in the present volume. On the other hand new fields have come up not yet mentioned in volume VI/2a. Such fields are added to the present volume by appropriate extension of the decimal numbering of sections, as is the case, for instance, for subsection 1.3.9 (Scientific charge-coupled devices (CCDs)).

The concept "extension and supplement" allows the volume to be shorter and published faster than a completely new edition.

My thank first of all is due to the authors of the various sections. They have had to do the scientific work proper in collecting the data and bear the final responsibility. They followed the suggestions of editor and publisher without grumbling.

I also want to thank the editorial staff of Landolt-Börnstein, especially Dr. W. Finger, who is responsible for the present LB volumes on astronomy and astrophysics. Thanks are also due to Springer, always being willing to follow our wishes as far as possible. Most of the authors have submitted the manuscripts on floppy disks. Although various kinds of software have been used the publisher has succeeded in preparing a uniform printout.

Göttingen, September 1993 **The Editor**

Contents for subvolumes VI/2a and VI/3a

The contents of the present subvolume VI/3a is indicated in roman. Sections of the former subvolume VI/2a which have not been supplemented in the present subvolume VI/3a are referred to in italic.

2 Positions and time determination, astronomical constants

3 The solar system

1 Astronomical instruments

1.1 Optical telescopes

1.1.1 – 1.1.5 see LB VI/2a

1.1.6 New and future developments

To enable ground-based, optical telescopes with entrance apertures larger than about 6 m, several new technologies have been introduced to telescope manufacturing and operation [88U,90B] (In the following, we will define an optical telescope as a collector of light with wavelengths between 0.3 and 20 μm). This includes segmented-mirror (e.g. Keck telescope), and multi-telescope designs where the telescopes may be mounted individually (like the VLT) or on a common mounting (like the MMT). For abbreviations see Table 3. Consequentely, it is necessary to develop telescope coupling methods (for incoherent as well as coherent beams) [89F, 88B]. If coherent coupling will be successful, interferometric methods will enable very high resolution observations at least in the near infrared [88M, 88U, 90B].

To reduce the weight of the single blanks, thin mirrors [89R] with diameters up to 8.2 m as well as honeycomb mirrors [85A] are under construction. To account for the figuring problems of the thin mirrors, the concept of active optics [91W2] has been developed based on wavefront-sensors [89W]. These sensors enable closed-loop technics as they are able to analyse the wavefront aberrations in a quasi-on-line manner. The concept of the closed-loop regulation of the optics will be further developed into the concept of adaptive optics which should be able to correct for seeing-induced image distortions [88U, 89B, 90B].

It is meanwhile common to place large mirrors on alt-azimuth mountings which again reduce the weight and the cost of the design [88U, 90B]. Therefore, many future instruments will be used in Nasmyth focus stations and their design has to account for field de-rotation. New technologies are in discussion also for the mountings like, e.g., optical gyro-control systems for the pointing and tracking [91M, 91S] or carbon-reinforced plastics [89K] instead of steel.

Further increase of the overall efficiency of (planned as well as of existing) telescopes were achieved by high-quantum-efficiency detectors [90C and subsect 1.3.9], improved understanding and operation of the local observing conditions ("dome seeing") [86M, 88U, 90B], and "high-throughput" instrumentations like, e.g., focal-reducers and multi-object spectrographs [88U, 90C].

For several of the future projects, remote observing is planned meaning that the observer operates the telescope from the home-institute [85M, 86S]. First such regular services are offered by ESO for some of their La Silla telescopes. The discussion of the design of the "post-VLT" telescope generation has already started and is related to the progress and success of space telescopes [90B, 91W1].

References for 1.1.6

85A Angel, J.R.P., Cheng, A.Y.S., Woolf, N.J.: SPIE Conf. Proc. **571** (1985) 123.

85M Martin, R., Hartley, K.: Vistas Astron. **28** (1985) 555.

86M Millis, R.L. (ed.): Identification, Optimization and Protection of Optical Telescope Sites, Proc. Flagstaff (1986).

86S Staffi, G., Ziebell, M.: ESO Conference and Workshop Proc. **24** (1986) 317.

88B Brown, D.S., Doel, A.P., Dunlop, C.N., Major, J.V., Myers, R.M., Purvis, A., Thompson, M.G.: ESO Conference and Workshop Proc. **30** (1988) 761.
88M Merkle, F. (ed.): High Resolution Imaging by Interferometry, ESO Conference and Workshop Proc. **29** (1988).
88U Ulrich, M.H. (ed.): Very Large Telescopes and Their Instrumentations, ESO Conference Proc. No. **30** (1988).
89B Beckers, J.M., Merkle, F.: New Technology for Astronomy, SPIE Conf. Proc. **1130** (1989) 10.
89F Faucherre, M., Greenaway, A.H., Merkle, F., Noordam, J.E., Perryman, M.A.C., Rousel, P., Vakili, F., Volonte, S., Weigelt, G.: SPIE Conf. Proc. **1130** (1989) 101.
89K Kärcher, H.J., Nicklas, H., Maurer, D., Czarnetzki, W.: The German Large Telescope Project, Göttingen (1989) 86.
89R Roddier, F.J. (ed.): Active Telescope Systems, SPIE Conf. Proc. **1114** (1989).
89W Wilson, R.N., Noethe, L.: SPIE Conf. Proc. **114** (1989) 290.
90B Barr, L.D. (ed.): Advanced Technology Optical Telescopes IV, SPIE Conf. Proc. **1236** (1990) Part I and II.
90C Crawford, D.L. (ed.): Instrumentation in Astronomy VII, SPIE Conf. Proc. **1235** (1990), I and II.
91M Merkle F., Ravenbergen, M.: ESO Messenger **65** (1991) 53
91S Schröder, W., Dahlmann, H., Huber, B., Schüssele, L., Merkle, F., Ravenbergen, M.: SPIE Conf. Proc. **1585** (1991) (preprint).
91W1 Wilson, R.N.: ESO Messenger **63** (1991) 15.
91W2 Wilson, R.N., Franza, F., Noethe, L., Andreoni, G.: Active Optics III, J. Mod. Optics **38** (1991) 219.

1.1.7 Updated list of large optical telescopes erected after 1960

Table 3. List of large optical telescopes erected after 1960. It includes reflectors with a minimum aperture of 1.2 m and wide-field cameras with a minimum aperture of 0.6 m. For telescopes erected before 1960, see [65B]. Included are some new generation telescope projects if they are in the phase of realization.

Abbreviations:

ATT	Advanced Technology Telescope	Nas	Nasmyth focus
alt-alt	alt-alt mounting	New	Newtonian focus
alt-az	alt-azimuth mounting	NTT	New Technology Telescope
Cas	Cassegrain focus	Pr	prime focus
CAT	Coudé Auxiliary Telescope	RC	Ritchey Chrétien focus
Cou	Coudé focus	Re	reflector
doubl.	doublet corrector	Sch	Schmidt camera
Gasc.	aspheric Gascoigne plate	SST	Spectroscopic Survey Telescope
IR	infrared telescope facility	tripl.	triplet corrector
MMT	Multiple Mirror Telescope	VLT	Very Large Telescope
mod.	modified		

Table 3

Location (Observatory)	Type	Mount	Aperture [m]	Optical system	Focal-length [m]	Year	Ref.
Abastumani	Re		1.25				
Apache Point (Astronomical Research Consortium)	Re	alt-az	3.5			1992	90S1
	Sch		2.5				
Asiago (Padua Univ.)	Sch	yoke	0.65/0.92	Pr	2.15	1964	65B
	Re	fork	1.82	Cas	16.4	1973	74B
Brazopolis (Brazilian Nat. Obs.)	Re	off-axis	1.6	RC	16.0		
				Cou	240.0/49.9		
Budapest (Budapest)	Sch	fork	0.6/0.9	Pr	1.8	1963	65B
Byurakan (Byurakan)	Sch	fork	1.0/1.5	Pr	2.13	1961	65B
	Re	fork	2.6	Pr	9.4	1975	
				Nas	41.6		
				Cou	104.0		
Calar Alto (Max-Planck-Inst. for Astronomy, German-Spanish Astronomical Centre)	Re	pole universal	1.23	mod.RC	9.86	1975	71S
				mod.RC/doubl.	9.81		
	Re	fork	2.2	RC	17.6	1979	73B1
				RC/doubl.	17.0		
				Cou	8.0		
	Sch	fork	0.8/1.2	Pr	2.4	1980	65B
	Re	horseshoe	3.5	Pr/doubl.	12.2	1984	75B
				Pr/tripl.	13.8		
				RC	35.0		
				Cou	122.5		
				IR	157.5		
Calar Alto (Spanish Nat. Obs. Madrid)	Re	cross axis	1.5	RC	12.0	1978	
				Cou	45.0		
Calern (CERGA, France)	Re		1.5			1981	
Calgary	Re		1.5			1988	
Cananea (Univ. of Mexico)	Re	fork	2.12	Cas	25.4	1986	87ST1
Castel Gandolfo (Specola Vaticana)	Sch	fork	0.64/0.98	Pr	2.4	1961	65B

Table 3 (continued)

Location (Observatory)	Type	Mount	Aperture [m]	Optical system	Focal-length [m]	Year	Ref.
Cerro Las Campanas (Carnegie Southern Obs. "Du Pont"	Re	fork	2.54	RC/Gasc.	19.05	1976	73B2
				Cou	76.2		74C
Cerro Las Campanas (Carnegie, Hopkins Univ., Univ. Arizona) "Magellan"	Re	alt-az	8.0	RC	43.2		90D
				IR	120.0		
Cerro La Silla (European Southern Obs.)	Re	cross axis	1.5	Cas	22.4	1968	89S
				Cou	46.9		
	Re	horseshoe and fork	3.6	Pr/Gasc.	10.9	1976	89S
				Pr/tripl.	11.3		
				RC	28.6		
				Cou	114.6		
	Sch	fork	1.0/1.6	Pr	3.06	1969	89S
"CAT"	Re	alt-alt	1.4	CAT		1980	
	Re	fork	2.2	RC	17.6	1984	
				RC/doubl.	17.0		
				IR	77.0		
"NTT"	NTT	alt-az	3.5	Nas	38.5	1988	89T
Cerro La Silla (Univ. Copenhagen/ESO)	Re	off-axis	1.5	RC	13.1	1979	89S
Cerro Pachon (NOAO, SRC, Canada) "Gemini" (south)	Re	alt-az	8.0	Cas/Nas	56.0		90O
				Nas	96.0		
				Nas	120.0		
				IR	280.0		
Cerro Pachon (NOAO, North Carolina, Columbia)	Re	alt-az	4.1	Nas	34.5	1995	91ST
Cerro Paranal (European Southern Obs.) "VLT"	VLT	alt-az	8.0 * 4 +2.0 * 6	Cas	107.0	1996	90E
				Nas	120.0		
				Cou (vis)	592.0		
				Cou (IR)	256.0		
				Cou/combined (vis)	208.0		
				(IR)	151.0		
Cerro Tololo (Inter-American Obs.)	Sch	cross axis	0.61/0.91	Pr	2.13	1967	65B
	Re	off-axis	1.52	RC/Gasc.	11.4		80W
				Cas	20.5		
				IR	45.6		
				Cou	47.4		

(continued)

Table 3 (continued)

Location (Observatory)	Type	Mount	Aperture [m]	Optical system	Focal-length [m]	Year	Ref.
	Re	horseshoe	4.0	Pr	10.6		80W
				RC	31.2		71C
Cloudcroft (New Mexico)	Re		1.2	New	7.8		80S
Coonabarabran (Siding Spring	Sch	fork	1.2/1.8	Pr	3.06	1973	72R
Obs., Anglo Australian Obs.) "AAT"	Re	horseshoe	3.9	Pr/Gasc.	12.7	1975	76M
				Pr./doubl.	12.7		
				Pr./tripl.	13.5		
				RC	30.8		
				Cas	57.9		
				Cou	140.2		
Coonabarabran (Siding Spring Obs., Australian Nat. Univ.)	ATT	alt-az	2.3			1983	85ST
Crimea (Crimean Astrophys. Obs.)	Re	fork	2.64	Pr	10.0	1961	65B
				Cas	43.0		
				Nas	41.0		
				Cou	105.0		
	Re		1.2				
Crimea (Sternberg South Station)	Re		1.25	Pr	5.0	1960	65B
				New	5.0		
				Cas	21.0		
Flagstaff (Perkins Obs.)	Re	cross axis	1.83	Cas	31.0	1961	65B 69H
Flagstaff (U.S. Naval Obs.)	Re	fork	1.55	Pr	15.0	1964	65B
Fort Davis (McDonald Obs., Univ. Texas)	Re	cross axis	2.7	RC	24.0	1969	68S
				Cas	48.6		
				Cou	89.1		
	SST		1.0 * 85 ≘ 9.22	Pr		1992	90R1
Gornergrat	Re	fork	1.5	IR	30.0	1980	81C
Hamburg-Bergedorf (Hamburg Obs.)	Re	fork	1.25	RC	15.6	1976	
Helwan (Helwan Obs.)	Re	cross axis	1.88	New	9.14	1963	65B
				Cas	34.0		
				Cou	56.0		

Table 3 (continued)

Location (Observatory)	Type	Mount	Aperture [m]	Optical system	Focal-length [m]	Year	Ref.
Hsing-lung (Peking Obs.)	Sch/Re	fork	0.6/0.9	Pr	1.8	1963	65B
				Cas (0.9m)	13.5		
Hyderabad (Nizamiah Obs.)	Re	cross axis	1.22	Pr	4.9	1963	65B
				Cas	18.0		
				Cou	37.0		
Japal Rangapur Obs. (Osmania Univ.)	Re		1.2				80ST2
Jelm Mt. (Univ. Wyoming)	IR	yoke	2.3	Pr	4.8	1977	78G
				Cas	62.1		
Jena (Jena Univ.)	Sch/Re	fork	0.6/0.9	Pr	1.8	1963	65B
				Cas (0.9m)	13.5		
Kavalur (Vainu Bappu Obs.)	Re	horseshoe	2.3	Pr	7.5	1985	87S
				Cas	29.9		
				Cou	103.5		
Kiaton (Univ. Athens)	Re	off-axis	1.2	Cas	15.6	1975	74K
Kiso Mt. (Kiso Obs.)	Sch	fork	1.05/1.5	Pr	3.25	1974	
				Cas	34.5		
Kitt Peak (Kitt Peak Nat. Obs.)	Re	fork	2.13	RC	16.2	1963	65B
				IR	57.2		82G
				Cou	66.5		
	Re	horseshoe	4.0	Pr/tripl.	11.1	1973	82G
				RC	30.8		71C
				Cou	652.0		
	Re		1.27	IR	18.8		82G
	Re	fork	2.4	RC	18.0	1985	82ST
				RC	32.4		
Kitt Peak (McGraw-Hill Obs., Univ. Michigan)	Re	cross axis	1.32	Cas	10.0	1975	71W
				Cas	17.8		
				Cou	44.2		
Kitt Peak (Steward Obs., Univ. Arizona)	Re	fork	2.28	RC	20.5	1969	69ST
				Cou	70.7		
Kitt Peak (Wisconsin, Indiana, Yale, NOAO) "WIYN"	Re	alt-az	3.5	Nas	22.0	1993	90J

Table 3 (continued)

Location (Observatory)	Type	Mount	Aperture [m]	Optical system	Focal-length [m]	Year	Ref.
Kvistaberg Station (Uppsala Univ.)	Sch	fork	1.0/1.35	Pr	3.0	1964	65B
Llano del Hato (Univ. of the Andes, Merida)	Sch	bent yoke	1.0/1.52	Pr	3.0	1978	65B
London, Ontario (Univ. Western Ontario)	Re		1.22			1968	
Mauna Kea (Canada,France, Hawaii) "CFHT"	Re	horseshoe	3.6	Pr/tripl. Cou	13.7 72.0	1979	77O
Mauna Kea (NASA) "IRTF"	IR	yoke	3.0	Cas Cou	105.0 360.0	1979	80N
Mauna Kea (United Kingdom) "UKIRT"	IR	yoke	3.8	Cas Cas Cou	34.2 133.0 76.0	1978	77B 77C
Mauna Kea (Univ. Hawaii)	Re	fork	2.2	RC Cou	22.0 72.6	1970	70ST
Mauna Kea (Keck Inst.) "Keck I"	Re	alt-az	1.8 * 36 ≘ 10.0	Pr RC/Nas IR	17.5 150.0 250.0		90N
"Keck II"	Re	alt-az	1.8 * 36 ≘ 10.0	Pr RC/Nas IR	17.5 150.0 250.0		
Mauna Kea (Japan) "Subaru"	Re	alt-az	8.0	Pr RC Nas	15.0 97.6 100.8	1999	90K
Mauna Kea (NOAO, SRC, Canada) "Gemini" (north)	Re	alt-az	8.0	Cas/Nas Nas Nas IR	56.0 96.0 120.0 280.0		90O
Mendoza (La Plata Obs.)	Re	fork	2.13	RC Cou	16.2 66.5		
Merate (Milan-Merate)	Re	fork	1.37	Cas	20.1	1972	72M
Mt. Abu (Rajasthan)	IR		1.2				87ST2

Table 3 (continued)

Location (Observatory)	Type	Mount	Aperture [m]	Optical system	Focal-length [m]	Year	Ref.
Mt. Chikurin (Okayama Obs.)	Re	cross axis	1.88	New Cas Cou	9.2 33.9 54.3	1960	65B
Mt. Graham (Specola Vaticana) "VATT"	Re	alt-az	1.8				90ST
Mt. Graham (Univ. Arizona, Italy, n.n.) "Columbus"	MMT	alt-az	8.4 * 2	Cas IR combined	43.7 126.0 277.2		90S2
Mt. Hopkins (Smithonian Astrophys. Obs.)	Re		1.52	Cas Cou	15.2 36.6	1970	73S
Mt. Hopkins	Re		1.2				
Mt. Hopkins (Smithonian Astrophys. Obs., Univ. Arizona)	MMT	alt-az	1.82 * 6 ≘4.46	Nas/Cas	57.7 49.9	1979	77H 77S
	will be changed to						
	Re	alt-az	6.5	Cas Cas IR	35.1 58.5 97.5		90C
Mt. Korek (Iraqui Nat. Astron.Obs.)	Re	pole universal	1.25	mod.RC		not operational	82B
	Re	horseshoe	3.5	Pr/doubl. Pr/tripl. RC Cou	12.2 13.8 35.0 122.5	not installed	
Mt. Megantic (Univ. Montreal)	Re	off-axis	1.6	Cas Cas	12.8 24.0	1978	
Ondrejov	Re	off-axis	2.0	Pr Cas Cou	9.0 29.6 72.0	1967	68JR 71G
Palomar Mt. (Hale Obs.)	Re	fork	1.52	RC Cou	13.3 45.6	1970	66B
Pic du Midi (Pic du Midi)	Re	horseshoe	2.0	Pr Cas	9.98 50.0	1979	80R
Pico Veleta (IAA-CSIC, Granada)	Re	fork	1.5	Nas			

Table 3 (continued)

Location (Observatory)	Type	Mount	Aperture [m]	Optical system	Focal-length [m]	Year	Ref.
Piszkéstetö (Konkoly Obs.)	Sch	fork	0.6/0.9	Pr	1.8	1962	
Rattlesnake Mt. (Penn State Univ.)	Re	yoke	1.52	Cas		1964	75Z
Roque de los Muchachos (IAC) "Isaac Newton"	Re	fork	2.49	Pr	7.5	1983	85L
				Pr/doubl.	8.2		
				Cas	36.8		
				Cou	82.0		
"William Herschel"	Re	alt-az	4.2	Pr		1988	85B
				Cas/Nas	46.2		90R2
Roque de los Muchachos (Nordic Optical Telescope Scientific Association) "Nordic Tel."	Re		2.56	RC	28.2	1988	90A 85A
Roque de los Muchachos (Italy) "Galileo"	NTT	alt-az	3.5	Nas	38.5	1994	90B
Rozhen (Bulgaria)	Re		2.0	RC	16.0	1980	80ST1
Saltsjöbaden (Stockholm Univ.)	Sch	fork	0.65/1.0	Pr	3.0	1964	65B
San Pedro Mt. Baja California (Univ. of Mexico)	IR		1.5			1971	80A
	Re		2.12	RC	15.9	1979	
				RC	28.6		
				IR	57.2		
Shemakha (Shemakha Astrophys. Obs.)	Re	off-axis	2.0	Pr	9.0	1967	68JR
				Cas	29.6		
				Cou	72.0		
Sutherland (South African Astronomical Obs.)	Re	cross axis	1.88	New	9.15	1974	65B
				Cas	34.0		
				Cou	53.0		
	NTT	alt-az	3.5	Nas			
Tautenburg (Landessternwarte Thüringen) "Karl-Schwarzschild-Obs."	Sch/Re	fork	1.34/2.0	Pr	4.0	1960	65B
				Cas(2.0)	21.0		

Table 3 (continued)

Location (Observatory)	Type	Mount	Aperture [m]	Optical system	Focal-length [m]	Year	Ref.
Torun	Sch/Re	fork	0.6/0.9	Pr	1.8	1962	65B
				Cas(0.9)	13.5		
Victoria, British Columbia (Dominion Astrophys. Obs.)	Re	off-axis	1.22	Pr	4.88	1962	65B
				Cas	22.0		
				Cou	36.5		
				Cou	177.0/36.5		77R
Vienna (Leopold-Figl-Astrophys. Obs.)	Re	fork	1.52	RC/doubl.	12.5	1969	69M
				Cas	22.5		
				Cou	45.0		
Zelenchuk (Special Astrophys. Obs.)	Re	alt-az	6.0	Pr	24.0	1976	77I
				Nas	180.0		

Abbreviations of institutions

AAT	Anglo-Australian Telescope
CERGA	Centre d'Etudes et de Recherches Géodynamiques et Astronomiques, Grasse France
CFHT	Canada, France, Hawaii Telescope
ESO	European Southern Observatory
IAA-CSIC	Instituto de Astrofisica de Andalucia, Consejo Superior de Investigaciones y Ciencias
IAC	Instituto de Astrofisica de Canarias
IRTF	Infrared Telescope Facility
NASA	National Aeronautics and Space Administration
NOAO	National Optical Astronomical Observatories
SRC	Scientific Research Council, United Kingdom
UKIRT	United Kingdom Infrared Telescope
VATT	Vatikan Advanced Technology Telescope
WIYN	Wisconsin, Indiana, Yale, NOAO

References for 1.1.7

65B Bahner, K.: Landolt-Börnstein, NS, Vol. VI/1 (1965) p. 1.
66B Bowen, I.S., Rule, B.H.: Sky Telesc. **32** (1966), 185.
68JR Jena Review (Jenaer Rundschau) **13** (1968).
68S Smith, H.J.: Sky Telesc. **36** (1968) 360.
69H Hall, J.S., Slettebak, A.: Sky Telesc. **37** (1969) 222.
69M Meurers, J.: Sterne Weltraum **8** (1969) 195.
69ST Sky Telesc. **38** (1969) 164.
70ST Sky Telesc. **40** (1970) 276.
71C Crawford, D.L.: J. Opt. Soc. Am. **61** (1971) 682.

71G Grygar, J., Koubsky, P.: Bull. d'Information, Ass. Dévelоppement Internat. Obs. Nice No. **8** (1971) 5.

71S Schlegelmilch, R.: Mitt. Astron. Ges. **30** (1971) 84.

71W Wehinger, P.A., Mohler, O.C.: Sky Telesc. **41** (1971) 72.

72M de Moltoni, G.: Sky Telesc. **43** (1972) 296.

72R Reddish, V.C.: Conf. on the Role of Schmidt Telescopes in Astronomy, Proc. ESO/SRC/Hamburger Sternwarte (Haug, U. ed.), (1972) p. 135.

73B1 Bahner, K.: Sterne Weltraum **12** (1973) 103.

73B2 Bowen, I.S., Vaughan, A.H.: Appl. Opt. **12** (1973) 1430.

73S Schild, R.E.: Smithonian Astrophysical Observatory Special Report No. **355**, 5 + 23 + A5 (1973).

74B Barbieri, C., Rosino, L., Stagni, R.: Sky Telesc. **47** (1974) 298.

74C Carnegie Inst. Washington, Year Book **74** (1974/75) 366.

74K Kotsakis, D.: In Honorem S. Plakidis (Kotsakis, D., ed.), Athens (1974) p. 161.

75B Bahner, K.: Mitt. Astron. Ges. **36** (1975) 57.

75Z Zabriskie, F.R.: Sky Telesc. **49** (1975) 219.

76M Morton, P.C. (ed.): Anglo Australian Telescope Observer's Guide (1976).

77B Brown, D.S., Humphries, C.M.: Proc. ESO Conf. Optical Telescopes of the Future (Pacini, F., Richter, W., Wilson, R.N., eds.), Geneva (1977) p. 55.

77C Carpenter, G.C., Ring, J., Long, J.F.: Proc. ESO Conf. Optical Telescopes of the Future (Pacini, F., Richter, W., Wilson, R.N., eds.), Geneva (1977) p. 47.

77H Hoffman, T.E.: Proc. ESO Conf. Optical Telescopes of the Future (Pacini, F., Richter, W., Wilson, R.N., eds.), Geneva (1977) p. 185.

77I Ioannisiani, B.K.: Sky Telesc. **54** (1977) 356.

77O Odgers,G.J., Richardson, E.H., Grundman, W.A.: Proc. ESO Conf. Optical Telescopes of the Future (Pacini, F., Richter, W., Wilson, R.N., eds.), Geneva (1977) p. 79.

77R Richardson, E.H.: Proc. ESO Conf. Optical Telescopes of the Future (Pacini, F., Richter, W., Wilson, R.N., eds.), Geneva (1977) p. 179.

77S Strittmatter, P.A.: Proc. ESO Conf. Optical Telescopes of the Future (Pacini, F., Richter, W., Wilson, R.N., eds.), Geneva (1977) p. 165.

78G Gehrz, R.D., Hackwell, J.A.: Sky Telesc. **55** (1978) 467.

80A Allen, C., de la Herran,J., Johnson, H.: Sky Telesc. **60** (1980) 270.

80N NASA, The Infrared Telescope Facility Observer's Manual, Hawaii (1980).

80R Rösch, J., Dragesco, J.: Sky Telesc. **59** (1980) 6.

80S Schneeberger, T.J., Worden, S.P., Africano, J.L., Tyson, E.: Sky Telesc. **59** (1980) 109.

80ST1 Sky Telesc. **60** (1980) 113.

80ST2 Sky Telesc. **59** (1980) 387.

80W Walker, A.R., Mu§oz, J.: The Facilities Book of Cerro Tololo Inter-American Observatory, La Serena (1980).

81C Citterio, O., Dilworth, C., Iucci, N.: Sky Telesc. **62** (1981) 17.

82B Bok, B.J.: Sky Telesc. **64** (1982) 33.

82G Goad, J.B.: The Facilities Book of Kitt Peak National Observatory, Tucson (1982).

82ST Sky Telesc. **64** (1982) 410.

85A Ardeberg, A.: Vistas Astron. **28** (1985) 561.

85B Boksenberg, A.: Vistas Astron. **28** (1985) 531.

85L Laing, R., Jones, D.: Vistas Astron. **28** (1985) 483.

85ST Sky Telesc. **69** (1985) 106.

87S Salwi, D.M.: Sky Telesc. **73** (1987) 375.

87ST1 Sky Telesc. **73** (1987) 367.

87ST2 Sky Telesc. **74** (1987) 12.

89S Schwarz, H.E.: User's Manual, European Southern Observatory, Garching (1989).

89T Tarenghi, M., Wilson, R.N.: SPIE Conf. Proc. **1114** (1989) 303.

90A Ardeberg, A., Andersen, T.: SPIE Conf. Proc. **1236** (1990) 543.
90B Barbieri, C.: Telescopio Nazionale Galileo, Technical Report No. **1** (1990).
90C Chaffee, F.H., et al.: SPIE Conf. Proc. **1236** (1990) 507.
90D Dressler, A.: SPIE Conf. Proc. **1236** (1990) 42.
90E Enard, D.: SPIE Conf. Proc. **1236** (1990) 63.
90J Johns, M.W., Pilachowski, C.: SPIE Conf. Proc. **1236** (1990) 2.
90K Kodaira, K.: SPIE Conf. Proc. **1236** (1990) 56.
90N Nelson, J.E., Mast, T.S.: SPIE Conf. Proc. **1236** (1990) 47.
90O Osmer, P.S.: SPIE Conf. Proc. **1236** (1990) 18.
90R1 Ray, F.B.: SPIE Conf. Proc. **1236** (1990) 790.
90R2 Ridpath, I.: Sky Telesc. **80** (1990), 136.
90S1 Siegmund, W.A.: SPIE Conf. Proc. **1236** (1990) 559.
90S2 Strittmatter, P.A.: SPIE Conf. Proc. **1236** (1990) 71.
90ST Sky Telesc. **79** (1990) 585.
91ST Sky Telesc. **82** (1991) 345.

1.2 Solar telescopes

Table 1. Solar telescopes with free aperture $\phi \geq 25$ cm. Update of Vol. VI/2a (stations with unchanged name and unchanged equipment are not included).

Explanation of columns:

1. Location of institution (name of institute), [location of field station]
2. Type of telescope
   A altazimuth mounting
   E equatorial mounting
   H horizontal system
   T tower telescope
3. Coel: Coelostat
   Diameter of coelostat and auxiliary separated by /
   Other systems denoted by H (heliostat), S (siderostat)
4. Opt: Optics
   L lens optics (refractor)
   M mirror optics (reflector)
5. ϕ: aperture of telescope objective
6. f_T: effective focal length(s) of telescope
7. Arr: arrangement
   C Cassegrain
   Cde coudé
   Cor coronograph
   G Gregorian
   N Newtonian
8. Equipment
   CPh coronal photometer
   CPol coronal polarimeter
   EGSp Echelle grating spectrograph (focal length of camera [cm])
   F( ) filter (type: Hα, K, UBF = universal birefringent filter)
   FTSp Fourier Transform spectrograph
   GSp grating spectrograph (focal length of camera [cm])
   IRSp infrared spectrograph
   Mag magnetograph
   MOF magneto-optic filter
   Shg spectroheliograph
   VMag vector magnetograph/Stokesmeter
9. Ref: reference

Table 1

Location	Type	Coel [cm]	Opt	ϕ [cm]	f_T [cm]	Arr	Equipment	Ref.
Abastumani (Abastumani Astrophys. Obs.)	E		L	53		Cor	GSp	
Bangalore (Indian Inst. of Astrophys.) [Kodaikanal]	T	61/61	M	30	1830	G,N	VMag	
[Kavalur]	Has no solar instruments anymore							
Beijing (Beijing Astron. Obs.) [Huairou]			L	35	470		VMag	27A
Crimea (Crimean Astrophys. Obs.)	T	60/50	M	45	2000/3500/ 1200		GSp(800,1600); EGSp(800); IRSp(1600)	
	E		L	53	1300/2000	Cor	GSp(800); F(Hα)	
Debrecen (Heliophysical Obs.)	E		L	53	1200	Cor,Cde	GSp(800); F(Hα)	82D
Freiburg (Kiepenheuer Inst.) [Schauinsland]	T	60/60	L	45	1300		GSp(800)	74M
[Anacapri]	No longer operating							
[Izaña]	See under Freiburg, Göttingen, Würzburg							
Freiburg (Kiepenheuer Inst.),	E		M	40	3700	N	F(Hα,K)	75C
	E		M	45	2500	G,Cde	GSp(1000); F(Hα, K)	85S2,87W
Göttingen (University Obs.), Würzburg (Inst. for Astron. Astrophys.) [Obs. del Teide]	T	80/80	M	70	4500		EGSp(1525); F(Hα,K,UBF)	85S2
Göttingen (University Obs.) [Hainberg]	T	65/65	M	50	2500	G	GSp(800); F(Hα)	
[Izaña]	See under Freiburg, Göttingen, Würzburg							
[Locarno]	See under Locarno							
[Muchachos]	No longer operating							
Herstmonceux (Royal Greenwich Obs.)	No longer operating							
Huntsville (NASA/Marshall Space Flight Center)	T		M	30	390	C	VMag	82H1
Hurbanovo (SÚAA)	H	60/60	M	50	3500		GSp(1000)	80A

(continued)

Table 1 (continued)

Location	Type	Coel [cm]	Opt	ϕ [cm]	f_T [cm]	Arr	Equipment	Ref.
Kiev (Main Astron. Obs. of the Ukrainian Acad. of Sci.) [Terskol Peak]	H	65/65	M	65	1800	N	GSp(800)	
Kunming	H	40	M,L	30	1636/4500		GSp; Mag	
(Yunnan Obs.)	E		L	26	93		F	90W
Kyoto								
(Kwasan Obs.)	H	70/70	M	50	2000		GSp(1500)	
(Hida Obs.)	A		M	60	3200	G	GSp(1400,1000); F(Hα)	
Liège	H	30/30	M	30	150		FTSp	
(Univ. of Liège) [Jungfraujoch]	E		M	76	3240	Cde	FTSp (mainly non-solar use)	
Locarno (Fond. Ist. Ric. Sol.) [Orselina]	E		M	45	2500	G,Cde	GSp(1000); F(Hα)	80W
Mt Wilson (Mt Wilson Inst.)	T	43/32	L	30	1830		MOF	86R
Naini Tal (Uttar Pradesh State Obs.)	Equipment unchanged							
Northridge (San Fernando Obs.) [Sylma]	E		M	61/28	1220/560	C,G,Cde	Mag; Shg; F(Hα,K)	89C
Ondřejov	H	60/60	M	50	3500		GSp(1000); Mag	80A
(Astron. Inst. of the Czech. Acad. of Sci).	H	60/60	M	50	3500		Mag	
Oslo (Inst. of Astrophys.)	No longer operating							
Paris	H	75(S)	M	40	2300	N,G	GSp(700); Mag	
(Obs. of Paris)	T	80/70	M	60	4500		GSp(1400)	77M
[Pic-du-Midi]	Instrument no longer operating							
Pasadena (Caltech)								
[Big Bear]	E		L	25	3600/10800		VMag; F(Hα,K)	
	E		M	65	3250	G	GSp(400); F(UBF); VMag	
[Mt Wilson]	See under Mt Wilson							
Pic-du-Midi	E		L	26	400	Cor	CPh	
(Obs. of Midi-Pyrénées)	E		L	25	250	Cor	CPol	

Table 1 (continued)

Location	Type	Coel [cm]	Opt	ϕ [cm]	f_T [cm]	Arr	Equipment	Ref.
Potsdam (Solar Obs. 'Einstein-turm')	Equipment unchanged							81S
Pulkovo								
(Pulkova Obs.)	E		L	53	800/1200	Cor,Cde	GSp(800); Mag; F(Hα,K)	
[Kislovodsk]	H	30	M	30	1700		GSp; Shg	
Stará Lesná (Astron. Inst. of the Slovak Acad. of Sci.)	H	60/60	M	50	3500		GSp(1000)	80A
Stockholm (Royal Swedish Acad. of Sci.) [La Palma]	T		L	48	2235		GSp(300); F(Hα); Mag	85S1
Sydney (CSIRO) [Culgoora]	No longer operating							
Tokyo (National Astron. Obs.)	Instruments at Norikura and Okayama unchanged							74M,85M
Tucson	T	203(H)	M	152	8262		GSp(1370); Mag;	91A
(National Solar Obs.)		91(H)	M	81	4037		FTSp/Pol;	
[Kitt Peak]		91(H)	M	81	3580		GSp(71) (solar-stellar work)	
	T	104/91	M	70	3639		GSp(1039); Mag	
[Sacramento Peak]	T	112/112	M	76	5500		EGSp(1214); F(UBF); Sp(152); F(Hα,K)	
	E		L	40	800	Cor	GSp(1300); Shg; GSp(152); CPh	
	H	41/41	L	30	1068		GSp(1300); Shg; GSp(152)	
	E		L	25	875	C	VMag (instrument of-Johns Hopkins)	
Würzburg (Inst. for Astron. Astrophys.)	See under Freiburg, Göttingen, Würzburg							
Zürich (Inst. of Astron.)								
[Zürich]	T	30	L	25	1070		GSp(466); VMag; F(Hα)	
[Arosa]	H	30	L	25	2950		GSp(1250); VMag; F(Hα)	
SOON (USAF Weather Service)	E		L	25	533		GSp(227); Mag; F(Hα)	
Identical equipment in: San Vito, Italy; Learmonth, Australia; Ramey, Puerto Rico; Holloman AFB, New Mexico; Palehua, Hawaii								

Abbreviations of institutions in the table and references:

CSIRO Commonwealth Scientific and Industrial Research Organisation (Australia)
JOSO Joint Organisation for Solar Observations
NASA National Aeronautics and Space Administration
NSO National Solar Observatory
SOON Solar Observing Optical Network
SUAA Slovak Union of Amateur Astronomers
USAF United States Air Force

References for 1.2

74M Mattig, W.: Sonnenforschung auf dem Schauinsland, Freiburger Universitätsblätter Heft 44 (1974).
75C Casanovas, J., Mattig, W.: JOSO Annual Report (1975) 18.
77M Mein, P.: Sol. Phys. **54** (1977) 45.
80A Ambrož, P., Bumba, V., Klvaňa, M., Macák, P.: Phys. Solari-Terr. **14** (1980) 107.
80W Wiehr, E., Wittmann, A., Wöhl, H.: Sol. Phys. **68** (1980) 207.
82D Dezsó, L.: Sol. Phys. **79** (1982) 125.
82H1 Hagyard, M.J., Cumings, N.P., West, E.A., Smith, J.E.: Sol. Phys. **80** (1982) 33.
82H2 Hiei, E.: Sol. Phys. **80** (1982) 113.
85M Makita, M., Hamana, S., Nishi, K., Shimizu, M., Koyano, H., Sakurai, T., Komatsu, H.: Publ. Astron. Soc. Jpn. **37** (1985) 561.
85S1 Scharmer, G.B., Brown, D.S., Pettersson, L., Rehn, J.: Applied Optics **24** (1985) 2558.
85S2 Schröter, E.H., Soltau, D., Wiehr, E.: Vistas in Astronomy **28** (1985) 519.
86R Rhodes, E.J., Cacciani, A., Tomczyk, S., Ulrich, R.K. in: Seismology of the Sun and Distant Stars (Gough, D.O., ed.), Dordrecht: Reidel (1986) 309.
87A Ai, G.: Publ. Beijing Astron. Obs. **10** (1987) 27.
87W Wiehr, E. in: The Role of Fine Scale Magnetic Fields on the Structure of the Solar Atmosphere (Schröter, E.H., Vázquez, M., Whyller, A., eds.), Cambridge: Cambridge University Press (1987) 354.
89C Chapman, G.A., Walton, S.R. in: High Spatial Resolution Solar Observations (van der Lühe, ed.), Sunspot, New Mexico: NSO/Sacramento Peak (1989) 402.
90W Wu, M., Li, Z., Li, Z., Liu, Z., Luan, D.: Publ. Yunnan Obs. **3** (1990) 1.
91A Anon.: National Solar Observatory Users' Manual. Tucson (1991).
91S Staude, J.: Rev. Mod. Astron. **4** (1991) 69.

1.3 Photoelectric photometry

1.3.0 – 1.3.8 see LB VI/2a

1.3.9 Scientific charge-coupled devices (CCDs)

This chapter deals only with optical CCDs in slow-scan scientific CCD cameras.
A CCD system consists of

- the CCD detector in its cooling system,
- the control electronics for operating the CCD and for signal processing and digitisation,
- the control computer for data handling, data display, control of housekeeping data and coordination with other instruments.

1.3.9.1 Functioning of a CCD detector

1.3.9.1.1 Technical description

A CCD is a photoelectronic imaging device which in principle can be described as an array of individual pixels (PICture ELementS) defined by potential wells in the silicon semiconductor material. In these pixels, the electrons which are generated by the absorption of photons are collected. By applying different voltages to the small semi-transparent electrodes on the CCD surface, the electric fields defining the potential wells can be changed. By this means, a transfer of the charge from pixel to pixel onto the output node is carried out. At the output capacitance the charge of each pixel can be measured as a small voltage (typically 1 $\mu V/e^-$). The CCD is operated by the camera electronics, which amplifies and converts the output voltages of the individual pixels into a digital number (DN), often referred to as "Counts" or "ADU" [86M, 89M].

1.3.9.1.2 A single CCD pixel as a metal-oxide-semiconductor (MOS) device

When photons of energy $E > 1.12$ eV are absorbed in the silicon semiconductor material, they can excite valence electrons to the conduction band, thus creating electron-hole pairs. The electrons are separated, collected and stored in a "potential well" generated by the structure of the MOS-type capacitor in combination with the applied gate-voltage. A MOS-type pixel consists of several layers (see Fig. 2a). If the polysilicon (or metal) gates are separated from the uniformly p-doped silicon by a thin SiO_2 insulator layer, one speaks of a *surface-channel* CCD. In *buried channel* devices, an additional n-doped layer is implanted between substrate and insulator.

When a positive voltage is applied to the gate electrode, the holes, which are the majority carriers, are repelled from the electrode and a region depleted of charge is generated. In one direction the borders of that *depletion region* are formed by neighbour electrodes with a less positive voltage and in the other direction by heavily p-doped narrow channels ("channel stops"). In this potential well, electrons can be collected and stored. If a triplet of electrodes defines the width of a pixel, the CCD is called a *three-phase* CCD. Since the minimum of the potential well in a *surface channel* CCD lies at the Si-SiO_2 interface, transfer of electrons during read-out is strongly disturbed by the irregularities at this crystal surface. With the n-layer implanted (*buried channel* CCD), the potential minimum is shifted away from that interface and the electrons are no longer transferred along the silicon surface.

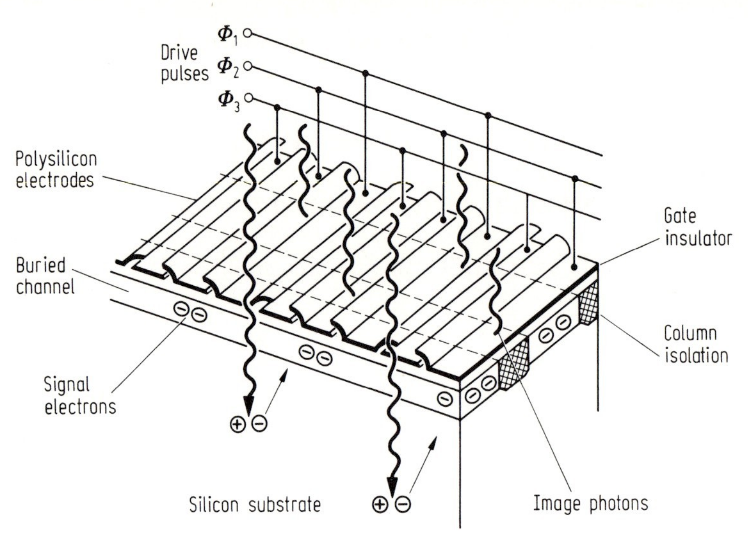

Fig. 1. Schematic view of the layout of a three-phase CCD. Courtesy EEV (see list of suppliers in Table 1)

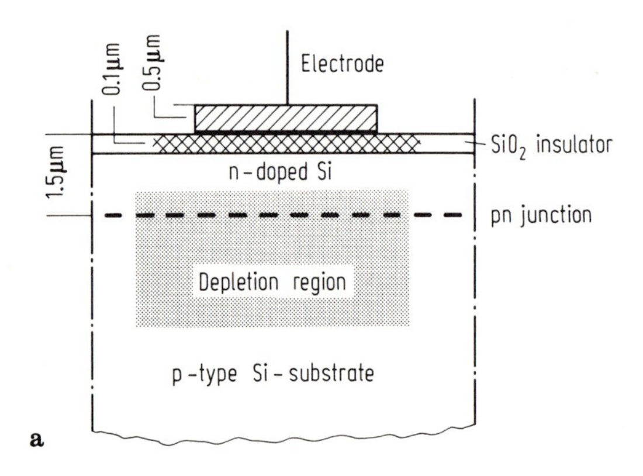

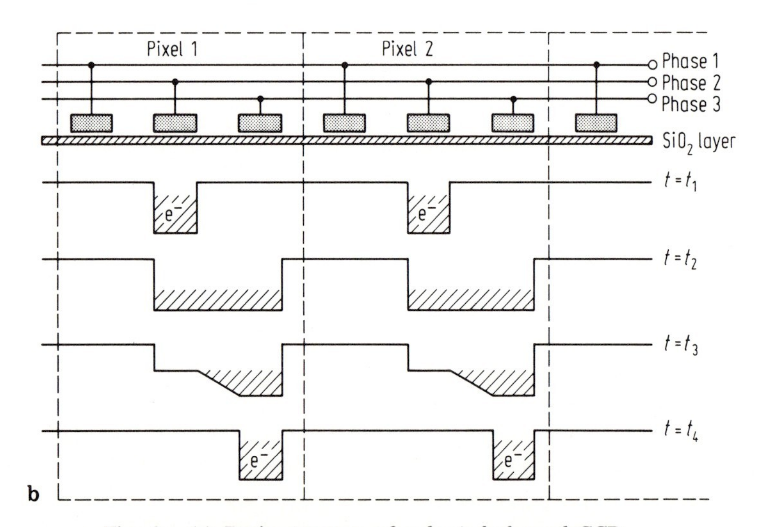

Fig. 2. (a) Basic structure of a *buried channel* CCD pixel and (b) principle of charge-transfer process for a 3-phase CCD.

1.3.9.1.3 The read-out process of a CCD

During the read-out of the CCD, the charges in all pixels in a row are simultaneously shifted, transferring the lowest row into the serial output register (Figs. 2b and 3). Before the next shift, all the pixels in the output register are read out one by one. This *charge coupling* from pixel to pixel is achieved by adequately modifying the voltages at the electrodes [86M, 89M]. Since CCDs used in ordinary TV-cameras normally work without a shutter, half of the area of these TV-CCDs is covered. The contents of the pixels in the *image area* is shifted very rapidly into the covered *storage area*. During read-out of the storage area the next integration takes place. The covered storage area can be either the lower half of the CCD (*frame transfer* CCD) or each second column (*interline transfer* CCD).

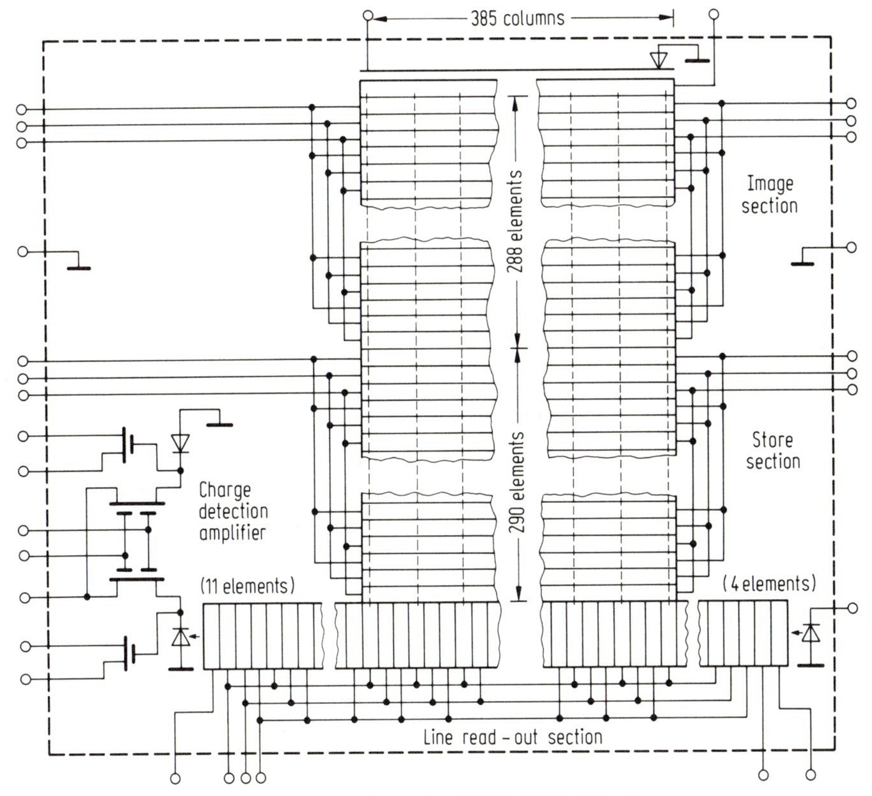

Fig. 3. Schematic view of a P8600 GEC CCD-detector. Courtesy EEV (see list of suppliers in Table 1)

1.3.9.1.4 Correlated double sampling in the CCD camera electronics

Since the read-out noise of the CCD system normally is dominated by the detector's output amplifier noise, the *correlated double sampling* (CDS) technique is used. Before the charge of a pixel is sampled, the output capacitance is loaded by the reset-transistor to a certain level. The signal then decreases this charge. To remove the uncertainty in the voltage to which the output node is reset *(reset noise)*, the output voltage is sampled directly before and after the signal charge has been dumped onto the output node. The difference between these two voltages then represents the true collected signal. (See Fig. 4). This procedure is possible because of the different RC time constants during reset ($R_{on} \approx 10^4\,\Omega$, $\rightarrow t > RC_0$) and after switch-off of the reset pulse ($R_{off} \approx 10^{12}\,\Omega$, $\rightarrow t \ll RC_0$). Because of the large $R_{off}C_0$ time constant, the level of the reset voltage is "frozen" and thus there is time enough for a double sampling.

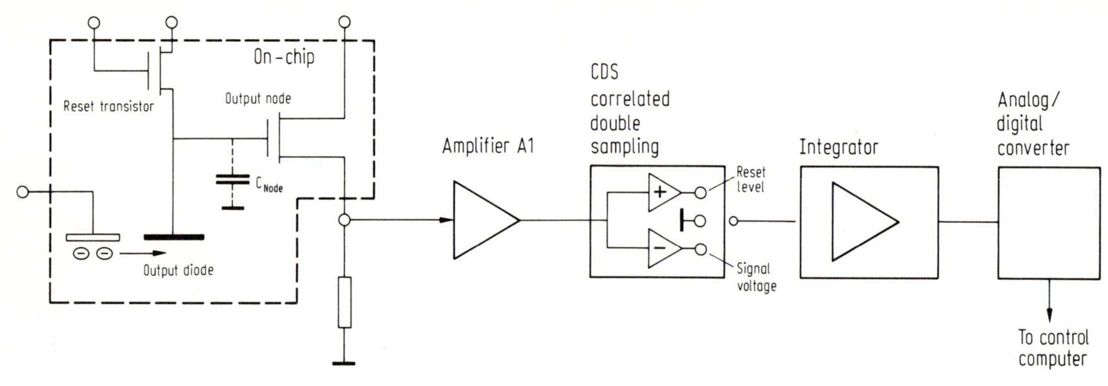

Fig. 4. Principle of CCD output circuit and schematic of signal chain in slow-scan CCD electronics.

This CDS and the final digitisation time determine the pixel read-out frequency. Since this is typically $5 \cdot 10^4$ pixel per second, such cameras are called *slow-scan cameras*.

1.3.9.2 Characteristic properties of modern CCD detectors

1.3.9.2.1 Nomenclature for conversion of incident photons into a digital number

In the CCD detector:
conversion of incident photons of number p_i into an equivalent output voltage v_s.

- Quantum efficiency $\{QE\}_i$: interactions per incident photon.
  → Number of generated electrons per interacting photon: η,
  ($\eta \cong 1$ for wavelengths $250\ \text{nm} < \lambda < 1106\ \text{nm}$).
- Quantum efficiency QE: number of generated electrons per incident photon, $QE = \eta \cdot \{QE\}_\text{i}$.
- Charge collection efficiency: CCE.
- Charge transfer efficiency: CTE.
- Effective quantum efficiency QE_eff: measured electrons per incident photons.
- Sensitivity S_ν [μV/e⁻]: volt per electron at the output node.

The signal consists of an **output voltage** v_s at the output node of the CCD.

In the camera electronics:
amplification of v_s and conversion into a digital number [DN].

- Gain A_1: amplification of CCD output voltage v_s.
- Addition of an electronic signal-offset *(bias)* b [mV].
- Digitisation A_ADC [DN/μV]: signal conversion into a digital number [DN].

The final result is the signal

$$S\,[\text{DN}] = (p_i QE_\text{eff} S_\nu A_1 + b) \cdot A_\text{ADC}, \quad or \quad S = g \cdot s + B,$$

with the signal s in electrons: $s[\text{e}^-] = p_\text{i} QE_\text{eff}$, the conversion factor: $g[\text{DN/e}^-] = S_\nu A_1 A_\text{ADC}$, and the digitized bias $B = b \cdot A_\text{ADC}$.

1.3.9.2.2 The measured quantum efficiency QE_{eff}

QE_{eff}: Ratio of measured to incident photons, i.e. number of electrons that are measured at the output of the CCD in relation to the number of incident photons. QE_{eff} is measured, e.g., by comparison with calibrated diodes [87M].

Photoelectric effect in silicon:
For photon energies E_i between 1.12 eV and 5 eV (i.e. wavelength λ between 1110 nm and 250 nm) a single electron-hole pair is created for each interacting photon. For $E_i > 5$ eV more than one electron-hole pair is generated.

Number of created electrons:

$$n(E_i) = \frac{E_i\ [\text{eV}]}{3.65} \qquad \text{with} \qquad \lambda_i\ [\text{nm}] = \frac{1239}{E_i\ [\text{eV}]}.$$

The final QE_{eff}
depends on the following parameters:

- The absorption probability in the photosensitive volume which depends
  a) on the *absorption length* $\tau(\lambda)$ [μm] in silicon and
  b) on the transparency, reflectance and surface conditions of the overlying layers.
- The charge collection efficiency (CCE), which is the probability that an electron is collected in that pixel where it was generated by the photon. This can be influenced by charge diffusion, recombination or trapping of electrons. When generated in the depletion region, nearly all electrons are collected. The CCE can be measured by using x-ray sources producing a definite number of electrons per absorbed photon, e.g., 1620 and 1270 e^- for the 6.5 and 5.9 keV lines of Mn from a ^{55}Fe source [87J1].
- The charge transfer efficiency (CTE), which – for the read-out process of the CCD – gives a measure for the completeness of the charge transfer from one potential well to the next. The CTE of modern CCDs, in the range of 0.99999 or better, is mainly limited by *spurious potential pockets*. Charge is lost in improper potential wells caused, e.g., by impurities or diffusion effects during the fabrication process. Often trapped charge is released at later times (*deferred charge*) [87J2, 88B]. Besides on the applied gate voltages the CTE also depends on the clock frequency, the rise and fall time of the clocking pulses and their overlap, and on the temperature of the CCD.

Methods for increasing the quantum efficiency:

In silicon, photons with wavelengths between 100 and 500 nm have a short absorption length τ (≤ 1 μm). Above $\lambda = 600$ nm, $\tau(\lambda)$ increases and the QE eventually becomes dependent on the depth of the photosensitive volume ($\tau \approx 2$μm for 600 nm, and $\tau \approx 10$μm for 800 nm, respectively). As in frontside-illuminated CCDs the 'blue photons' are absorbed within the overlying poly-silicon or metal gate electrodes two methods have been developed to overcome this lack of sensitivity in the blue: frontside-illuminated CCDs are coated with a thin layer (about 0.3 ... 3 μm) of organic phosphor material (e.g. lumogen or coronene) which converts photons of $\lambda < 420$nm into light of about 530 nm [86D]. Another solution is the illumination of the CCD from the backside. First, the thick silicon substrate has to be thinned to about 10 μm [90L2]. Then a native SiO_2 layer forms on the backside when the silicon oxidizes. With this depth photons of all wavelengths can be absorbed in the CCDs depletion region [90L2]. Yet an additional treatment is necessary in order to avoid instabilities like the quantum efficiency hysteresis (QEH) or charge loss due to, e.g., diffusion or recombination in so-called "backside wells". To direct the created charge from the backside to the front pixel wells, different techniques like ion-implantations, backside charging by UV-flashing, an additional thin metal film (*flash gate* [86J]) or an highly doped ultra-thin boron layer is applied. See e.g. [86R, 87L2, 91J2].

Frontside-illuminated virtual-phase CCDs (like the Texas Instrument CCDs) achieve an acceptable blue responsivity because only part of the front side is covered by electrodes [89J1].

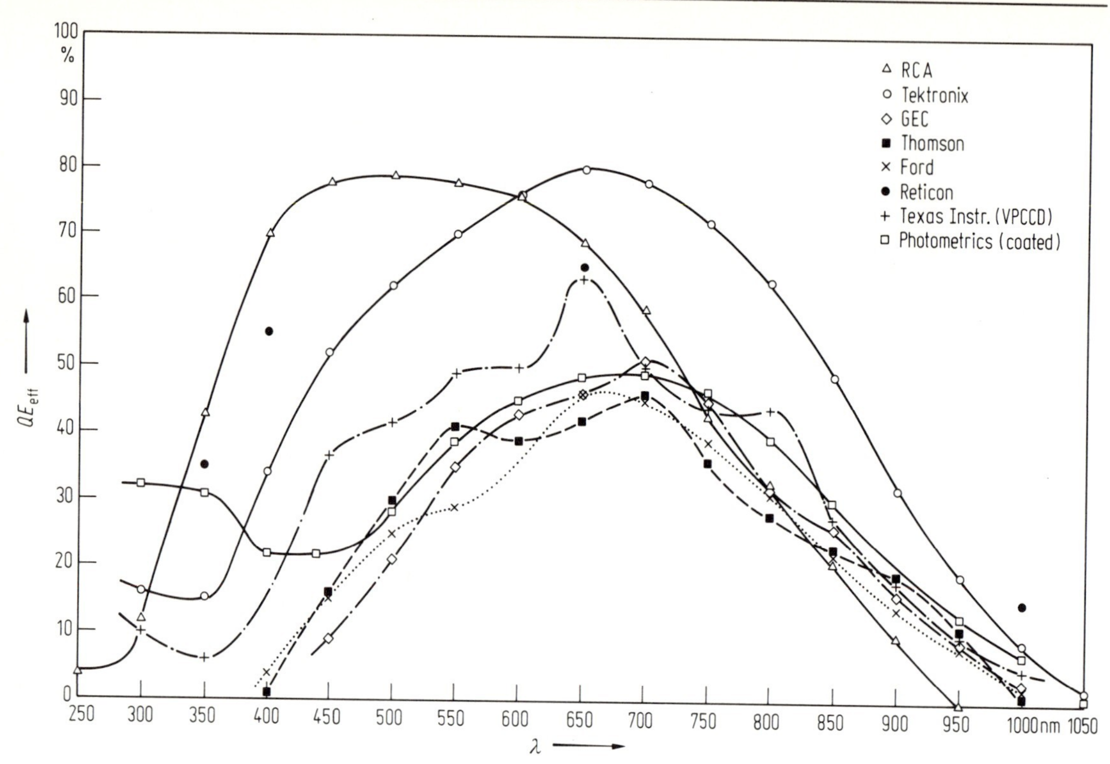

Fig. 5. Typical QE_{eff}-curves of scientific CCD detectors.

As the reflectivity of silicon in the visible is nearly 40%, an additional antireflective (AR) coating is applied in order to achieve a quantum efficiency of up to 90% with backside illuminated CCDs [87L1].

UV-coated frontside-illuminated CCDs and thinned, backside-illuminated CCDs show an interference fringe pattern when illuminated with nearly monochromatic light.

As in the near infrared (NIR) the photon absorption length increases with decreasing temperature, warming up the CCD by some degrees leads to an enhanced quantum efficiency in the NIR.

1.3.9.2.3 Sensitivity [μV/e$^-$]

The sensitivity or output voltage per electron has typical values of $0.1 \cdots 4$ μV/e$^-$, with values of the output capacitance of $0.1 \cdots 0.6$ pF, and can be measured indirectly via the photon transfer curve (see subsection 1.3.9.2.9), when A_1 and A_{ADC} are known.

1.3.9.2.4 Read-out-noise *RON* [e$^-$/pixel]

The main sources for this noise term caused by the electronic read-out process are [87J2, 91J2]:

- the output amplifier noise: noise of the on-chip output amplifier, mainly consisting of a white noise and a $1/f$ -noise component.
- the reset noise: uncertainty of the voltage to which the output node is reset between the read-out of two pixels. Most of the reset noise ($\sim \frac{1}{e} \cdot \sqrt{kT/C}$) is eliminated by the *correlated double sampling* read-out procedure in the camera electronics (see 1.3.9.1.4).
- the charge transfer noise: [87J2].

The read-out noise in [DN] can be determined from an image without light exposure or from the overscan areas (see 1.3.9.4). In [e^-] it is derived by applying the conversion factor g [DN/e^-] determined from the photon transfer curve. Typical values of the read-out noise are $3 \cdots 30$ e^-. A specially designed *skipper* CCD allows for nondestructive read-out of the pixel charge. By measuring the charge n times, the error is reduced by a factor of $\sqrt{n}$, thus achieving values of < 1 [e^-/pixel] [90C, 90J].

1.3.9.2.5 Dark current at a temperature T[K] in [e^-/min] per pixel

To reduce the thermally generated electrons, CCDs normally are operated at temperatures between 145 and 165 K. A drop in temperature of 10 K reduces the dark current by about a factor of three. Lower temperatures are limited by the decreasing CTE performance and by a poor quantum efficiency in the near infrared. The dark current at any temperature can be derived from the formula:

$$\text{dark current } [(e^-/s)/\text{pixel}] = R_0 C_1 T^{1.5} e^{-E_g/2kT} \qquad [91J2]$$

with R_0 = dark current in [nA/cm^2] at room temperature,
$C_1 = 5.86 \cdot 10^7 \cdot ps^2$, with ps = pixel size in [μm],
$k = 8.62 \cdot 10^{-5}$ eV K^{-1}. (Boltzmann constant), and
E_g = bandgap = $(1.11557 - (7.021 \cdot 10^{-4} T^2)/(1108 + T))$ eV.

Typical values are 1 e^- per pixel and minute at a temperature of 150 K. The signal in a long-integrated but not illuminated frame (e.g. with a closed shutter) gives a measure for the dark current (when corrected for *cosmics*, see subsections 1.3.9.2.7 and 1.3.9.4).

For dark current there are mainly three source regions: the silicon substrate, the depletion region, and the surface states at the Si-SiO_2 interface. The latter is by far the dominating one. Disturbances in the crystal lattice at the Si-SiO_2 interface can have energy levels between the valence and the conduction band, and thus allow electrons to first hop to this mid-band state and eventually to the conduction band. In order to populate these interface states with free carriers, CCDs are operated in the *partially inversion* mode. During integration the collecting electrodes are set to about 0 V (instead of, e.g., 10 V) and the confining ("barrier"-) electrodes to a negative voltage (e.g. -8 V). Then holes from the channel stops are attracted to the Si-SiO_2 interface and thus "fill" the interface states. Modern CCDs are manufactured with MPP (*multi-pinned phase*) technology, which allows the CCD alternatively to be operated totally inverted at all times [89J2, 91J2]. With MPP technique dark current values of < 0.05 e^- per pixel and minute are reached.

1.3.9.2.6 Linearity and charge capacity [e^-/pixel]

The charge capacity or "full well" is the maximum number of electrons a single pixel can hold before spilling over. Since the p-doped channel stops are a much higher potential barrier, excess charge normally flows only into the vertical neighbour pixels. The effect of overflow of the saturated pixel is called *blooming*. Full well is reached when the linearity curve of the CCD (signal [DN] versus illumination level) shows a cut-off. Typical values are from $5 \cdot 10^4$ to $6 \cdot 10^5$ [e^-/pixel]. See also subsect. 1.3.9.2.9.

Up to their saturation level, which can be slightly different from pixel to pixel, CCDs are linear devices (typically $< 0.5\%$). The *dynamic range* of a CCD is defined as the ratio of it's full well capacity to the read-out noise. Typical values are about 10^4 to 10^5.

1.3.9.2.7 On-chip binning, overscan, bias, cosmics, luminescence

1. On-chip binning:
   The addition of charges of adjacent pixels on the chip. Binning in vertical direction occurs when rows

are added in the serial register with the horizontal clocking sequence omitted. By summing up charge in the output node without a reset, pixels are binned in horizontal direction. To allow for binning the horizontal register and the output node of modern CCDs have distinctly higher charge capacity as the normal pixels. Binning on the one hand leads to a lower spatial resolution and a higher "cosmics-per-pixel" value, but on the other hand to an improved signal-to-noise ratio and to a much faster read-out time.

2. Overscan and bias:
   Overscan: read-out of virtual pixels, i.e. clocking more horizontal and/or vertical transfers than the number of physical pixels in that direction. The resulting overscan area in the image then only contains the "bias"-signal, – a voltage level added to each pixel signal in the camera electronics before digitisation. The bias signal should be adjusted to a level which excludes negative voltage values (caused by the read-out noise) at the analog-to-digital converter. The overscan area allows to determine the bias value and the read-out noise directly from the frame. Also bad CTE can be detected when parts of the charge collected in the last real pixel are to be seen in the adjacent overscan pixels.
3. Cosmics:
   Events of up to some thousand electrons in one or a few adjacent pixels. In most cases they are caused by the energy release of a muon (about 80 e^- per μm of silicon) when passing through the CCD. As the muons have been generated in nuclear interactions of cosmic rays with atoms of the earth's atmosphere these events have been called "cosmics".
4. Luminescence:
   In some CCDs (e.g. RCA-CCDs) the output transistor amplifier can act like a faint LED. This effect is eliminated by switching off the output transistor during integration. Sometimes, the gate voltages of the pixels also have to be reduced. During the read-out of the CCD these voltages are reset to their nominal values.
5. Residual images:
   When the CCD is over-exposed to light, electrons can be trapped at the Si-SiO_2 interface. In some CCDs release of the trapped charge can take several days. Then this *residual image* can be seen on a very faint level in the following exposures.

1.3.9.2.8 CCD formats

CCDs used in astronomy normally have square pixels. Pixel-sizes range from 7.5 μm to 30 μm (see Table 1). At the moment CCDs with pixel-numbers of up to 4096 × 4096 are available [90J]. Often these big CCDs are equipped with two or four output nodes [90G]. Since blemish-free big CCDs are very difficult to produce, mosaics of CCDs are used alternatively [87G2]. For this reason special *buttable* CCDs have been developped [89R].

1.3.9.2.9 Cosmetics

Visible blemishes in CCDs, known as, e.g., "bad pixels" or "blocked columns", occur in many devices [88B, 91J2], see also subsection 1.3.9.4. Usually they are introduced during the manufacturing process. Mostly they are stable and sometimes can be modified to a certain extent by the gate voltages.

1.3.9.2.10 Measuring technique: the photon transfer curve

From the photon transfer curve [87J2] the conversion factor g [DN/e^-], the read-out noise *RON*, and the full-well capacity can be derived directly. For this purpose, the noise of a subset of n pixels is plotted versus the mean signal at different levels of uniform illumination (after subtraction of the electronic bias from each frame). The dark current is neglegible at exposure times of less than a minute. In a plot of noise versus signal, one can distinguish four regimes (Fig. 6): at the very lowest signal levels, the constant read-out noise is the main contribution; at low and medium signal levels, the shot noise (the statistical fluctuation in the photon detection rate) dominates and at high signal levels, the *fixed-pattern noise* dominates the total noise. Eventually the examined pixels go into saturation and the photon transfer curve shows a distinct cut-off.

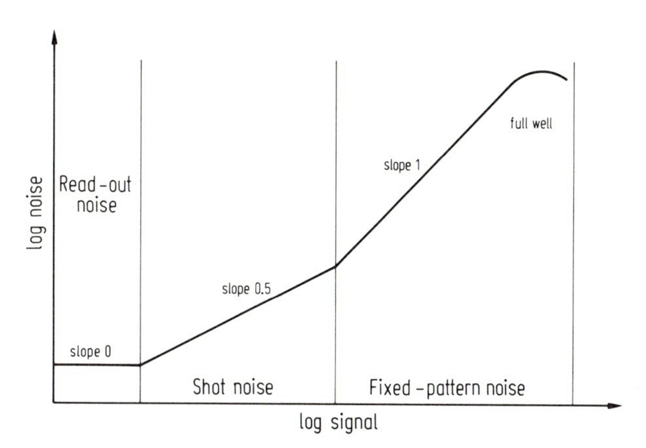

Fig. 6. Schematic photon transfer curve of a CCD detector.

The *fixed-pattern noise* is proportional to the signal and is a measure for the nonuniformity of the different single pixels. The effective quantum efficiency varies from pixel to pixel by typically some percent. If one eliminates the fixed-pattern contribution, either by comparing a pixel value in several identical exposures or by calibrating *(flatfielding)* each of the N pixels of the subarray, the following formulae for the mean signal $\bar{S}$ and the variance $\bar{V}$ apply:

$$\bar{S}[\mathrm{DN}] = 1/\mathrm{N}\sum_{\mathrm{ij}}^{\mathrm{N}} S_{\mathrm{ij}}, \qquad \text{and} \qquad \bar{V}[\mathrm{DN}] = 1/(\mathrm{N}-1)\sum_{\mathrm{ij}}^{\mathrm{N}}(S_{\mathrm{ij}}-\bar{S})^2 .$$

On the other hand the total noise in $[e^-]$ is given by

$$\sigma^2 = (RON_{\mathrm{e^-}})^2 + (\Delta s_{\mathrm{e^-}})^2,$$

where $RON_{\mathrm{e^-}}$ is the read-out noise in [e$^-$] and $\Delta s_{\mathrm{e^-}}$ is the shot noise of the signal s [e$^-$].

By multiplying each term with g [DN/e$^-$] and setting $\Delta s_{\mathrm{e^-}}^2 = \bar{S}/g$, one yields

$$(g\,\sigma)^2 = (g\,RON_{\mathrm{e^-}})^2 + g^2\bar{S}/g, \qquad \text{which is} \qquad \bar{V} = RON_{\mathrm{DN}}^2 + g\,\bar{S}.$$

g and RON can be either determined from the plot or by computation. Since RON_{DN} can be directly measured (e.g. from the overscan areas of the images), g is evaluated according to

$$g = (\bar{V} - RON_{\mathrm{DN}}^2)/\bar{S}.$$

Usually a mean $\bar{g}$ is derived from a set of subarrays.

In practice there are three methods of eliminating the fixed-pattern noise. In the first method always k exposures are made per signal level. Then from these k frames $\bar{S}_{\mathrm{ij}}$ and $\bar{V}_{\mathrm{ij}}$ are determined for each single pixel and eventually averaged over the N pixels of the subarray. In this case the illumination device must be extremely stable. Another, much faster, way is to make only two identical exposures E_1 and E_2 per signal level and to compute a fixed-pattern-free third frame $E_3 = E_1 - E_2$. Then the following equations hold:

$$\bar{S} = (S_{\mathrm{E_1}} + S_{\mathrm{E_2}})/2 \qquad \text{and} \qquad \bar{V} = V_3/2 \qquad (\sigma = \sigma_3/\sqrt{2})\,.$$

If the light source is not constant enough even for two exposures (i.e. at the telescope) then a single but *flatfielded* (see subsection 1.3.9.4) image can be used for each signal level. Since with this testing method it can not be excluded that there is still a rest of fixed-pattern noise C in the image, one has to fit in the following parabolic equation the values for g and $(g\,C)^2$:

$$\bar{V} = RON_{\mathrm{DN}}^2 + g\,\bar{S} + g^2 C^2 \bar{S}^2 .$$

1.3.9.3 CCD systems

1.3.9.3.1 The cooling system for the CCD

To keep CCDs cold, they are housed in vacuum dewars and cooled by liquid nitrogen via a copper finger [83L]. Thermoelectric cooling with, e.g., Peltier elements is used for CCD-TV cameras or similar applications, yet there are plans to use it for detectors with MPP technique, too. To regulate the temperature, a resistor and a diode (as temperature sensor) are embedded in the copper socket. Vacuum pumps are used to achieve the operating pressure (typically 10 Pa) and ion getter pumps or sorbent material like zeolite for maintaining this pressure over long periods.

1.3.9.3.2 Slow-scan scientific CCD cameras

Modern CCD cameras should have the following characteristics:

- Optimal correlated double sampling.
- The read-out noise of the whole system is determined by the read-noise of the detector.
- Different CCDs can be operated.
- The clock pulses are easily adjustable in amplitude, shape and frequency.
- On-chip pixel binning in both axes is possible (to reduce read-out noise per mm^2).
- A subarea of the whole image can be read out by dropping of the signal before the preamplifier.
- Overscanning and control of CCDs with nondestructive readout (skipper).
- Control of CCDs with MPP option.

For CCD camera systems see e.g. [87G1, 88L, 89C, 90L1, 90W].

1.3.9.3.3 The control computer

Often personal computers are used for controlling the CCD electronics and the environment (e.g. shutter, filterwheel, and other light-selecting devices). In addition to data handling and storing they supply more or less complex image procession and display options. A minimum "quick look" system must be available in order to decide whether the exposure had been successful or not [87R].

Table 1. Typical characteristics of main astronomical CCDs.

	Texas Instrument	RCA	GEC (EEV)	Tektronix
Format	800 ×800	1024 ×640	1152 ×770	1024 ×1024
Pixel size [μm]	15	15	22.5	24
Read noise [e^-/pixel]	25 ···100	30 ···60	3 ···10	4 ···15
Technology	virtual phase	3-phase	3-phase	3-phase
Illumination	frontside	backside	frontside	backside
Full well [e^-/pixel]	$1 \cdot 10^5$	$1.1 \cdot 10^5$	$2 \cdot 10^5$	$5 \cdot 10^5$
Dark current at 150 K [(e^-/min)/pixel]	< 0.07 (inverted)	0.1 ···1	< 0.1	0.001 (MPP)
Other formats: format/pixelsize	(1024×1024)/18 (390×584)/22	(512×320)/30	(576×385)/22 (1242×1152)/22.5	(512×512)/27 (2048×2048)/24

Table 1. (continued)

	Thomson	Reticon	Ford Aerospace (LORAL)	Photometrics
Format	1024 ×1024	1200 ×400	2048 ×2048	512 ×512
Pixel size [μm]	18	27	15	20
Read noise [e^-/pixel]	3 ···15	< 10	< 10	6 ···18
Technology	4-phase	4-phase	3-phase	4-phase
Illumination	frontside	backside	frontside	frontside
Full well [e^-/pixel]	$1.7 \cdot 10^5$	$5 \cdot 10^5$	$1 \cdot 10^5$	$3.2 \cdot 10^5$
Dark current at 150 K [(e^-/min)/pixel]	1.5	0.05 (MPP)	< 0.01 (MPP)	0.05 (MPP)
Other formats: format/pixelsize	(576×384)	(512×512)/27	(4096×4096)/7.5 (2048×2048)/7.5	

Suppliers of scientific CCD detectors:

(a) As listed in Table 1:
1. Texas Instrument: Custom device for NASA.
2. RCA: Withdrawn from market. Succeeded by (b) 1.
3. GEC: English Electric Valve Co. (EEV), 106 Waterhouse Lane, Chelmsford, Essex CM1 2QU, United Kingdom
4. Tektronix: P.O. Box 500, Mail Sta. 59-420, Beaverton, OR 97077, USA
5. Thomson: Thomson CSF, 38 Rue Vauthier, 92102 Bologne-Billancourt CEDEX, FRANCE
6. Reticon: EG & G Reticon, 345 Potrero Ave, Sunnyvale, CA 94086, USA
7. Ford Aerospace: LORAL, Fairchild Imaging Sensors, 1801 McCarthy Blvd., Milpitas, California 95035, USA
8. Photometrics: Photometrics Ltd., 2010 N.Forbes Boulevard, Suite 103, Tucson, Arizona 85745-1418, USA

(b) Additional suppliers:
1. David Sarnoff Research Center: CN533300,M/S SW 331A, 201 Washington Road, Princeton, NJ. 08543-5300, USA
2. Dalsa Inc., 605 McMurray Road, Waterloo, Ontario, Canada N2V2E9
3. Eastman Kodak, KP1-34, Rochester, NY. 14652 3708, USA

References for 1.3.9.1 – 1.3.9.3

Review articles and books
86M, 87J2, 88R, 89M, 91J1, 91J2

Special References

83L Lauer, T.R., et al: CCD use at Lick Observatory, SPIE Conf. Proc. **445** (1983) 132.

86D Deiries, S.: CCD-coating at ESO, in: ESO Conference and Workshop Proc. **25** (Baluteau, U.P., D'Odorico, S., eds.) (1986) p. 73.

86J Janiseck, J., et al.: The CCD flash gate, SPIE Conf. Proc. **627** (1986) 54.

86M Mackay, C.D.: Ann. Rev. Astron. Astrophys. **24** (1986) 255.

86R Robinson, L.B.: A Review of Surface Treatment for CCD, in: ESO Conference and Workshop Proc. **25** (Baluteau, U.P., D'Odorico, S., eds.) (1986) p. 53.

87G1 Gunn, J.E., et al.: The Palomar Observatory CCD camera, Publ. Astron. Soc. Pac. **99** (1987) 518.

87G2 Gunn, J.E., et al.: Four-shooter: a large format Charge-coupled device camera for the Hale telescope, Optical Eng. **26** (1987) 779.

87J1 Janiseck, J.R., Klaasen, K., Elliott, T.: CCD charge collection efficiency and the photon transfer technique, Optical Eng. **26** (1987) 972.

87J2 Janiseck, J.R., Elliott, T., Collins, S., Blouke, M.M., Freeman, J.: Scientific charge-coupled devices, Optical Eng. **26** (1987) 692.

87L1 Lesser, M.: Antireflective coatings for silicon charge-coupled devices, Optical Eng. **26** (1987) 9.

87L2 Leach, R.W., Lesser, M.: Improving the blue and UV response of silicon CCD detectors, Publ. Astron. Soc. Pac. **99** (1987) 668.

87M Marien, K.H., Pitz, E.: Measurement of the absolute quantum efficiency of a charge-coupled device in the ultraviolet, Optical Eng. **26** (1987) 742.

87R Robinson, L.B., et al.: Lick Observatory charge-coupled device data acquisition system, Optical Eng. **26** (1987) 795.

88B Blouke, M., et al: Traps and deferred charge in CCDs, in: Instrumentation for ground-based optical astronomy (Robinson, L.B., ed.). Berlin, Heidelberg, New York: Springer-Verlag (1988) p. 462.

88L Leach, R.W.: Design of a CCD controller optimized for mosaics, Publ. Astron. Soc. Pac. **100** (1988) 1287.

88R Robinson, L.B.(ed.): Instrumentation for ground-based optical astronomy: present and future, 9th Santa Cruz Summer Workshop in Astronomy and Astrophysics. Berlin, Heidelberg, New York: Springer-Verlag (1988).

89C Chen, P.C., Novello, J.: A general purpose CCD controller, Publ. Astron. Soc. Pac. **101** (1989) 940.

89J1 Janiseck, J.: Open pinned-Phase CCD Technology, SPIE Conf. Proc. **1159** (1989) 153.

89J2 Janiseck, J., et al.: Charge-coupled device pinning technologies, SPIE Conf. Proc. **1071** (1989) 153.

89M McLean, I.S.: Electronic and Computer-Aided Astronomy. Chichester: Ellis Horwood Limited, John Wiley & Sons (1989).

89R Reiss, R., et al.: Buttable optical CCD Mosaics: Concept and first results at the European Southern Observatory, SPIE Conf. Proc. **1130** (1989) 152.

89Y Yang, F.H., et al.: Trap characteristics in large area CCD image sensors, SPIE Conf. Proc. **1071** (1989) 213.

90C Chandler, C.E., et al: Sub-electron noise charged-coupled devices, SPIE Conf. Proc. **1242** (1990) 238.

90G Geary, J.C., et al.: Development of a 2048 × 2048 imager for scientific applications, SPIE Conf. Proc. **1242** (1990) 38.

90J Janiseck, J., et al.: New advancements in CCD technology – sub-electron noise and 4096 × 4096 pixel CCDs, SPIE Conf. Proc. **1242** (1990) 223.

90L1 Leach, R.W., Beale, F.L.: Design and operation of a multiple readout CCD camera controller, SPIE Conf. Proc. **1235** (1990) 284.
90L2 Lesser, M.: Recent Charge-Coupled Device Optimization, Results at Steward Observatory, SPIE Conf. Proc. **1242** (1990) 164.
90W Waltham, N.R., van Breda, I.G., Newton, G.M.: Simple Transputer based CCD camera controller, SPIE Conf. Proc. **1235** (1990) 328.
91J1 Jacoby, G.H.(ed.): CCDs in Astronomy, Astronomical Society of the Pacific Conference Series **8**. San Francisco: (1991).
91J2 Janiseck, J.R.: Scientific Charge-Coupled Devices, Short Course Notes SC22, SPIE Symposium on Electronic Imaging, San Jose (1991).

1.3.9.4 CCD data reduction

Data reduction has to correct for the above mentioned detector properties in order to produce an astronomically usable frame. In the following we describe each step in the reduction procedure in the order it may conveniently be performed. Main emphasis will be on broad band continuum imaging. A table will list all calibration exposures to be taken at the telescope to enable a successful data reduction (see [89M] and [87B]).

- *Bias subtraction*: information on the bias level (subsect. 1.3.9.2.7) is obtained from dark exposures with zero exposure time (plain CCD read out). Ideally the bias is a constant to be subtracted from the count level of every pixel in the science frame. If the bias frame does show structure averaging and smoothing of several frames may be neccessary in case of low exposure levels before applying bias correction to science frames. With some CCD cameras the bias level may also vary with time. Its value is then best extracted from the overscan area (subsect. 1.3.9.2.7) for each frame individually. Bias level may also vary with binning.
- *Dark exposure*: several dark exposures should be used to derive the amount and distribution of charge accumulated during an exposure without detector illumination (subsect. 1.3.9.2.5). Averaging of dark frames (to eliminate cosmic ray events) is done following the procedure outlined below for derivation of the fringe pattern. The resulting average dark frame should be smoothed or modelled to reduce noise before scaling to and subtracting from the science frame. Smoothing is not possible if the dark frame contains correctable small-scale structure. In this case dark exposure times have to be several times the typical exposure time of the science frames in order to avoid deterioration of the signal-to-noise-ratio in the corrected frame.
- *Corrupted pixels/columns* are identified via a map containing their positions. In case these pixels are irreversably lost, their count levels are replaced by suitable interpolation between neighbouring pixels or transformed and substituted from another frame, for which the telescope has been shifted slightly. Some CCDs (e.g. RCA's) show column offsets, which depend on the count level in the respective pixel according to the empirical law

$$S_{\text{offset}} = S_{\infty}(1 - \exp(-S_{\text{observed}}/S_{\text{critical}})).$$

The constants S_{∞} and S_{critical} are valid for a specific column and have to be derived from a series of flatfields with increasing exposure level. For applications with sufficiently high background levels only knowledge of S_{∞} is required, which can be derived from a single flatfield frame exposed to a level slightly higher than $\max(S_{\infty}, 5 \cdot S_{\text{critical}})$.
- *Flatfield* corrects for sensitivity variations over large scales and from pixel to pixel (subsect. 1.3.9.2.9) as well as vignetting effects (e.g. due to dust on the dewar window) in a multiplicative way. Sensitivity variations could be colour-dependent and thus require separate flatfields for each filter. To correct vignetting effects it is important to have a homogeneous illumination of the entrance aperture identical to that in the science frame. This is best accomplished with exposures of the twilight sky.

– *Sky-background fringe pattern* may be present on thinned back-illuminated or coated detectors (subsect. 1.3.9.2.2). Fringes are due to line emission from the night sky so they are present only in the signal of the background and not of the objects. As such the fringe pattern is highly wavelength-dependent and strictly speaking can be derived only from the science frames themselves. The amplitude of these modulations has to be scaled to each science frame according to the exposure level and then subtracted. Obviously contributions to the background level which do not affect the amplitude of the fringe modulations (e.g. moon light) have to be eliminated from both the modelling and the scaling process.
  The fringe pattern value for each pixel (i,j) is calculated from a suitably selected batch of normalized science frames k according to

$$S(\mathrm{i},\mathrm{j}) = \mathrm{median}\,[S_{\mathrm{k}}(\mathrm{i},\mathrm{j})],$$

  thus eliminating objects and cosmic ray events. This procedure requires that the object(s) of interest must not be placed at the same pixel position in all the frames but the telescope be moved by small amounts between exposures in order to illuminate every pixel with pure night sky in several frames. This is not possible if mainly science frames of extended objects or crowded fields are obtained. In these cases it is mandatory to also expose empty fields to facilitate creation of the fringe pattern.
– *Cosmic rays* or other energetic particles produce a signal in the CCD detector which is either confined to one or a few isolated pixels (thinned chips) or is spread out forming bright spots and tracks (thick chips) (subsect. 1.3.9.2.7). In the case of isolated events and a proper sampling of the astronomical data (i.e. seeing FWHM > 2 pixels) methods employing a search for pixel-to-pixel gradients which are steeper than permitted by the seeing profile succeed to detect the cosmics ray events on a single frame. In all other cases a reliable detection requires at least two exposures with identical instrument setting. Pixels affected by cosmic ray events are treated as described above for corrupted pixels.
– *Conversion to detected number of photons* s_{ij} should be done before archiving and further processing (details see subsect. 1.3.9.2.1).

Table of calibration frames needed for data reduction:

Type	Mode	Illumination	Exposure level/time	Number of exposures	Remarks
bias	dark	none	0 sec	several scattered over run	overscan preferred
dark	dark	none	1000 sec $\approx t_{\mathrm{integ}}$*)	> 3 > 3	smooth dark frame small-scale structure
flat	dome	arbitrary no line emission	$0 \cdots$ saturation	≈ 25	derivation of chip parameters (subsect. 1.3.9.2.9)
flat	dome	arbitrary no line emission	$0 \cdots S_{\infty}$	1 per level	evaluation of column offsets
flat	twilight	homogeneous	$\approx 1/2$ saturation	> 2	for every filter
flat	empty field	homogeneous	$\approx t_{\mathrm{integ}}$	> 4	for every filter in case of extended objects or crowded fields

*) t_{integ} is the typical exposure time of the science frames.

1.3.9.5 Photometry

Compared with the standard, single-aperture photometry using photomultipliers the 2-dimensional images obtained with a CCD have several advantages:

- Position and optimum width of the integration aperture can be chosen *a posteriori*. For objects much smaller than the CCD field the actual sky background can be determined very accurately on the same frame without beam switching. This results in an improved photometric accuracy and a deeper limiting magnitude in the *photometry of faint objects* (background limited application).
- Various methods of image analysis (e.g. image deconvolution) are able to disentangle individual components in *crowded fields*.
- *Surface photometry* of extended objects needs one single exposure and allows a seeing-limited spatial resolution.
- For *variability studies* the object(s) of interest and several reference stars can be recorded simultaneously on a single frame. Such a *relative photometry* is little affected by rapid transmission variations in the atmosphere.

For recent references see [89R], who provide a summary of references to CCDs and their use in stellar photometry (see also [90P, 90J].

With DNs converted to photon counts (subsect. 1.3.9.2.1) one has to follow a similar reduction (and observing) strategy as used with single-aperture photometers [74Y] in order to derive magnitude or flux of an object outside the atmosphere.

1.3.9.5.1 List of symbols

The symbols are arranged in the order of occurrence in subsects. 1.3.9.5.2 and 1.3.9.5.3.

A	area of aperture [pixels]
$\langle b \rangle$	mean background level in aperture A [e^-]
s_{ij}	signal in pixel (i,j) [e^-]
s_{obj}	integrated signal of an object [e^-]
A_{obj}	area around an object [pixels] which contains 95% of the light. For stars and a Gaussian seeing profile $A_{obj} = \pi(\mathrm{FWHM})^2$
RON_{e^-}	read-out-noise [e^-]
$\sigma_{th}(\mathrm{obj})$	rms noise from pure photon statistics and read-out-noise
σ	best guess of the true (rms) error of s_{obj}
$\eta, \eta'(\mathrm{i,j})$	relative accuracy σ_{FF}/FF of the adopted flat field level FF at pixel (i,j). It depends on photon statistics and uncertainties in reproducing the appropriate illumination of the detector in the flatfield exposures. η' refers to the flatfield correction of the sky background (fringe pattern)
$\langle \eta \rangle, \langle \eta' \rangle$	mean flatfield error within aperture A
f_{ij}	fraction of pixel (i,j) lying within A
w_{ij}	value of weighting function at pixel (i,j)
q_w	normalization constant
A_{eff}	$= \sum_{(i,j)}\{w_{ij}\}$, effective aperture of weighting function
w	e-folding width of Gaussian w_{ij}

1.3.9.5.2 Simulated aperture photometry

Using a CCD detector as an integrating photon counter requires to sum up the detected photons in an aperture A. The signal from an object is then given by

$$s_{\text{obj}} = \sum_{(\text{i,j})\in A} \{s_{\text{ij}}\} - A\langle b \rangle. \tag{1}$$

The theoretical (rms) noise of s_{obj} is

$$\sigma_{\text{th}}(\text{obj}) = [\sum_{(\text{i,j})\in A} \{s_{\text{ij}}\} + A(RON_{\text{e}^-})^2]^{\frac{1}{2}}. \tag{2}$$

In practice the true error σ will depend also upon the relative accuracy of the flatfield correction $\eta(i,j)$ and the digitization error introduced by the finite pixel size. For large rectangular apertures $A \gg$ seeing FWHM the true error is estimated as

$$\sigma(\text{obj}) = [\sigma_{\text{th}}^2(1 + \langle\eta\rangle)^2 + (A\langle b \rangle\langle\eta'\rangle)^2]^{\frac{1}{2}}. \tag{3}$$

While the first term reflects errors in the flatfield correction of the object's signal the latter term results from uncertainties in the background level due to the subtraction of the fringe pattern.

If a small (circular) integration aperture $A \leq 30$ pixels is required to avoid confusion and to minimize the background noise, counts have to be derived within the aperture – to be centered on the object to a fraction of the pixelsize (≤ 0.1 FWHM of the objects profile) – according to

$$s_{\text{obj}} = \sum_{(\text{i,j})} \{f_{\text{ij}} s_{\text{ij}}\} - A\langle b \rangle, \tag{4}$$

where f_{ij} is the fraction of pixel ij within A and the summation has to be carried out over all pixels with $f_{\text{ij}} > 0$. The error will be estimated accordingly following equ. (3).

1.3.9.5.3 Weighted summation and profile fitting

For faint (isolated) stellar objects (background-limited case) the best photometric accuracy is obtained by employing a weighted summation in a large aperture,

$$s_{\text{obj}} = q_w \sum_{(\text{i,j})\in A} w_{\text{ij}}(s_{\text{ij}} - \langle b \rangle), \tag{5}$$

where $w_{\text{ij}} \in [0,1]$ is a peaked weighting function centered on the object and $q_w > 1$ is a normalization constant. The optimum width of w_{ij} is about the width of the point-spread-function (PSF). For every single CCD frame q_w should be determined by adjusting the weighted summation result of brighter field stars to the value obtained with a large simulated aperture. The error estimate in this case is then

$$\sigma(\text{obj}) = ([s_{\text{obj}} + q_w A_{\text{eff}}(\langle b \rangle + (RON_{\text{e}^-})^2)][1 + (\langle\eta\rangle)^2] + [q_w A_{\text{eff}}\langle b \rangle\langle\eta'\rangle]^2)^{\frac{1}{2}}, \tag{6}$$

where $A_{\text{eff}} = \sum_{\text{ij}} w_{\text{ij}}$. If one uses the observed PSF as a weighting function, the weighted summation is roughly equivalent to the methode of profile fitting [87S]. Assuming a Gaussian PSF and an identical weighting function $w_{\text{ij}} = \exp(-r_{\text{ij}}^2/w^2)$ one obtains $q_w = 2$ and $A_{\text{eff}} = \pi w^2$. Great care in terms of image alignment and FWHM determination has to be taken if one applies these methods to extended objects [91R].

1.3.9.5.4 Problems in using a CCD as a photometric detector

CCD photometry with respect to photometric standard stars calibrated by conventional photometry has to account for the fact that the spectral sensitivity curves of CCDs (subsect. 1.3.9.2) differ considerably from those of photomultiplier cathodes (subsect. 1.3.2 in LB VI/2a). Bessel [90B] and Beckert and Newberry [89B], e.g., show how the standard response functions can be approximated with CCD detectors and which color equations enable the transformation into standard photometric systems.

Having available an increasing list of spectrophotometric standard stars, e.g. [90M], it is alternatively possible to employ methods of synthetic photometry [86B], which rely only on a good knowledge of the system response function. This method has the additional advantage that the flux is directly measured in physical units ($WHz^{-1}m^{-2}$) which is mandatory if one has to compare optical photometry with flux measurements at other wavelengths. The latter method is also of interest if one wants to measure nonthermal radiation sources.

A second problem is inherent in the use of sensitive CCD detectors: bright standard stars ($m_V \leq 12^{mag}$) have to be exposed for only a few seconds to avoid non-linearity effects or saturation. Thus shutter errors may limit the photometric accuracy. Two types of shutter errors are commonly encountered:

- position-dependent exposure time due to the finite opening/closing time of an iris shutter (in front of the CCD camera)
- timing errors which are caused by variations in the actual shutter performance.

It is recommended to test the shutter function before photometric CCD observations.

1.3.9.5.5 Software packages available for CCD data reduction

Several software packages are available to facilitate data reduction as outlined above, IRAF [86T] and MIDAS [88B] being widely distributed systems. For extracting photometric information we also refer to these systems and especially to packages like e.g. DAOPHOT [87S] or ROMAFOT [83B, 87B] implemented there.

References for 1.3.9.4 and 1.3.9.5

74Y Young, A.T., in: Methods of Experimental Physics – Astrophysics Part A, Optical and Infrared (Carleton, N. ed.). New York, London: Academic Press (1974) p. 123.

83B Buonanno, R., Buscema, G., Corsi, C.E., Ferraro, I., Iannicola, G.: Astron. Astrophys. **126** (1983) 278.

86B Buser, R., in: Synthetic Photometry in Highlights of Astronomy **7** (J.-P. Swings, ed.). Dordrecht, Boston, Lancaster, Tokyo: D. Reidel Publishing Company (1986) p. 799.

86T Tody, D., in: Instrumentation in Astronomy VI (D.L. Crawford, ed.) SPIE Conf. Proc. **627** (1986) 733.

87B Baluteau, J.-P., D'Odorico, S., ed.: The Optimization of the Use of CCD Detectors in Astronomy, ESO Conference and Workshop Proc. **25**. European Southern Observatory, D8046 Garching (1987).

87S Stetson, P.B.: Publ.Astron.Soc.Pac. **99** (1987) 191.

88B Banse, K., Ponz, D., Ounnas, Ch., Grosbol, P., Warmels, R., in: Instrumentation for Ground-Based Optical Astronomy (L.B. Robinson, ed.). Berlin, Heidelberg, New York: Springer-Verlag (1988) p. 431.

89B Beckert, D.,C., Newberry, M.V.: Publ.Astron.Soc.Pac. **101** (1989) 849.

89M McLean, I.S.: Electronic and Computer-Aided Astronomy. Chichester: Ellis Horwood Limited, John Wiley & Sons (1989).

89R Rufener, F.: Stellar Photometry with Modern Array Detectors in Highlights of Astronomy **8** (D. Mc-Nally, ed.). Dordrecht, Boston, London: Kluwer Academic Publishers (1989) p. 615.

90B Bessel, M.S.: Publ.Astron.Soc.Pac. **102** (1990) 1181.

90J G.H. Jacoby, ed.: CCDs in Astronomy, Astronomical Society of the Pacifif Conference Series **8** (1990), San Fransisco.
90M Massey, P., Gronwall, C.: Astrophys.J. **358** (1990) 344.
90P Philip, A.G.D., Hayes, D.S., Adelman, S.J., eds.: CCDs in Astronomy II. Schenectady, N.Y.: L. Davis Press (1990).
91R Röser, H.-J., Meisenheimer, K.: Astron.Astrophys. **252** (1991) 458.

1.4 see LB VI/2a

1.5 Spectrometers and spectrographs

1.5.1 - 1.5.5 see LB VI/2a

1.5.6 New and future developments

During the last decade, the overall efficiency and versatility of grating spectrographs have been considerably improved [86D, 88D, 88O] through the realisation of new spectrograph design concepts, optimization of the spectrograph components, utilization of modern CCD detectors (see 1.3.9), and application of optical fibers. CCD detector arrays provide high quantum detective efficiency, linear and reproducible response, and excellent geometric stability. The pixel size and the two-dimensional format of the detector arrays largely govern the layout of modern spectrographs [85W, 86S]. Fibers are used as an optical link between the spectrograph and the telescope [88B].

The following new design concepts have been adopted for present and projected future grating spectrographs:

Fiber-optic spectrograph
The fiber-optic spectrograph is operated at a remote location off the moving telescope. The light from a stellar object in the focal plane of the telescope is picked up by an optical fiber attached to the entrance aperture of the spectrograph [83L, 88G].

Multi-object spectrograph
The multi-object spectrograph permits observations of more than one object in a single exposure. The entrance slit is replaced by (1) an aperture plate with punched holes matching the proper positions of the objects to be observed [82B, 83G, 91D], (2) movable slitlets centered on the objects [88D], or (3) the exit ends of a bundle of fibers which have their movable entrance ends located in the focal plane of the telescope [80H, 83L, 86H]. The latter concepts provides access to virtually all objects in field of view of the telescope [90H]. Rather sophisticated positioning devices are required for remote control of the fiber ends or slitlets [88P, 90P].

Dual-beam spectrograph
The dual-beam spectrograph ("double spectrograph") permits simultaneous observations of the same object in two separate spectral regions (e.g. "blue" and "red" ranges) [79A, 82O, 86D]. The instrument consists of two separate spectrograph channels (including the detectors) fed by a dichroic beam divider below the entrance slit. Each channel has been optimized for its own spectral domain.

Multi-order spectrograph

The multi-order spectrograph concept largely extends the spectral range covered in a single exposure. High-spectral-order gratings (echelle oder echellette gratings) are used in combination with cross-dispersers (gratings, prisms) to obtain a two-dimensional "echellogram" which ideally matches the format of modern array detectors [76C, 86W, 90W]. The wavelength coverage depends on the actual spectral resolution and the number of available detector elements. Echelle spectrographs provide high spectral resolution at moderate wavelength coverage [76C]. Echellette spectrographs provide full coverage of wavelength from 0.3 to 1.1 μm at moderate resolution [90S].

Focal reducer – low/medium-resolution spectrograph

The focal reducer – low/medium-resolution spectrograph is a multi-mode instrument permitting both direct imaging and low-to-medium-resolution spectroscopy of faint sources [83E, 86D, 90D]. The instrument utilizes all-refractive optics and straight-line dispersing elements (e.g. a transmission grating mounted on a prism: "grism"). The design allows quick change-over from imaging to spectroscopic mode and can incorporate both dual-beam and multi-object facilities [86D, 91D].

References for 1.5.6

76C Chaffee, F.H., Schroeder, D.J.: Annu. Rev. Astron. Astrophys. **14** (1976) 23.

79A Angel, J.R.P., Hilliard, R.L., Weymann, R.J.: Smithson. Astron. Obs. Spec. Rep. No. **385** (1979) 87.

80H Hill, J.M., Angel, J.R.P., Scott, J.S., Lindley, D., Hintzen, P.: Astrophys. J. **242** (1980) L69.

82B Butcher, H.: SPIE Conf. Proc. **331** (1982) 296.

82O Oke, J.B., Gunn, J.E.: Publ. Astron. Soc. Pacific **94** (1982) 586.

83E Enard, D., Delabre, B.: SPIE Conf. Proc. **445** (1983) 522.

83G Gray, P.M.: SPIE Conf. Proc. **445** (1983) 57.

83L Lund, G., Enard, D.: SPIE Conf. Proc. **445** (1983) 65.

85W Walker, D.D., Diego, F.: Mon. Not. R. Astron. Soc. **217** (1985) 355.

86D Dekker, H., Delabre, B., D'Odorico, S.: SPIE Conf. Proc. **627** (1986) 339.

86H Hill, J.M., Lesser, M.P.: SPIE Conf. Proc. **627** (1986) 303.

86S Solf, J.: ESO Conf. & Workshop Proc. **24** (1986) 107.

86W Walker, D.D., Diego, F., Charalambous, A., Hirst, C.J., Fish, A.C.: SPIE Conf. Proc. **627** (1986) 291.

88B Barden, S.C. (ed.): Fiber Optics in Astronomy, Astron. Soc. Pacific Conf. Ser. **3** (1988).

88D D'Odorico, S., Dekker, H., Delabre, B.: ESO Conf. & Workshop Proc. **30** (1988) 1003.

88G Guérin, J. Felenbok, P.: ESO Conf. & Workshop Proc. **30** (1988) 1131.

88O Oke, J.B.: ESO Conf. & Workshop Proc. **30** (1988) 1037.

88P Pitz, E., in: [88B], p. 163.

90D D'Odorico, S.: The Messenger (ESO) **61** (1990) 51.

90H Hamilton, D.: SPIE Conf. Proc. **1235** (1990) 673.

90P Parry, I.R., Lewis, I.J.: SPIE Conf. Proc. **1235** (1990) 681.

90S Solf, J., Eislöffel, J.: Astron. Astrophys. **234** (1990) 583.

90W Walker, D.D., Bingham, R.G., Diego, F.: SPIE Conf. Proc. **1235** (1990) 535.

91D Dekker, H. et al.: The Messenger (ESO) **63** (1991) 73.

1.6 Optical high-resolution methods

1.6.1 – 1.6.4 see LB VI/2a

1.6.5 New developments

During the last few years many new high-resolution methods have been developed and many astronomical results have been obtained. The new methods for speckle imaging, speckle spectroscopy and optical long-baseline interferometry are discussed in the next sections. In spite of atmospheric image degradation, speckle imaging and speckle spectroscopy yield diffraction-limited resolution, for example, 0.02 arcsec for a 5-m telescope and wavelength 400 nm. The resolution of optical long-baseline interferometry with 100 m baseline is about 1 milli-arcsec.

1.6.5.1 Reconstruction of diffraction-limited images

Speckle interferometry [70L] yields the diffraction-limited autocorrelation of astronomical objects. True images are not obtained since speckle interferometry cannot measure the phase of the object Fourier transform. This problem has stimulated the development of the following methods which can reconstruct diffraction-limited images since they measure both the amplitude and the phase of the object Fourier transform:

(1) Knox-Thompson method [74K],
(2) shift-and-add method [80B],
(3) non-redundant mask method [58J, 86B, 88R1], and
(4) speckle masking method (bispectrum or triple correlation method) [77W, 83L, 83W].

The first image processing step in the Knox-Thompson method is the calculation of the ensemble average cross-spectrum of many speckle interferograms. From the average cross-spectrum the object cross-spectrum and an image of the object can be reconstructed.

The shift-and-add analysis is mainly used for point-source-dominated objects and at IR wavelengths. In the shift-and-add method each speckle interferogram is shifted in such a way that the brightest speckle is moved to the center of the coordinate system. Then the shifted speckle interferograms are simply averaged.

In the non-redundant mask method, a mask with several small holes is inserted into the telescope pupil to obtain Michelson interferograms. The position of the holes is chosen such that each baseline appears only once. From the Michelson interferograms images can be reconstructed by the Radio phase closure method and related methods.

The first image processing step in speckle masking is the calculation of the ensemble average bispectrum or triple correlation of all speckle interferograms. After the compensation of the speckle masking transfer function the bispectrum of the object is obtained. From the object bispectrum a diffraction-limited image can be reconstructed by a recursive method or by various least-squares methods (see [90H] and several papers in [92B1]).

The non-redundant mask technique and speckle masking have the advantage that they can measure closure phases which are very important for optical long-baseline interferometry. Various modifications of the above methods and references for other methods can be found in the proceedings of the ESO conference on High Resolution Imaging by Interferometry II [92B1] and in review papers [88R2, 91W]. Many different types of objects have already been observed by high-resolution techniques, for example, solar surface structures, asteroids, moons of planets, surface structures on stars, star clusters, and Seyfert galaxies (see [92B1]).

1.6.5.2 Speckle spectroscopy

Speckle spectroscopy methods [86W, 87R, 92G, 92W] can yield simultaneously diffraction-limited angular resolution and spectral information. For example, objective prism speckle spectroscopy [86W] reconstructs diffraction-limited objective prism spectra, or projection speckle spectroscopy [92G] (and references herein) can reconstruct spectrally dispersed one-dimensional projections of the object.

1.6.5.3 Optical long-baseline interferometry

The great advantage of optical long-baseline interferometry with three or more telescopes is that images with 1 milli-arcsec resolution can be reconstructed with a baseline of 100 m at a wavelength of 500 nm. Possible image reconstruction methods are the phase closure or non-redundant mask method [58J, 86B, 88R1] and the speckle masking method (see [92R] and references herein).

Several optical interferometers have already been built (after the Michelson and the intensity interferometer; see 1.6.1 and 1.6.2 in LB VI/2a p. 33), for example, I2T at CERGA [76L], GI2T at CERGA [86L], Multi-Mirror Telescope (MMT) [88B], Mark III Interferometer at Mt. Wilson [88S], Berkeley IR Heterodyne Interferometer [77S], Sydney University Stellar Interferometer (SUSI) [92D], COAST [92B2], and the Infrared Michelson Array (IRMA) [91B]. Many impressive astronomical results have already been obtained, for example orbits of very close binaries and stellar diameters [92B1].

References for 1.6.5

58J Jennison, R.C.: Mon. Not. R. Astron. Soc. **118** (1958) 276.

70L Labeyrie, A.: Astron. Astrophys. **6** (1970) 85.

74K Knox, K.T., Thompson, B.J.: Astrophys. J. **193** (1974) L45.

76L Labeyrie, A., High-Resolution Techniques in Optical Astronomy, in: Progress in Optics (Wolf, E., ed.), Vol. 24, Elsevier Science Publishers B.V. (1976) Ch. 2.

77S Sutton, E.C., Storey, J.W.V., Betz, A.L., Townes, C.H., Spears, D.L.: Astrophys. J. **217** (1977) L97.

77W Weigelt, G.: Opt. Commun. **21** (1977) 55.

80B Bates, R.H.T., Cady, F.M.: Opt. Commun. **32** (1980) 356.

83L Lohmann, A.W., Weigelt, G., Wirnitzer, B.: Appl. Opt. **22** (1983) 4028.

83W Weigelt, G., Wirnitzer, B.: Opt. Lett. **8** (1983) 389.

86B Baldwin, J.E., Haniff, C.A., Mackay, C.D., Warner, P.J.: Nature **320** (1986) 595.

86L Labeyrie, A., Schumacher, G., Dugue, M., Thom, C., Bourlon, P., Foy, F., Bonneau, D., Foy, R.: Astron. Astrophys. **162** (1986) 359.

86W Weigelt, G., Baier, G., Ebersberger, J., Fleischmann, F., Hofmann, K.-H., Ladebeck, R.: Opt. Eng. **25** (1986) 706.

87R Ridgway, S.T., Mariotti, J.-M.: A Method for Multispectral Infrared Interferometry, in: Proc. of the ESO/NOAO Conf. on Interferometric Imaging in Astronomy (Goad, J.W., ed.), ESO/NOAO (1987), p. 93.

88B Beckers, J.M.: Interferometry with the MMT and NNTT, in: Proc. Conf. High-resolution Imaging by Interferometry (Merkle, F., ed), Garching: ESO (1988), p. 879.

88R1 Readhead, A.C.S., Nakajima, T.S., Pearson, T.J., Neugebauer, G., Oke, J.B., Sargent, W.L.W.: Astron. J. **95** (1988) 1278.

88R2 Roddier, F.: Phys. Rep. **170** (1988) 97.

88S Shao, M., Colavita, M.M., Hines, B.E., Staelin, D.H., Hutter, D.J., Johnston, K.J., Mozurkewich, D., Simon, R.S., Hershey, J.L., Hughes, J.A., Kaplan, G.H.: Astron. Astrophys. **193** (1988) 357.

90H Hofmann, K.-H., Weigelt, G.: Proc. Soc. Photo-Opt. Instrum. Eng. **1351** (1990) 522.

91B Benson, J.A., Dyck, H.M., Ridgway, S.T., Dixon, D.J., Mason, W.L., Howell, R.R.: Astron. J. **102** (1991) 2091.

91W Weigelt, G.: Triple-Correlation Imaging in Optical Astronomy, in: Progress in Optics (Wolf, E., ed.), Vol. 29, Elsevier Science Publishers B.V. (1991) Ch. 4.

92B1 ESO Conf. Proc. High Resolution Imaging by Interferometry II (Beckers, J., Merkle, F., eds.), Garching: ESO (1992)

92B2 Baldwin, J.: COAST: The Current Status, in: ESO Conf. Proc. High Resolution Imaging by Interferometry II (Beckers, J., Merkle, F., eds.), Garching: ESO (1992) p. 747.

92D Davis, J., Tango, W.J., Booth, A.J., Minard, R.A., ten Brummelar, T., Shobbrook, R.R.: An Update on SUSI, in: Conf. Proc. High Resolution Imaging by Interferometry II (Beckers, J., Merkle, F., eds.), Garching: ESO (1992) p. 741.

92G Grieger, F., Weigelt, G.: Objective Prism Speckle Spektroscopy and Wideband Projection Speckle Spectroscopy, in: ESO Conf. Proc. High Resolution Imaging by Interferometry II (Beckers, J., Merkle, F., eds.), Garching: ESO (1992) p. 481.

92R Reinheimer, T., Hofmann, K.-H., Weigelt, G.: Computer Simulations of Interferometric Imaging with the VLT Interferometer, in: ESO Conf. Proc. High Resolution Imaging by Interferometry II (Beckers, J., Merkle, F., eds.), Garching: ESO (1992) p. 827.

92W Weigelt, G., Grieger, F., Hofmann, K.-H., Pausenberger, R.: Slit Speckle Spektroscopy, in: ESO Conf. Proc. High Resolution Imaging by Interferometry II (Beckers, J., Merkle, F., eds.), Garching: ESO (1992) p. 471.

1.7 X-ray and γ-ray instruments

1.7.1 and 1.7.2 see LB VI/2a

The principles of the instruments have not changed since the last edition. Therefore only the list of X-and γ-ray satellites in 1.7.3 is updated.

1.7.3 X- and γ-ray satellites

Table 2. X- and γ-ray satellites (continuation of Table 2 in LB VI/2a, p.40).

Code for instrumentation:

1 = gas proportional counter	10 = polarimeter
2 = scintillation counter	11 = spark chamber
3 = solid state detector	12 = Cerenkov counter
4 = collecting mirror	13 = gas scintillation proportional counter
5 = dispersive spectrometer	14 = filter spectroscopy
6 = modulation collimator	15 = all sky monitor
7 = pin hole camera	16 = coded mask telescope
8 = imaging telescope	17 = double Compton telescope
9 = focal plane image detector	

Table 2. Continuation of Table 2 in LB VI/2a, p. 40.

Name	Mission	Major scientific objectives	Instrumentation	Energy range [keV]	Ref.
HAKUCHO	1979-83	scanning and pointed observations of the X-ray sky	1,2,6	$0.1 \cdots 100$	81H,81K
SMM (Solar Maximum Mission)	1980-89	solar X- and γ-ray observations	1,2,5,16	$20 \cdots 9000$ ($10 \cdots 100$ MeV)	80A,80B,80F,80O, 80V
HINOTORI	1981	solar X- and γ-ray studies, γ-ray bursts	2	$20 \cdots 6000$	83Y
TENMA	1983-84	medium resolution X-ray spectroscopy	4,13,15	$0.1 \cdots 60$	83F
EXOSAT	1983-1986	low and medium energy X-ray observations	1,5,8,9,13,14	$0.05 \cdots 20$	81D,81P,81T,88W
Spacelab 2	1985 (shuttle)	medium energy X-ray imaging observations	1,16	$2.5 \cdots 25$	84W,88S
MIR-KVANT	1987-	medium to high X-ray energy pointed observations	1,2,13,16	$2 \cdots 2000$	83B,83R,83S,86S
GINGA	1987-1991	medium energy, high sensitivity pointed X-ray observations, γ-ray bursts.	1,2,15	$1.5 \cdots 400$	87M
GRANAT	1989-	X-ray to low energy γ-ray imaging observations, γ-ray bursts	1,2,16	$3 \cdots 1500$ ($0.1 \cdots 100$ MeV)	84L,84M
GRO (Gamma Ray Observatory)	1991-	low to high energy γ-ray observations, γ-ray bursts	2,11,17	$50 \cdots 30 \cdot 10^6$	83K,88D,89F,89J, 89K1,89K2,91S
ROSAT (Röntgen Satellite)	1990-	first all sky survey with imaging X-ray and XUV telescopes, pointed observations	1,8,9,14	$0.01 \cdots 2.4$	83T,84T
ASTRO	1990 (shuttle)	high resolution X-ray spectroscopy	3,8	$0.5 \cdots 10$	90B,91M

References for 1.7.3

80A Acton, L.W., et al.: Solar Physics **65** (1980) 53.
80B Bohlin, J.D., Frost, K.J., Burr, P.T., Gruha, A.K., Withbroe, G.L.: Solar Physics **65** (1980) 5.
80F Forrest, D.J., et al.: Solar Physics **65** (1980) 15.
80O Orwig, L.E., Frost, K.J., Dennis, B.R.: Solar Physics **65** (1980) 25.
80V Van Beek, H.F., Hoyng, P., Lafleur, B., Simnett, G.M.: Solar Physics **65** (1980) 39.
81D De Korte, P.A.J., et al.: Space Sci. Rev. **30** (1981) 495.
81H Hayakawa, S.: Space Sci. Rev. **29** (1981) 221.
81K Kondo, I., et al.: Space Sci. Instrum. **5** (1981) 211.
81P Peacock, A., et al.: Space Sci. Rev. **30** (1981) 525.
81T Turner, M.J.L., Smith, A., Zimmermann, H.U.: Space Sci. Rev. **30** (1981) 513.
83B Brinkman, A.C., Dam, J., Mels, W.A., Skinner, G.K., Willmore, W.P., in: Proc. Rome Workshop on Non-thermal and Very High Temperature Phenomena in X-Ray Astronomy (eds. Perola, Salvator), 1983, p.263.
83F Fujii, M., et al.: ISAS preprint No. **215**, 1983.
83K Kurfess, J.D., et al.: Adv. Space Res. **3** (1983) 109.
83R Reppin, C., Pietsch, W., Trümper, J., Kendziorra, E., Staubert, R., in: Proc. Rome Workshop on Non-thermal and Very High Temperature Phenomena in X-Ray Astronomy (eds. Perola, Salvator), 1983, p.279.
83S Smith, A., in: Proc. Rome Workshop on Non-thermal and Very High Temperature Phenomena in X-Ray Astronomy (eds. Perola, Salvator), 1983, p.271.
83T Trümper, J.: Adv. Space Res. **2** (1983) 241.
83Y Yoshimori, M., Okudaira, K., Hirasima, Y.: Nucl. Instrum. Methods **215** (1983) 255.
84L Lund, N., in: Proc. 2nd Int. Symp. on Optical and Electrooptical Applied Science and Engineering, Cannes, Dec. 1984, p.55.
84M Mandrou, P.: Adv. Space Res. **3** (1984) 525.
84T Trümper, J.: Phys. Scr. **T7** (1984) 209.
84W Willmore, A.P., Skinner, G.K., Eyles, C.J., Ramsey, B.: Nucl. Instrum. Methods **221** (1984) 284.
86S Sagdeev, R.Z., Academy of Sciences of the USSR, Space Research Institute Preprint **1171** (1986).
87M Makino, F.: Astron. Lett. and Commun. **25** (1987) 223.
88D Diehl, R.: Space Sci. Rev. **49** (1988) 85.
88S Skinner, G.K., et al.: Astron. Lett. and Commun. **27** (1988) 199.
88W White, N.E., Peacock, A., in: X-Ray Astronomy with EXOSAT (eds. R. Pallavicini, N.E. White), Mem. Soc. Astron. Ital. **59** (1988) 7.
89F Fishman, G.J., in: Proc. Gamma Ray Observatory Science Workshop, GSFC, Greenbelt, Maryland, USA, April 1989, p.2-39.
89J Johnson, W.N., in: Proc. Gamma Ray Observatory Science Workshop, GSFC, Greenbelt, Maryland, USA, April 1989, p.2-22.
89K1 Kniffen, D.A., in: Proc. Gamma Ray Observatory Science Workshop, GSFC, Greenbelt, Maryland, USA, April 1989, p.1-1.
89K2 Kanbach, G., et al., in: Proc. Gamma Ray Observatory Science Workshop, GSFC, Greenbelt, Maryland, USA, April 1989, p.2-1.
90B Blair, W.P., Gull, T. R.: Sky and Telescope **79** (1990) 591.
91M Maran, S.P.: Sky and Telescope **81** (1991) 591.
91S Schönfelder, V.: Adv. Space Res. **11** (1991) 313.

1.8 Infrared technique

The infrared (IR) region can be divided into the near IR (NIR: 0.8 ⋯ 3 μm), the middle IR (MIR: 3 ⋯ 30 μm) and the far IR (FIR: 30 ⋯ 1000 μm). Occasionally 5 μm is used as a boundary between the NIR and MIR. In ground-based astronomy, the range $\lambda > 3$ μm is called thermal IR; in radio astronomy the range $\lambda < 1000$ μm is called the submillimeter region. A comprehensive introduction to NIR astronomical techniques is given in [93B3].

1.8.1 Infrared detectors

(See also sect. 1.3)

Beyond the usual desires (high quantum efficiency, low dark current, etc.) the requirements on IR astronomy detectors can be quite different: for $\lambda < 5$ μm large arrays with low readout noise (< 10e) of the associated electronics are desired. For ground-based MIR observations high well capacity and fast readout are required in order to deal with the large background flux. In space IR observations the background is 10^6 times smaller and the emphasis is on the FIR range where detectors with low drift and memory effects and long on-chip integration times are required [92L1]. There is no "universal" detector for these diverse applications.

Several optical detectors are also sensitive in the near IR. Photoelectric and thermal (bolometric) detectors are used in the middle and far IR. Two-dimensional monolithic arrays with hybridized readout electronics are available for the whole middle IR [87W2, 89W]. In the far IR, arrays are often bundled single detectors. All IR detectors require cooling. For coherent detectors see also sect. 1.9.

1.8.1.1 Detector types

Photographic plates and photocathodes: see LB VI/2a, p. 42.

Silicon charge-coupled devices (CCDs)
Large CCDs (≈ 2048 × 2048 pixels) are the most common near IR-detectors. This intrinsic silicon photodetector is sensitive up to 1.2 μm with a quantum efficiency of 15% at 1 μm [92B2]. Operational temperature is −90°C. For CCDs see sect. 1.3.

Photodetectors
Examples of modern IR detectors are given in Table 1, an updated version of Table 1 in LB VI/2a, p. 43.

Radiation-hardened detectors
Bulk material detectors react by spikes, dark current and responsivity increases when exposed to ionizing radiation as encountered in space observatories (passage through earth's radiation belts) [93L2, 92P]. Blocked impurity band (BIB) detectors have a highly doped but very thin IR-sensitive layer. They are less vulnerable to ionizing particle hits, have a fast and linear response and show an increased long-wavelength response as compared with the bulk material. Si:As- and Si:Sb-BIB detectors and arrays can be used to $\lambda < 30$ or 40 μm respectively [89W, 92H3]. Ge:Ga-BIBs are under development for $\lambda < 160$ μm [88W].

Thermal detectors: see LB VI/2a, p. 43.

Arrays
Monolithic arrays of 256 × 256 pixels are available for the 0.8 ⋯ 5.5 μm range. The detector chip is usually bonded by indium balls to a Si readout chip. The selection of a suitable IR array is influenced by

its pixel size, wavelength coverage, cooling requirement, flat fielding requirement, quantum efficiency, readout noise and background flux; guidance for the popular HgCdTe, PtSi, InSb array is given in [92F2]. Si:Ga arrays with 58 × 62 pixels cover the range $5 \cdots 18$ μm [92G1]. Si:As arrays with 20 × 64 pixels operate in the $2 \cdots 26$ μm range [93H]. Larger arrays (128 × 128 Rockwell, 256 × 256 Hughes) have already been developed but are not yet available for general export. Far IR arrays of Ge:Ga ($\lambda < 110$ μm) and stressed Ge:Ga ($\lambda < 240$ μm) are reviewed by [92W2]. Bolometer arrays of up to 40 pixels were used for airborne spectrometers [92K2].

Table 1. Detector materials. PV: photovoltaic, PC: photoconductive.

Material	Type	Wavelength range [μm]	Operating temperature [K]
InSb	PV	0.8 ··· 5.5	35
HgCdTe	PV,PC	1 ··· 14	60
PtSi	Schottky barrier	1 ··· 5.6	40
Si:Ga	PC	2 ··· 17	4
Si:As	PC	2 ··· 25	4
Si:Sb	PC	2 ··· 29	4
Ge:Be	PC	25 ··· 55	3
Ge:Ga	PC	50 ··· 120	2.5
Stressed Ge:Ga	PC	100 ··· 240	< 1.8
Ge:Bolometer	thermal	1 ··· 1000	0.1 ··· 2

1.8.1.2 Detector parameters

Detector parameters characterize the detector's practical properties, operational conditions and figures of merit. The measurement of these parameters requires great care in order to be comparable between different laboratories [89V]. Important parameters are:

Spectral range

Photodetectors have a limited sensitivity range with a steep long-wavelength cutoff $\lambda_c[\mu m] = 1.24/E_{min}$ [eV] corresponding to a minimal photon energy E_{min} required to excite electrical carriers. While the quantum efficiency vs. wavelength curve is usually flat, the commonly used responsivity curve (e.g. photocurrent per incident IR flux [A/W]) increases with λ and peaks at λ_c.

Sensitivity: see Lb VI/2a, p. 43.

The noise equivalent power *(NEP)* is a measure of detector sensitivity, quoting the minimum incident flux giving a signal equal to the rms noise in a 1 Hz electrical bandwidth and 1 s integration time. The *NEP* is expressed in units of [W Hz$^{-1/2}$]. Typical *NEP* of a detector operated in a photometric/imaging mode on a ground-based telescope is $\approx 10^{-16 \cdots -14}$ WHz$^{-1/2}$ (increasing with λ), at a cooled space telescope > 10^{-18} WHz$^{-1/2}$.

Noise

IR detectors and their preamplifiers or readout electronics suffer from different noise sources; (i) Johnson noise, (ii) $1/f$ noise, (iii) generation-recombination noise, (iv) phonon noise (bolometer), (v) electronic readout noise [84D]. In practice, the (vi) photon background noise, due to the statistical arrival of background photons, or (vii) the "sky noise" at ground or airborne telescopes, or readout noise in speckle application may limit a detection system [74L, 77B, 89V]. The dark current of certain detectors can practically be cancelled by sufficiently low operational temperatures. Because no detector can suffer from less noise than that generated by incident photon flux Q, a detector under these conditions is called a background-limited photon detector (BLIP). The BLIP-*NEP* (PC) $= hc/\lambda\sqrt{4Q/\eta}B$ (η-quantum efficiency; B-Bose factor, which becomes > 1 for long λ and high T) is the optimum a detector can achieve [89V].

Time Constant

This is a measure of the response speed of the detector/readout electronics. If too large, fast readout as required by several applications (speckle) may be restricted and the chopping frequency may also be limited, thus preventing the elimination of low-frequency noise components ($1/f$-noise, sky noise). Usually, thermal detectors have larger time constants ($> 10^{-2}$s) than photodetectors. When operated under low-background conditions several detectors respond very slowly (minutes) or show memory effects [92F1, 92W1, 93L2].

Operating temperature and cooling

All IR detectors used in astronomy require cooling in order to prevent the thermal excitation of carriers. Usually, the better the sensitivity and the longer the wavelength, the lower the required operational temperature. Open cooling systems offer large cooling power and temperature stability, but usually need replenishment every few days: (i) dry ice, 193 K; (ii) liquid nitrogen, 77 K; (iii) pumped nitrogen, > 45 K; (vi) liquid helium-4, 4.2 K, (v) pumped helium-4, 1.5 K. For $T \geq 20$ K, closed-cycle cooling machines become common at ground [78D, 90E] and are being developed for space applications [92W3]. Temperatures in the 0.1-K-range, as required by sensitive bolometers, are available with helium-3 cryostats or by adiabatic demagnetization [92H2].

1.8.2 See LB VI/2a

1.8.3 Infrared telescopes

Ground-based optical mirror telescopes can be used for IR observations, provided the precipitable water vapour column of the atmosphere above is small enough, and some measures have been taken to reduce the telescope's thermal emission. An IR instrument at a telescope normally includes a chopper, a cooled focal plane and the data acquisition and control system (Fig. 1).

1.8.3.1 Chopper

See LB VI/2a, p. 45

Sky subtraction for NIR arrays is usually by slow nodding of the telescope, MIR arrays use fast chopping [93B3].

1.8.3.2 Optics of the IR telescope

See LB VI/2a, p. 45

An existing ground-based Cassegrain telescope is normally modified for use in the IR, as described in [74L]. Because these requirements often compromise the use of the telescopes in the optical range, large dedicated IR telescopes have been built (see subsect. 1.8.3.3).

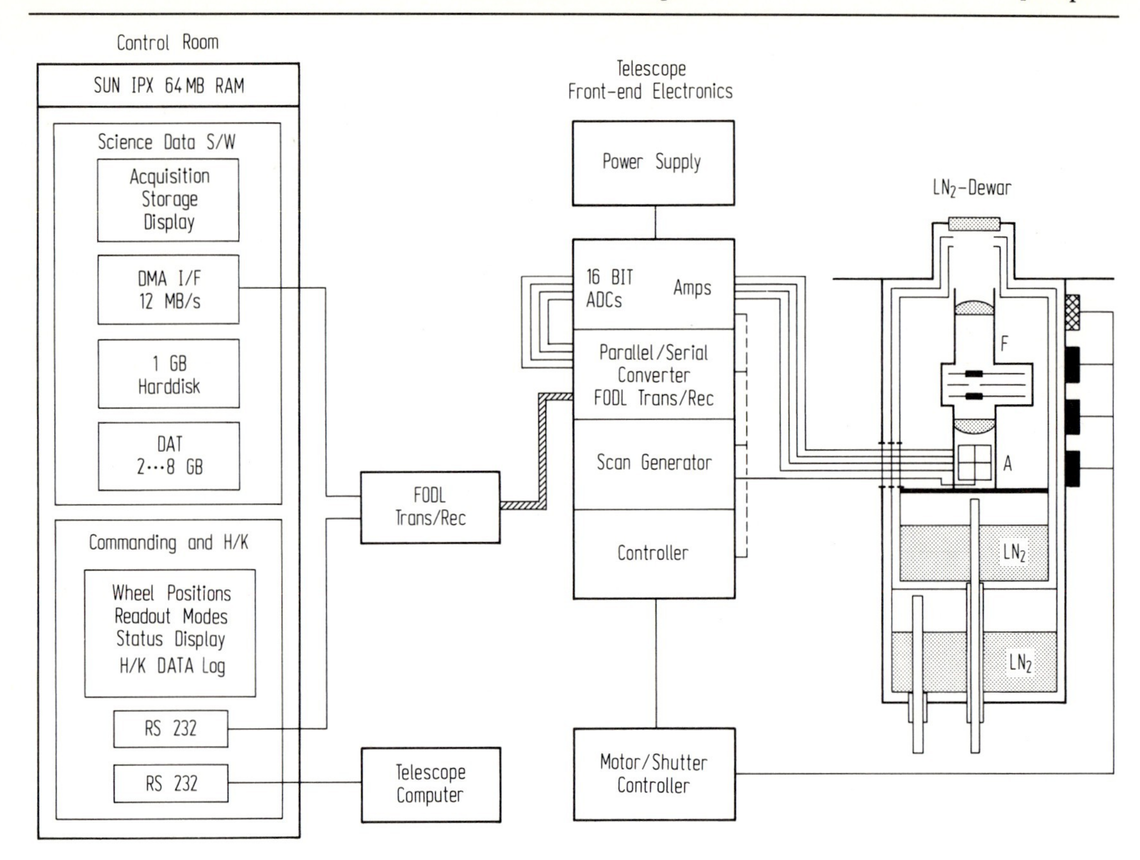

Fig. 1. Block diagram of a NIR camera as described in [93B2]. The four-quadrant HgCdTe-array (A) is driven and read out by a front-end electronics which also controls filter and aperture wheels (F) in the dewar. The data are transmitted via a fibre optics data link (FODL) to the control room, where a host computer system (SUN) displays the images and allows the astronomer to command the observations.

1.8.3.3 Telescope platforms

In order to access the whole IR region, ground-based telescopes have to be complemented by instruments operating in the upper atmosphere or in space.

Ground-based

High mountains located in arid climates or on islands offer high atmospheric transmission because the precipitable water column above is often $\approx 1 \cdots 3$ mm [83C, 83G]. Many major optical telescopes are routinely used in the IR after temporary modifications, often including a particular IR frontring (e.g. the ESO 3.6-m-telescope and the MPIA 2.2-m-telescope at La Silla). Large dedicated telescopes for almost exclusive use in the IR include the 3.8 m UKIRT and the 3.2 m NASA telescopes on Mauna Kea, and the Multi-Mirror Telescope (MMT) on Mt. Hopkins. The four Very-Large-Telescope (VLT) dishes of ESO to be located at Paranal site will be optimized for the IR and equipped with 1.1 m chopping secondaries [91K1]. The VLT capabilities in the 10 and 20μm atmospheric windows are assessed in [93K].

The 30-m-IRAM antenna is routinely used in the far IR [89Q] and the same is planned with the 10-m-dish on Mt. Graham [90B1]. Because of its cold and dry climate, the Antarctica offers good IR potentials with precipitable water vapor of $100 \cdots 500$ μm during Austral winter [89P, 92H1].

Aircraft
Routine observations up to 6 hours at an altitude of $\approx$ 12 km are made with the 90-cm-telescope on board the NASA G.P. Kuiper Airborne Observatory (KAO) [91K3]. A Stratospheric Observatory for the Infrared (SOFIA) with a 2.5-m-telescope on board a B747SP aircraft is in preparation [92E2].

Balloon-borne: see LB VI/2a, p. 45.

Rockets: see LB VI/2a, p. 45.
Measurements of the cosmic background radiation [90M] were obtained at $\approx$ 150 km altitude within a $\approx$ 5 min observing time.

Satellites
The Infrared Astronomical Satellite IRAS surveyed 95% of the sky with its liquid-helium-cooled 60-cm-telescope in four wavelength bands at 12, 25, 60 and 100 μm during a 300-day-mission [84N]. The Cosmic Background Explorer (COBE), another helium-cooled satellite, mapped the sky with a 0.7° field-of-view between 2 and 300 μm, and also in the mm range [92B3]. Under construction is the ISO satellite, a versatile IR Space Observatory equipped with four focal-plane-experiments described in [92E1]. In order to reduce the thermal emission of the satellite telescope, the mirrors are gold-coated [65H]. Future IR space observatories will include larger but radiatively cooled telescopes for long-lifetime missions [93T].

1.8.3.4 Adaptive optics

This technique reduces the image distortion caused by atmospheric turbulence. It includes sensing the wave front errors and their real-time compensation by deformable mirrors [89M, 90B2, 92C2, 92R1, 93B1]. The goal of near-diffraction-limited imaging is achieved more easily in the IR than in the visible, because coherence length and coherence time of the atmosphere vary as $\lambda^{6/5}$. The larger isoplanatic angle in the IR allows bright enough guide stars near the object to be found for most of the sky. First-order corrections by a fast tip-tilt mirror allowing only image stabilization achieved improvements in peak brightness and reduction of the image diameter by a factor of $\approx$ 3 at 3 μm [93G2]. Full atmospheric correction can be expected for $\lambda > 2$ μm with the 8m-class-telescopes, their focal plane instrument development therefore requires diffraction limited instrumentation for the $1 \cdots 5$ μm region [93L3].

1.8.4 Infrared photometry and imaging

See LB VI/2a, p. 45

A few bright IR standard stars are given there in Table 4. Fainter standard stars are listed in [82E, 87Z, 92C1]. See also LB VI/3b, sect. 4.2.

1.8.4.4 Filters and windows

See LB VI/2a, p. 47

High-quality metallic grid structures for high transmission and steep band-passes have been made by X-ray lithography [91R]. These filters often have to be combined with crystalline and other blockers outside the bands [93L1]. Neutral density filters attenuating incident IR radiation (low-background simulation) over a wide spectral range are made by depositing Ni–Cr films on Si, Ge or mylar substances

[90I]. Entrance windows for cryostats or cooled IR instruments are often CaF_2 ($\lambda < 5$ μm), KRS-5 ($\lambda < 50$ μm), polyethylene ($\lambda > 30$ μm), polymethylpenten TPX ($\lambda > 50$ μm), quartz ($\lambda > 50$ μm) [78W].

1.8.5 Infrared spectroscopy

The recent availability of large two-dimensional detector arrays allows the application of spectroscopic techniques developed in the visible now also to the infrared ranges.

Medium resolution ($\lambda/\Delta\lambda \approx$ few 10^2) can be achieved by circular variable filters (CVF) and by grisms, a combination of a transmission grating mounted on a prism [88B2]. These small and simple devices are used like interference filters in a cold filter wheel [93B2, 92K1].

For high spatial and high spectral resolution ($\geq 10^3$) imaging Fabry-Perot interferometers (FPI) can deliver monochromatic images of extended objects [92G2]. FPIs have been built for ground-based applications in the near and middle IR [88B1, 92R2] as well as for airborne applications in the far IR [91P].

High resolution ($\approx 10^4$) grating/echelle spectrometers can be used in a long-slit mode with 2D arrays. Cooled grating spectrometers are used for $\lambda < 5$ μm [92M, 93W], advanced systems for the 10 and 20 μm windows are under development [93G1]. Long-slit spectrometers allow obtaining spectral line images of extended sources by slit scanning.

Flexible spectroscopic capabilities to be combined with imaging, photometry and polarimetry are contained in a new generation of focal plane instruments for the Very-Large-Telescope [92M].

In order to reduce the strong OH-airglow emission to 1/30 of the natural background an OH line suppressor spectrograph for the J- and H-band ($1.1 \cdots 1.8$ μm) has been developed [93M].

1.8.6 Infrared polarimetry

See LB VI/2a, p. 48

Polarization foils are used for $\lambda < 2.5$ μm, for $\lambda < 5$ μm MgF_2-Wollaston prisms can be applied, while for $\lambda > 1.5$ μm throughout the far IR wire-grid polarizers are used [88J, 89N].

1.8.7 High-resolution imaging

(See also subsect. 1.6.5)

Interferometers involving several telescopes separated by several meters have been used in the near and middle IR; ongoing projects are described in [88M1, 92B1]. ESO's Very-Large-Telescope (VLT) is planned to be used in an interferometric mode in the $3.5 \cdots 20$ μm range with distances up to ≈ 100 m [88M2, 92L2]. The most common single-aperture technique is speckle interferometry originally using slit-scans [87L]. Most speckle instruments are available in the $\lambda < 5$ μm range [88P], some in the 10 μm band [87D]. With the advent of arrays, 2-dimensional speckle cameras for $\lambda < 5$ μm have been built [88B3, several papers in 92B1]. Speckle data reduction techniques are reviewed in [91C]. Lunar occultation allows for angular resolution of 10^{-3} arcsec [87W1, 89R, 92S]. Because the angular information is contained in the amplitude of the diffraction fringes, the size of the telescopes affects primarily the signal-to-noise ratio. Best results are obtained at $\lambda = 2.2$ μm ($K \approx 9.4$ mag) [93R]. For adaptive optics see subsect. 1.8.3.4.

References for 1.8

65H Hass, G.: Mirror coatings. In: Applied Optics and Optical Engineering, (R. Kinslake, ed.). New York: Academic Press (1965).

74L Low, F.J., Rieke, G.H.: The instrumentation and technique of infrared photometry. In: Methods of Experimental Physics, (N. Carleton, ed.). New York: Academic Press (1974).

77B Bratt, P.R.: Impurity germanium and silicon infrared detectors. In: Semiconductors and semimetals 12, (R.K. Williardson, A.C. Beer, eds.). New York: Academic Press (1977).

78D Donabedian, M.: Chapter 15. In: The Infrared Handbook [78W].

78W Woolf, W.J., Zissis, G.J., eds.: The Infrared Handbook. Office of Naval Research, Dep. of the Navy, Washington, D.C. (1978).

82E Elias, J.H., et al.: Astron. J. **87** (1982) 1029.

83C Chandrasekhar, T., Sahu, K.C., Desai, J.N.: Infrared Phys. **23** (1983) 119.

83G Greve, A.: Infrared Phys. **23** (1983) 59.

84D Dereniak, E.L., Crowe, D.G.: Optical radiation detectors. New York: Wiley (1984).

84N Neugebauer, G., et al.: Astrophys. J. **278** (1984) L1. Special issue of Astrophys. J. Letters (March 1984) on IRAS and early results.

87D Dyck, H.M., Zuckerman, B., Howell, R.R., Beckwith, S.: Publ. Astron. Soc. Pac. **99** (1987) 99.

87L Leinert, Ch., Haas, M.: Mitt Astron. Ges. **68** (1987) 223.

87W1 White, N.M.: Vistas in Astron. **30** (1987) 13.

87W2 Wynn-Williams, C.G., Becklin, E.E., eds.: Proc. Workshop Infrared Astronomy with Arrays, Univ. Hawaii, Hilo, March 1987.

87Z Zuckerman, B., Becklin, E.E.: Astrophys. J. **319** (1987) L 99.

88B1 Baily, J. et al.: Publ. Astron. Soc. Pac. **100** (1988) 1178.

88B2 Beckers, J.M., Gatley, I, in: Proc. ESO Conf. VLT and their Instrumentation, (M.H. Ulrich, ed.), Proc. ESO Conf. **30** (1988) 1093.

88B3 Beckers, J.M., Christou, J.C., Probst, R.G., Ridgway, S.T., von der Lühe, O.: P. 393 in [88M1].

88J Jones, T.J., Klebe, D.: Publ. Astron. Soc. Pac. **100** (1988) 1158.

88M1 Merkle, F., ed.: High-Resolution Imaging by Interferometry. Proc. NOAO-ESO Conference, Garching (1988).

88M2 Merkle, F.: Current Concept of the VLT Interferometric Mode, p. 909 in [88M1].

88P Perrier, C.: p. 113 in [88M1].

88W Watson, D.M., Huffman, J.E.: Appl. Phys. Lett. **52** (1988) 1602.

89M Merkle, F., Beckers, J.M.: Proc. SPIE **1114** (1989) 36.

89N Novak, G., Perinc. R.J., Sundwall, J.L.: Appl. Optics **28** (1989) 3425.

89P Peterson, J.B., in: Am. Inst. Phys. Conf. Proc. Astrophysics in Antarctica, (D.J. Mullan et al., eds.), (1989) 116.

89Q Quesada, J.A.: Publ. Astron. Soc. Pac. **100** (1989) 441.

89R Richichi, A.: Astron. Astrophys. **226** (1989) 366.

89V Vincent, J.D.: Fundamentals of infrared detector operation and testing. New York: Wiley (1989).

89W Wolf, J., Grözinger, U., Burgdorf, M., Salama, A., Lemke, D., in: Proc. Third IR Detector Technology Workshop, (C.R. McCreight, ed.), NASA-TM 102209 (1989).

90B1 Baars, J.W.M., Krügel, E., Martin, R.N.: The SMT: a joint submillimeter telescope project of the Max-Planck-Institut für Radioastronomie and Steward Observatory. In: Submillimeter Astronomy, (G.D. Wall and A.S. Webster, eds.). The Netherlands: Kluwer Academic Publishers (1990).

90B2 Babcock, H.W.: Science **249** (1990) 253.

90E Ellis, T., Little, J.K., Christeler, J.C., Riedy, R.C.: Proc. SPIE **1235** (1990) 36.

90I Infrared Laboratories Inc.: Product Brochure, Tucson, AZ 85719, (1990).

90M Matsumoto, T., in: Proc. IAU Symp. No. 139, (S. Bowyer, Ch. Leinert, eds.). Dordrecht: Kluwer Acad. Publ. (1990) 317.

91C Christou, J.C.: Exp. Astron. **2** (1991) 27.

91K1 Käufl, H.U., Bouchet, P., van Dijsseldonk, A., Weilenmann, U.: Exp. Astron. **2** (1991) 115.

91K2 Krabbe, A., Genzel, R., Drapatz, S., Rotaciuc, V.: Astrophys. J. **382** (1991) L19.

91K3 G.P. Kuiper Airborne Observatory: Publication list, available from NASA Ames Res. Center, Moffet Field, Calif. 94035 (1991).

91P Poglitsch, A., Beeman, J.W., Geis, N., Genzel, R., Haggerty, M., Haller, E.E., Jackson, J., Rumitz, M., Stacey, G.J., Townes, C.H.: Int. J. Infrared and Millimeter Waves **12** (1991) 859.

91R Ruprecht, R., Bacher, W.: Report KfK 4825, Kernforschungszentrum Karlsruhe (1991).

92B1 Beckers, J.M., Merkle, F., eds.: High resolution imaging by interferometry II. Proc. ESO Conf. No. 39 (1992).

92B2 Blouke, M.M., et al.: SPIE **1656** (1992) 497.

92B3 Bogess, N., et al.: Astrophys. J. **397** (1992) 420.

92C1 Casali, M., et al.: JCMT – UKIRT Newsletter **4** (1992) 33.

92C2 Collins, G.P.: Physics Today, Febr. 1992, p. 17.

92E1 Encrenaz, Th., Kessler, M.F., eds.: Infrared Astronomy with ISO. Nova Science Publ. ISBN-1-56072-078-6, several papers in this volume (1992).

92E2 Erickson, E.F.: Space Sci. Rev. **61** (1992) 6.

92F1 Fouks, B.I.: Proc. ESA Symp. Photon Detectors for Space Instrumentation. ESA SP-356 (1992) p. 167.

92F2 Fowler, A.M.: Proc. ESA Symp. Photon Detectors for Space Instrumentation. ESA SP-356 (1992) p. 129.

92G1 Gezari, D.Y., Folz, W.C., Woods, L.A., Varosi, F.: Publ. Astron. Soc. Pac. **104** (1992) 191.

92G2 Gredel, R., Weilenmann, U.: ESO Messenger **70** (1992) 62.

92H1 Harper, D.A.: Report Center for Astrophys. Res. in Antarctica, No. 1, Univ. Chicago, Yerkes Obs., Williams Bay, WI, USA (1992).

92H2 Hepburn, I.D., Ade, P.A.R., Davenport, I., Smith, A., Summer, T.J.: Proc. ESA Symp. Photon Detectors for Space Instrumentation. ESA SP-356 (1992) p. 317; see also Davenport, I.J., et al: p. 275 in same ref.

92H3 Huffman, J.E., Crouse, A.G., Halleck, B.L., Downes, T.V., Herter, T.L.: J. Appl. Phys. **72** (1992) 273.

92K1 Käufl, H.U., Jouan, R., Lagage, P.O., Masse, P., Mestreau, P., Tarrius, A.: ESO Messenger **70** (1992) 67.

92K2 Kreysa, E.: Proc. ESA Symp. Photon Detectors for Space Instrumentation. ESA SP-356 (1992) p. 207; see also Moseley, S.H. et al.: p. 13 in same ref.

92L1 Low, F.J.: Proc. SPIE **1684** (1992) 168.

92L2 Lühe, O.v.d., Beckers, J.M., Braun, R.: p. 959 in [92B1]

92M Moorwood, A.: ESO Messenger **70** (1992) 10; also: Proc. SPIE **1946** (1993) 461.

92P Price, M.C., Griffin, M.J., Church, S.E., Murray, A.G., Ade, P.A.R.: Proc. ESA Symp. Photon Detectors for Space Instrumentation. ESA SP-356 (1992) p. 309.

92R1 Roddier, F.: p. 571 in [92B1].

92R2 Rotaciuc, V.: Max-Planck-Inst. Extraterr. Phys. Report 239, Dissertation Univ. München (1992); see also Krabbe, A., et. al.: Publ. Astron. Soc. Pac., in print (1994).

92S Simon M., Leinert, Ch.: Sterne und Weltraum **31** (1992) 380.

92W1 Wensink, J.W., Luinge, W., Beintema, D., Valentijn, E.A., de Graauw, Th., Kattenloher, R., Barl, L., Young, E.T.: Proc. ESA Symp. Photon Detectors for Space Instrumentation. ESA SP-356 (1992) p. 339.

92W2 Wolf, J.: Proc. ESA Symp. Photon Detectors for Space Instrumentation. ESA SP-356 (1992) p. 137.

92W3 Wanner, M.: Proc. ESA Symp. Photon Detectors for Space Instrumentation . ESA SP-356 (1992) p. 115.

93B1 Beckers, J.M.: Annu. Rev. Astron. Astrophys. **31** (1993).

93B2 Beckwith, S.V.W., Birk, Ch., Herbst, T.M., Hippler, S., McCaughrean, M.J., Mannucci, F., Wolf, J.: Proc. SPIE **1946** (1993) 605.
93B3 Beckwith, S.V.W.: Proc. 5th EADN Summer School Star Formation and Techniques in Infrared and mm-Wave Astronomy. In: Lecture Notes in Physics, (T. Ray and S. Beckwith, eds.). Heidelberg: Springer Verlag (1994).
93G1 Glasse, A.C.H., Atad, E.E.: Proc. SPIE **1946** (1993) 629.
93G2 Glindemann, A., Rees, N.: Proc. ICO-16 Satellite Conf. Active and Adaptive optics, (F. Merkle, ed.), Garching, (1993).
93H Hoffmann, W.F., Fazio, G.G., Shivanandan, K., Hora, J.L., Deutsch, L.K.: Proc. SPIE **1946** (1993) 449.
93K Käufl, H.U.: ESO Messenger **73** (1993) 8.
93L1 Lemke, D., Garzon, F., Gemünd, H.P., Grözinger, U., Heinrichsen, I., Klaas, U., Krätzschmer, W., Kreysa, E., Lützow-Wentzky, P., Schubert, J., Wells, M., Wolf, J.: Proc. SPIE **2019** (1993).
93L2 Lemke, D., Wolf, J., Schubert, J., Patraschin, M.: Proc. SPIE **1946** (1993) 261.
93L3 Lenzen, R.: Proc. SPIE **1946** (1993) 635.
93M Maihara, T., Iwamuro, F., Hall, D.N.B., Cowie, L.L., Tokunaga, A.T., Pickles, A.J.: Proc. SPIE **1946** (1993) 581.
93R Richichi, A.: IAU Symp. **158,** Sydney (1993).
93T Thronson, H.A., Hawarden, T.G., Bally, J., Bradshaw, T.W., Davies, J.K., Greenhouse, M., Orlowska, A.H., Stern, A.: Proc. SPIE **1945** (1993).
93W Wright, G.S., Mountain, G.M., Bridges, A., Daly, P.N., Griffin, J.L., Ramsay, S.K.: Proc. SPIE **1946** (1993) 547.

1.9 Radio astronomy instrumentation

1.9.0 Introduction

Notable advances in the decade since the publication of LB VI/2a in 1981 have occurred in the areas of large telescopes for millimeter (mm) and submillimeter (submm) wavelengths, including interferometer arrays, the creation of dedicated VLBI-networks, low-noise receivers over the entire mm- and cm-wavelength range and the processing and display of images, obtained with non-ideal instruments, in particular synthesis arrays. On the other hand radio astronomy has come of age and several of the telescopes listed in Vol. 2 have been taken out of operation.

Recent original books and reprint collections on the subject of the techniques of radio astronomy are:

- Christiansen, W.N., Högbom, J.A.: "Radiotelescopes" (2nd ed.) [85C].
- Rohlfs, K.: "Tools of Radio Astronomy" [86R].
- Thompson, A.R., Moran, J.M., Swenson, G.W.: "Interferometry and Synthesis in Radio Astronomy" [86T3].
- Goldsmith, P.F. (editor): "Instrumentation and Techniques for Radio Astronomy" [86G1]. This is a very useful collection of reprints covering all areas of technical radio astronomy.

1.9.1 Radio astronomy receivers

Developments in solid state devices, also actively pursued at or supported by a number of radio observatories, have led to a further reduction of the receiver noise temperature over the entire cm- and mm-wavelength band. The state of art is illustrated in Fig. 1. Most significant is the application of HEMTs (high electron mobility transistor), cooled to cryogenic temperature as low as 4 K, as amplifier element with a wide (about one octave) bandwidth [88P,91P3]. It is foreseen that HEMTs will be available at frequencies of 100 GHz in the near future. Consequently the parametric amplifier has essentially disappeared and masers are being phased out.

The advance of mm-astronomy has spurned the development of special mm- and submm-Schottky diodes, used as (cooled) mixers [85A, 87M, 90E]. At the same time the technology of the fabrication of SIS-diodes (superconductor-insulator-superconductor) [81P, 82G2, 90K1] and their practical application in mm-mixer circuits has resulted in sensitivity improvements of about a factor $2 \cdots 3$ with respect to Schottky-diodes for frequencies up to 350 GHz [82P, 88B2]. Efforts are being made to extend the applicability of SIS-elements to the submm region [89E, 90S]. As local oscillators for the submm region one uses carcinotrons [86K], IR-pumped gas lasers [87H, 87R] or a solid state oscillator with frequency multiplier [84A]. A 15-channel receiver for the 3 mm band is reported in [92E] and an all-solid-state radiometer at 700 GHz in [92Z].

For observations of continuum radiation at (sub)mm-wavelengths increasing use is being made of wideband bolometers [90K2]. Multichannel receivers of this type are under development, the biggest one being a 90-element submillimeter common-user bolometer array (SCUBA) for the James Clerk Maxwell Telescope (JCMT) [90G]. Noteworthy also is a 25-channel Fabry-Perot spectrometer for the far-infrared [91P2].

Because waveguide components become increasingly difficult to fabricate at wavelengths as short as one millimeter, so-called "quasi-optical" techniques, using lenses and mirrors to guide the signals, find increasing use in mm-wave technology. The principles and applications of these are described in [82G1, 85E].

Special custom-designed large-scale-integrated circuits have been used in the correlation section of synthesis arrays and also in spectrometer backends of single telescopes. Special mention deserves the correlator chip, developed at the Dwingeloo Observatory [91B1], which is being used in a number of observatories around the world.

Extragalactic spectroscopy at mm-wavelengths requires an analysing bandwidth of the order of 1 GHz. These have been realized with "hybrid" systems, consisting of a number of filter channels of a few hundred MHz, each followed by a digital autocorrelation section [85W]. An alternative approach is the application of acousto-optical Bragg-cells to achieve some 1000 spectral resolution points over an instantaneous bandwidth of 1 GHz [89S].

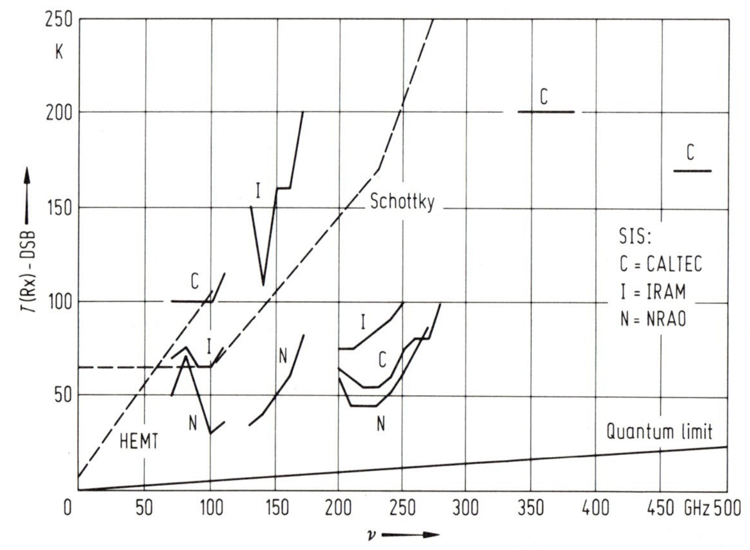

Fig. 1. Present double sideband (DSB) noise temperatures (T(Rx)) of the best radio astronomy receivers in the frequency range up to 500 GHz. The broken lines indicate the best performance of cryogenically cooled HEMT-amplifiers and Schottky-diode mixers. The connected points are results for SIS-mixers, obtained at three major development laboratories. The figure is adapted from Kerr and Pan [90K1] and Pospieszalski [91P3], where more details can be found. HEMT-amplifiers will soon reach a frequency of 100 GHz, Schottky-mixers are used up to 2 THz and SIS-mixers are being pushed to 1 THz.

1.9.2 Radio telescopes

For ease of comparison with Vol. VI/2a we maintain the classification of radio telescopes in instruments for meter (m), decimeter/centimeter (cm) and millimeter/submillimeter (mm) wavelengths. The following tables contain the necessary amendments to those in Vol. VI/2a plus the new instruments. The tables are not exhaustive; only the most important instruments have been listed, including those spaceborne telescopes, which are under construction or definitely planned. Since most telescopes are used for the entire range of astrophysical problems which can be tackled with radio techniques, the column "Programs" has been dropped from the tables to make more space for technical information. Each table begins with a listing of those telescopes which have ceased operation since the appearance of Vol. VI/2a.

1.9.2.1 Radio telescopes for m wavelengths

See Table 1 on next page

1.9.2.2 Radio telescopes for cm wavelengths

In this section we deal with both single-dish telescopes, limiting ourselves to diameters in excess of 30 m, and aperture synthesis arrays for the wavelengths range between roughly 1 and 90 cm. In the single-dish category the most dramatic event since the appearance of Vol. VI/2a is the collapse of the Green Bank 92-m-telescope (No. 2 in Table 2) in October 1988. This instrument will be replaced by a 100-m-diameter "clear aperture" antenna, the "Green Bank Telescope (GBT)", which should come into operation in 1995. Note that a number of telescopes have been taken out of operation. The three NASA deep-space tracking stations (Table 2, Nos.5, 6 and 7) have been enlarged to 70 m diameter and are still parttime available for radio astronomy, mainly geodetic VLBI. The large, fixed, spherical reflector at Arecibo will be further improved with a dual-subreflector system, and will be used for wavelengths as short as 3 cm. Several antennas in the $25 \cdots 30$ m class have been built specifically for VLBI work; these are listed seperately in subsection 1.9.2.4.

In the category of aperture synthesis telescopes the most important advance is the Australia Telescope National Facility, which provides a powerful, high-resolution instrument for detailed mapping of southern hemisphere objects. The large instrument in India (Table 1, No.9 and Table 3) will be used to a shortest wavelength of 20 cm, where it will be the most sensitive telescope on earth. Mention should also be made of MERLIN (Multi-Element Radio Link-Interferometer) in Britain, a set of 7 antennas, connected by phase-stable radiolink, which provides angular resolutions between those obtained at the VLA and those typical of VLBI. Some smaller interferometers are now used for restricted purposes; e.g. the Cambridge 5-km-array for cosmic background research and the Green Bank interferometer (not listed in Vol VI/2a) for astrometric work. The characteristics of the major synthesis telescopes have been summarized in Table 3. More details can be obtained from the references. A new instrument has been proposed for the Penticton observatory. It is called the Radio Schmidt Telescope and it is to have 100 antennas of 12 m diameter, observing at several frequencies between 0.4 and 26 GHz. At the time of writing funding has not been secured and international participation is being sought.

Table 1. Radio telescopes for m wavelengths – update of Table 1 in Vol. VI/2a.

Type: A.S. = aperture synthesis A=collecting area
P.A. = phased array ν=frequency

No.	Institution/Observatory	Location	Type	Characteristics	A [m^2]	ν [MHz]	Ref.
3	Clark Lake		out of operation				
4	Univ. Iowa		out of operation				
7	CSIRO Solar array		out of operation				
8	Tata Institute, Pune, India	Ooty, India	P.A.	cylindrical paraboloid along meridian	8000	326	71S
			A.S.	plus 7 outstations	+1000	326	88S
9	Tata Institute	Pune, India	A.S.	30 × 45 m paraboloids	30000	38 ⋯ 1420	Tab.3
10	Univ. Cambridge, UK	Cambridge, UK	A.S.	50-Yagi-dipole arrays		151	85B

Table 2. Single-reflector telescopes for cm wavelengths ($D > 30$ m) – update of Table 2 in Vol. VI/2a.

D = diameter,
sa = surface accuracy,
ν = operational frequency range

No.	Institution	Location	Characteristics	D [m]	sa [mm]	ν [GHz]	Ref.
1	MPIfR, Bonn, Germany	Effelsberg	improved reflector surface	100	0.4	$1 \cdots 86$	86G
2	NRAO, USA	Green Bank	not existent anymore				
8	NRC, Canada	Algonquin Park	out of operation				
9	AFCRL, USA	Sagamore Hill	out of operation				
10	Stanford Univ. USA	Stanford	out of operation				
13	Univ. Illinois, USA	Vermilion River	out of operation				
16	Nuffield Lab., England	Wardle	part of MERLIN (Table 3)				
5	NASA/JPL	Goldstone,USA	improved and larger				
6	NASA/INTA	Robledo,Spain	surface-primarily VLBI	70	1.2	$2 \cdots 22$	86C
7	NASA/JPL	Tidbinbilla,Aus.	and deep space tracking				89B1
17	Nat.Ionos.Astron.Center Cornell Univ. USA	Arecibo Puerto Rico	spherical reflector with focal carriage	300	2	$0.3 \cdots 8$	86T1
18	RATAN 600/Rus.Ac.Sci.	Arkhiz, USSR	variable profile antenna	600	1	$1.4 \cdots 35$	79K
18	IAR/Argentina	Villa Elisa	2 antennas, mesh surface	30	$5 \cdots 8$	$1.4 \cdots 2.7$	
19	RT70/Rus. Ac.Sci.(1994?)	Suffa, Uzbek SSR	active surface	70		$1.6 \cdots 115$	87I
20	NRAO, USA (1995)	Green Bank, W.Va.	clear aperture, active surface	100	1(0.25)	$1 \cdots 50(86)$	91V

Table 3. Aperture synthesis telescopes for cm wavelengths – replaces Table 3 in Vol. VI/2a.

Characteristic | Name	VLA	WSRT	Cambridge	MERLIN	AT	DRAO ST	GMRT
Full name of telescope	Very Large Array	Westerbork Synthesis Radio Telescope	5-km-Telescope	MERLIN	Australia Telescope	DRAO Synthesis Telescope	Giant Metre Radio Telescope
Institution [1]	NRAO USA	NFRA Netherlands	Cambridge Univ. UK	Nuffield Labs. UK	CSIRO Austr.	DRAO(NRC) Canada	Tata Inst. India
Location Latitude	Socorro, NM +34°05′	Westerbork +52°55′	Cambridge +52°10′	England +53°	Culgoora -30°19′	Penticton +49°19′	Pune +19°06′
Element antennas							
- diam. [m]	25	25	13	25,32,76	20	8.5	45
- total no. of elements	27	14	8	7	7	7	30
- nos. of moveable elements	27	4	4	0	6	3	0
Type of array	"Y" in 4 configur.	E-W	E-W	irregular	EW+out-stations	E-W	12 cluster + "Y" (3 × 6)
Max. baseline length [m]	36400	2800	4600	233000	2000 +	604	23000
Min./max. frequency [GHz]	0.3/23	0.3/5	2.7/15	0.15/5	1.4/23	0.4/1.4	0.038/1.4
Field of view [arcmin]	200/2	240/15	45/9	120/10	60/4	490/156	900/25
Max. Resolution [arcsec]	6/0.1	55/3.7	5/1	3/0.1	20/1.4	210/60	75/2
Ref.	83N	73B	72R	86T2	88N	73R,85V	91S

[1] NRAO = National Radio Astronomy Observatory, NFRA = Netherlands Foundation for Research in Astronomy,
CSIRO = Commonwealth Scientific and Industrial Research Organisation, DRAO = Dominion Radio Astrophysical Observatory.

1.9.2.3 Radio telescopes for mm and submm wavelengths

Great advances have been made over the last decade in the area of telescope design and construction for the short millimeter and submillimeter wavelengths (0.3 ⋯ 3 mm), further on designated mm wavelengths. The method of homology design, originally proposed by von Hoerner [67H], has been put to practice with the aid of modern structural analysis programs, leading to designs in which residual gravitational deformations have been reduced by an order of magnitude. At the same time the significance of thermally induced deformations has been recognized and several successful solutions to this problem have been found. For a survey of these aspects see [83B].

Particular mention should be made of the application of carbon-fiber reinforced plastic (CFRP) in some of the most accurate telescopes. This lightweight, but strong material has a very small coefficient of thermal expansion, which makes it ideal for the minimisation of thermal effects. A quick view of the state of art is presented in Fig.2, where a number of telescopes have been plotted on a "precision versus diameter" plot. The lines indicate limits due to thermal gradients and gravity deformations in a "classic non-homologous" design for the materials steel and CFRP. The most accurate telescopes are well above the gravitational limit and on the "day"-line for thermals, indicating that they perform properly during operation in sunshine.

In Vol. VI/2a, Table 4, 14 telescopes were listed with four being "under construction". Of these, three have been in operation for several years, while the 25-m-telescope planned by NRAO was never built. In addition 7 dedicated mm telescopes have been constructed in the meantime, while four powerful mm interferometers have come into operation. These are the extended versions of the Owens Valley and Hat Creek interferometers, as well as the new Japanese and IRAM arrays. We summarize in Table 4 the parameters of the most important single mm telescopes (extending the numbering of Vol. VI/2a), while the data on the mm interferometers are presented in Table 5. It should be noted that most new mm telescopes have been located at high and dry mountain sites to exploit the better atmospheric transparency.

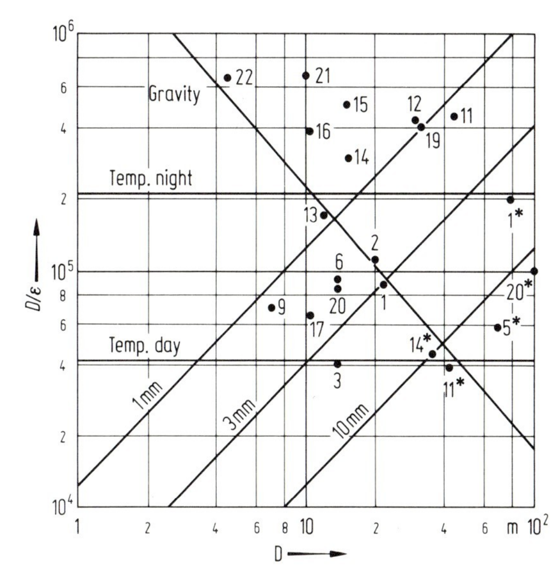

Fig. 2. Precision of radio telescopes, defined as the ratio of reflector diameter D to surface rms deviation ϵ, as a function of diameter D. Numbers refer to the entries in Tables 2 (with *) and 4. The "natural limits" for temperature gradients and gravitational deformation of a steel structure are indicated. All telescopes above the line "Gravity" employ a homologous design of the reflector structure. Temperature limits for CFRP material lie almost an order of magnitude higher. The three parallel lines indicate minimum wavelengths of 1,3 and 10 mm, respectively. All telescopes to the left of a line will perform well at this wavelength. Minimum wavelength is defined as 12.5 times reflector rms error ϵ. Adapted from Baars [83B].

Table 4. Radio telescopes for mm wavelengths ($D > 10$ m) – update of Table 4 in Vol. VI/2a.

protection: o.a. = open air, rad = closed radome, ast = astrodome with doors;
D = diameter, λ_{min} = minimum wavelength

No.	Name/Institution [1)]	Location Latitude	Year	Altitude [m]	D [m]	λ min [mm]	Remarks	Ref
7	Nat. Radio Astr. Obs.	Kitt Peak, Arizona	reflector replaced, now listed as No.13					
8	Calif. Inst. Tech. USA	Owens Valley, Calif.	extended to interferometer, listed in Table 5 (Caltech)					
10	Univ. Calif., Berkeley	Hat Creek, Calif.	extended to interferometer, listed in Table 5 (BIMA)					
13	NRAO	Mauna Kea, Hawaii	not realized					
11	Nobeyama Radio Observatory Nobeyama, Japan	Nobeyama 35°56′	1983	1350	45	1.2	o.a., CFRP panels thermal insulation	83A
12	IRAM – Germany/France/Spain Granada, Spain	Pico Veleta, Spain 37°04′	1985	2850	30	0.8	o.a., with active thermal control, CFRP subreflector	87B 88B1
13	Nat. Radio Astr. Obs. Tucson, AZ, USA	Kitt Peak, Arizona 31°57′	1983	1940	12	1.0	ast, new reflector replaces old 11 m reflector	84G
14	SEST- ESO/Onsala Obs. Sweden Santiago, Chile	La Silla, Chile −29°15′	1987	2400	15	0.6	o.a., IRAM design CFRP panels/backup	89B2
15	JCMT- UK/Netherlands/Canada Hilo, Hawaii, USA	Mauna Kea, Hawaii +19°50′	1988	4000	15	0.4	ast, with teflon membrane	

[1)] IRAM = Institute for Radio Astronomy in Millimeter range, French/German/Spanish cooperation; headquarter Grenoble, France;
SEST = Swedish-ESO Submillimeter Telescope;
JCMT = James Clerk Maxwell Telescope;

Table 4. (continued)

protection: o.a. = open air, rad = closed radome, ast = astrodome with doors;
D = diameter, λ_{min} = minimum wavelength

No.	Name/Institution [1)]	Location Latitude	Year	Altitude [m]	D [m]	λ min [mm]	Remarks	Ref
16	Calif. Inst. Tech. USA Hilo, Hawaii, USA	Mauna Kea +19°50′	1989	4000	10.4	0.4	ast	87L
17	Raman Research Institute Bangalore, India	Bangalore +13°	1989	930	10.4	2.0	o.a., Caltech design	
18	Deaduk Radio Astronomy Obs. Daejeon, Korea	Daejeon City +36°24′	1989	120	13.7	2.0	rad	
19	ROT-32/54/2.6; VNIIRI Obs. Armenian Acad. Science	Erevan, Armenia +40°20′	1989	1700	32	1.0	spherical refl.; integrated 2.6 m optical telescope	89H
20	Purple Mountain Observatory Nanjing, China	Delinghai, Qinghai +37°21′	1990	3200	13.7	2.0	rad	
21	SMT- MPIfR/Steward Obs. Tucson, AZ, USA	Mt.Graham, Arizona +32°42′	1992	3200	10	0.2	ast, CFRP panels/backup	90B 90M1
22	FIRST-European Space Agency	Space orbit	2003(?)		4.5	0.1	orbiting observatory	90F2

[1)] ROT = Radio-Optical Telescope;
SMT = Sub-Millimeter Telescope, collaboration of Max Planck Institute für Radioastronomie, Bonn and Steward Observatory, Univ. of Arizona, Tucson
FIRST = Far Infrared Space Telescope.

Table 5. Interferometers for mm wavelengths.

Characteristic|Name [1]	NRO	Caltech	BIMA	IRAM	SMA	MMA
Institution	Nobeyama Radio Obs., Japan	Californian Inst.Techn.	BIMA consortium	IRAM	Smithonian Astr.Obs.	NRAO, USA
Location	Nobeyama	Owens Valley	Hat Creek	Pl.de Bure	Mauna Kea	Arizona(?)
Latitude	36°	37°	41°	44°	19°	33°
Altitude [m]	1350	1236	1043	2552	4000	2800
Element antennas						
- diameter [m]	10	10.4	6	15	6	8
- total no. of elements	5 (6)	3 (5)	6 (9)	3 (4)	6	40
- nos. of moveable elements	5 (6)	3 (5)	6 (9)	3 (4)	6	40
Type of array	“T”	“T”	“T”	“T”	“ring’”	4 ellipses
Max. baseline length [m]	560	220	300 (750)	288	460	3000
Wavelength coverage [mm]	13,7,3	3,1.2	3,(1.3)	3,(1.3)	1.3 ⋯ 0.35	7 ⋯ 0.8
Field of view [arcsec]	320/60	70/35	120/(55)	55/(25)	54 ⋯ 15	220 ⋯ 25
Max. resolution [arcsec]	4.2/0.8	3.5/1.5	2.3/(0.3)	2.7/(1)	0.5 ⋯ 0.12	0.6 ⋯ 0.07
Status	operational	operational	operational	operational	under constr.	proposed to
-remarks	ext.ready 93	ext.ready 92	ext.ready 94	ext.ready 93	submm array	NSF,90;2005(?)
Ref.	90I	91P	84W,85U	92G	90M2,92M	91B2

[1] NRO = Nobeyama Radio observatory,
IRAM = Institute for Radio Astronomy in Millimeter range,
MMA = MilliMeter Array,
BIMA = Berkeley-Illinois-Maryland-Array,
SMA = Sub Millimeter Array,
NRAO = National Radio Astronomy Observatory

1.9.2.4 Very-long-baseline interferometers (VLBI)

Next to the mm-wavelength developments, the strongest activity in the last decade has been in the field of very-long-baseline interferometry (VLBI). This is a direct result of the success of the original VLBI experiments, which produced a host of new results, particularly in the areas of galactic nuclei, quasars and Galactic maser sources. Developments in the early eigthies went along two lines. Firstly, the use of more than two, sometimes up to ten, telescopes for an observing run was achieved by extensive international cooperation, which started on an ad hoc basis, but worked very well [89F]. Secondly, the development of recorders for larger bandwidths, from the original 2 MHz up to 56 MHz (Mark III, [81R]) and the construction of multi-station processing units [89A] increased the sensitivity and the mapping capabilities enormously. Once more telescopes were equipped with hydrogen maser frequency standards [87P], the integration time of an observation could be increased, opening the way to real aperture synthesis over intercontinental baselines.

Using existing telescopes, two semi-formal "networks" were set up, one in Europe (EVN = European VLBI network) and one in the USA, each with five consortium members plus a number of associate members. About 8 telescopes constitute each network, providing some 25% of observing time for VLBI. A global network is made by combination of the two networks. For a good and practical introduction to VLBI see [89F].

In 1983 the construction of a dedicated VLB-Array was started in the USA, composed of 10 antennas of 25 m diameter, placed from Hawaii to the Virgin Islands [85K]. Its characteristics are summarized in Table 6. It is being operated by the NRAO from Socorro, NM, and is approaching a full operational state. The locations of the elements have been carefully chosen to provide a good aperture plane coverage over a number of hours integration time. Thus this instrument will be able to produce radio pictures of high reliability. It is expected that for special projects the VLA and the new Green Bank Telescope, once completed, and possibly other antennas, will be incorporated into the VLBA. The correlator in Socorro will be able to handle 20 stations simultaneously.

In Europe, the EVN is growing through the addition of new telescopes due to the initiative of individual institutes. Its composition in the early nineties is presented in Table 7. The sensitivity of EVN is several times larger than that of the VLBA, and it too provides a good mapping capability. With support of the EC, a European VLBI data processing institute with a 20-station correlator will be established in the Netherlands.

With the maturity of both VLBI- and millimeter-techniques, the next logical step constitutes VLBI observations at mm wavelengths. From the first pilot projects at 7, 3 and even 1.3 mm it has emerged that such observations are technically feasible and astronomically highly significant. Thus a growing number of the larger mm-telescopes is being equipped for VLBI (Table 8). Three "millimeter-networks" of great potential emerge from this list:

(i) European, including SEST and Suffa telescope, – 6 elements
(ii) Western USA, including Hawaii, if and when equiped, – 5 ⋯ 6 elements
(iii) Asian-Australian, – 4 ⋯ 5 elements.

Additional VLBI arrays or single stations are listed in Table 9.

Noteworthy are the dedicated 6 element array across the former USSR and the three element array in China, both of which are partially operational. In the southern hemisphere there is the relatively short Australian array, which can be used with the station in South Africa. Finally, Table 9 also lists two space-borne antennas, dedicated to VLBI in combination with earth-bound arrays. Radioastron is an effort by the Russian Academy of Sciences under participation of the EVN, Finland, Australia, Hungary and Canada. VSOP is a project of the Japanese Institute for Space and Astronautical Science.

Table 6. The US Very-Long-Baseline Array (VLBA) [85K,91V].

Location	Lat. [° ′]	W Longit. [° ′]	Alt. [m]	Operational [yy/mm]	Frequency [GHz]	Max.Resolution [10^{-3} arcsec]
Pie Town, NM	34 18	108 07	2371	88/06	0.33	24
Kitt Peak, AZ	31 57	111 37	1916	89/06	0.61	13
Los Alamos, NM	35 46	106 15	1967	91/01	1.5	5.4
Fort Davis, TX	30 38	103 57	1615	91/06	2.3	3.5
N. Liberty, IA	41 46	91 34	241	91/09	4.8	1.6
Owens Val., CA	37 14	118 17	1207	91/12	8.4	0.9
Brewster, WA	48 08	119 41	255	91/12	15	0.5
St. Croix, VI	17 46	64 35	16	92/06	23	0.4
Hancock, NH	42 56	71 59	309	93/01	43	0.2
Mauna Kea, HI	19 49	155 28	3725	93/01		

Table 7. The European Very-Long-Baseline Network (EVN) [89F].

Location	Lat.	E Long.	Alt.	Diam.	Frequencies [GHz]								Remarks
	[° ']	[° ']	[m]	[m]	0.33	0.61	1.4	1.6	4.8	8.4	10.7	23	(i.o. = in operation)
Jodrell Bank, UK	53 14	-2 18	78	76	x	x	x	x	x				i.o.
Jodrell Bank, UK	53 14	-2 18	78	25			x	x				x	i.o.
Cambridge, UK	52 10	00 03	17	32		x	x	x	x	x	x	x	i.o.
Westerbork, Neth	52 55	06 36	16	94*)	x	x	x	x	x				i.o.
Effelsberg, Ger	50 32	06 53	369	100		x	x	x	x	x	x	x	i.o.
Wetzell, Germany	49 06	12 54	400	20						x			i.o. (geodesy)
Onsala, Sweden	57 24	11 55	24	26			x	x	x				i.o.
Onsala, Sweden	57 24	11 55	24	20						x		x	i.o.
Medicina, Italy	44 31	11 39	41	32			x	x	x	x	x	x	i.o.
Noto, Sicil, Italy	36 54	15 00	30	32			x	x	x	x	x	x	i.o.
Torun, Poland	53 06	18 36	90	15	x	x	x	x	x				i.o.
Torun, Poland	53 06	18 36	90	32	x	x	x	x	x				under construction
Nançay, France	47 23	02 12	81	94*)			x	x					i.o.
Simeiz, Russia	44 32	34 01	676	22	x	x		x			x	x	i.o.

*) Effective size

Table 8. Antennas equipped for millimeter wavelength VLBI [89F].

Location	Lat.	E Long.	Alt.	Diam.	Frequencies [GHz]				Remarks
	[°]	[°]	[m]	[m]	23	43	86	230	(i.o.= in operation)
Nobeyama, Japan	36	138	1350	45	x	x	x	x	i.o.
Pico Veleta, Spain	37	-3	2850	30		x	x	x	i.o.
Onsala, Sweden	57	12	24	20	x	x	x		i.o.
SEST, La Silla, Chile	-29	-71	2347	15			x	x	i.o.
BIMA, HatCreek, Calif.	41	-121	1043	15*)			x		i.o.
Owens Valley, Calif.	37	-118	1236	18*)			x	x	i.o.
Bangalore, India	13	77	930	10.4	x		x		i.o.
Pl.de Bure, France	45	06	2552	26*)			x	(x)	planned 1994
Suffa, Uzbek SSR	38	66	2200	70	x	x	x	(x)	time uncertain
Yebes, Spain	41	-3	914	14		x	x		i.o.
Metsahovi, Finland	60	24	60	14	x	x	x		i.o.
SMT, Mt.Graham, USA	32	-110	3200	10				x	planned
Mopra, Australia	-31	149	1149	22	x	x	x		(1993)
Qinghai, China	37	97	3200	14	x		x	(x)	(1992)
Daejeon, Korea	36	127	120	14		x	x		(1994)

*) Effective size

Table 9. Other Very Long Baseline Networks and individual stations [89F].

Location	Lat. [°]	E Long. [°]	Alt. [m]	Diam. [m]	Frequencies [GHz] 0.33	0.61	1.4	2.3	4.8	8.4	10.7	23	43	Remarks (i.o.= in operation)
Southern hemisphere Array														
Culgoora, Austral.	-30	150	217	54*)			x		x			x		Australia Telescope
Mopra, Australia	-31	149	1149	22								x	x	
Parkes	-33	148	392	64			x		x		x	x		i.o.
Tidbinbilla	-35	149	656	70				x		x				mainly geodetic
Hobart,Tasmania	-43	148	300	26			x		x					i.o.
Hartebeeshoek,SAfr.	-26	28	1391	26				x	x	x				i.o.
China VLBI Network (CVN)														
Shanghai	31	121	10	25	x	x	x	x	x	x	x	(x)		i.o.
Urumchi, Xinjiang	44	87	2000	25	x	x	x	x	x	x	x	x		under constr.(1992)
Kunming, Yunnan	25	103	2000	25	x	x	x	x	x	x	x	x		planned for 1994
Miyun, Bejing	41	118	90	47*)	x									parttime (1992)
Delinghai, Qinghai	37	97	3200	14								x		parttime (1992)
"former USSR" Array "Quasar" [90F1]														operational 1994/5(?)
Leningrad, Svetloe	61	30	45	32			x	x	x	x		x	(x)	under construction
Zelenchukskaya	44	42	1100	32			x	x	x	x		x	(x)	under construction
Irkutsk	52	104	830	32			x	x	x	x		x	(x)	under construction
Ashchbad	38	58	640	32			x	x	x	x		x	(x)	under construction
Odessa	47	30	110	32			x	x	x	x		x	(x)	under construction
Petropavlovsk	53	158	20	32			x	x	x	x		x	(x)	under construction
Spaceborne VLBI elements														
Radioastron, Russia + others			[88K]	10			x		x			x		launch 1994/95
VSOP (VLBI Space Obs.Proj.) Japan			[88H2]	10			x		x			x		launch 1994/95

*) Effective size

1.9.3 Data analysis and image processing

While this subject was not even mentioned in LB VI/2a, it is by now of an importance entirely comparable to that of the telescope and its associated receiver systems. Large aperture synthesis telescopes and VLBI arrays deliver a dense data stream, which must be subjected to a number of mathematical procedures before any meaningful astronomical interpretation is possible. Since all these telescopes are inherently imperfect, i.e. the instrumental characteristics have systematic and significant errors, the correction of the data for these errors becomes one of the most essential, difficult and time-consuming chores for the observer.

A number of observatories have put a considerable effort in the development of software packages for the reduction of observations and the construction of the best possible images of the observed source. In some cases these have developed into generally accepted, relatively easy to use and well supported programs, used in a larger number of institutes. By their nature, these programs are under continuous development and the actual performance can only be learned from studying the handbooks and using the program itself. Some conferences have been devoted to this subject [85G]. In the following we give a very short description of the major features of a number of the most widely used packages. Beacause these have been described in the above conference proceedings [85G], we refrain from giving further references.

An important aspect of the exchange of data is the format in which the astronomical and instrumental data are recorded. The FITS (Flexible Image Transport System) format [81W,88G2] enjoys a growing use among observatories. It is supported by the NASA Office of Standards and Technology (NOST).

AIPS (Astronomical Image Processing Software)
AIPS was developed at the NRAO for the analysis and display of synthesis observations made with the VLA. It runs under VMS on VAX-computers, and it has been adopted by a number of observatories. A new, extended version of the program, known as AIPS++, is being prepared under the leadership of NRAO with the active participation of several other groups. This version will run under UNIX and is scheduled to be available in 1994. AIPS is very versatile and can be used also for non-radio-astronomical imagery.

CLASS (Continuum and Line Analysis Single-dish Software)
This package was developed at CERMO and IRAM in Grenoble, primarily for the reduction of millimeter wave observations. It runs on a VAX-computer and allows post-real-time reduction and inspection of observations, a feature particularly useful at millimeter wavelengths, where short-term changes in atmospheric transmission are frequent. The program is used at several millimeter observatories. A version running under UNIX is in preparation.

DWARF (Dwingeloo Westerbork Astronomical Reduction Facility)
This is the name of an environment for the running of the reduction packages for the Westerbork Synthesis Radio Telescope. It runs under VMS or UNIX in a number of Dutch observatories. It is particularly useful for the reduction of synthesis data with baseline redundancy and in applying self-calibration, both of which improve the dynamic range of the observations.

MIDAS (Munich Interactive Data Analysis System)
MIDAS is a general-purpose data analysis package, developed and supported by ESO. It is implemented on VAX computers and is being continuously updated and improved. It is possible to use the system, located at the ESO headquarters near Munich, remotely over a data line.

NOD2
NOD2 is a package for the reduction and display of brightness distribution maps, obtained by scanning an area of sky with a single telescope. The program was developed at MPIfR and has been adopted

at several other observatories, and runs in a number of environments (VMS, UNIX). It is particularly suitable for the restauration of "dual beam" observations and for the removal of baseline effects in maps of large dimensions, obtained with the "basket weaving" technique [74H,79E].

STARLINK
This is a hardware and software environment for astronomical computing and data analysis. It is based on VAX/VMS machines; a transition to UNIX is being prepared. Starlink is installed at 22 sites within the UK and several abroad. It is centrally managed by the Rutherford Appleton Laboratory. A large number of computational routines and data handling and display packages are available, which normally are contributed by the participating sites.

UNIPOPS (UNIX People-Oriented Parcer Service)
UNIPOPS is a single-dish analysis package for all NRAO telescopes. It runs under UNIX, primarily on workstations. It is comparable in its features with CLASS. It replaces POPS, which has been used at NRAO for several years.

References for 1.9

67H Hoerner, S. von: Astron. J. **72** (1967) 35.
71S Swarup, G., Sarma, N.V.G., Joshi, M.N., Kapahi, V.K., Bagri, D.S., Damle, S.H., Ananthakrishnan, S., Balasubramanian, V., Bhave, S.S., Sinha, R.P.: Nature Phys.Sci. **230** (1971) 185.
72R Ryle, M.: Nature **239** (1972) 435.
73B Baars, J.W.M., Van der Brugge, J.F., Casse, J.L., Hamaker, J.P., Sondaar, L.H., Visser, J.J., Wellington, K.J.: Proc. IEEE **61** (1973) 1258.
73R Roger, R.S., Costain, C.H., Lacey, J.D., Landecker, T.L.: Proc. IEEE **61** (1973) 1270.
74H Haslam, C.G.T.: Astron. Astrophys. Suppl. **15** (1974) 333.
79E Emerson, D., Klein, U., Haslam, C.G.T.: Astron. Astrophys. **76** (1979) 92.
79K Korolkov, D.V., Pariskii Y.N.: Sky & Telescope **57** (1979) 324.
81P Phillips, T.G., Woody, D.P., Dolan, G.J., Miller, R.E., Linke, R.A.: IEEE Trans. Magn. **MAG-17** (1981) 684.
81R Rogers, A.E.E., Moran, J.M.: IEEE Trans. Instrum. Meas. **IM-30** (1981) 283.
81W Wells, D.C., Greisen, E.W., Harten, R.H.: Astron. Astrophys. Suppl. **44** (1981) 363.
82G1 Goldsmith, P.F., in: Infrared and Millimeter Waves, Vol.6 (K. Button, ed.) Academic Press (1982) p.277.
82G2 Gundlach, K.H., Takada, S., Zahn, M., Hartfuss, H.J.: Appl. Phys. Lett. **41** (1982) 294.
82P Phillips, T.G., Woody, D.P.: Annu. Rev. Astron. Astrophys. **20** (1982) 285.
83A Akabane, K., Morimoto, M., Kaifu, N., Ishiguro, M.: Sky & Telescope **66** (1983) 495.
83B Baars, J.W.M., in: Infrared and Millimeter Waves, Vol.9 (K. Button, ed.) Academic Press (1983) p.241.
83N Napier, P.J., Thompson, A.R., Ekers, R.D.: Proc. IEEE **71** (1983) 1295.
84A Archer, J.W.: IEEE Trans. Microwave Theory Tech. **MTT-32** (1984), 421.
84G Gordon, M.A.: Sky & Telescope **67** (1984) 326.
84W Welch, W.J., Thornton, D.D., in: Proc. Int. Symp. on Millimeter and Submillimeter Radioastronomy, Granada (1984) p.53.
85A Archer, J.W.: Proc. IEEE **73** (1985) 109.
85B Baldwin, J.E., Boysen, R.C., Hales, S.E.G., Jennings, J.E., Waggett, R.C., Warner, P.J., Wilson, D.M.A.: Mon. Not. R. Astron. Soc. **217** (1985) 217.
85C Christiansen, W.N., Högbom, J.A.: Radiotelescopes, 2nd ed., Cambridge University Press (1985).
85E Ediss, G.A., Wang, S-J., Keen, N.J.: IEE Proc., Part H: Microwaves, Opt. Antennas **H-132** (1985) 99.

85G Gesu, V. di, Scarsi, L., Crane, P., Friedman, J.H., Levialdi, S. (eds.): Data analysis in astronomy, New York-London: Plenum Press (1985).

85K Kellermann, K.I., Thompson, A.R.: Science **229** (1985) 123.

85U Ury, W.L., Thornton, D.D., Hudson, J.A.: Publ. Astron. Soc. Pacific **97** (1985) 745.

85V Veldt, B.G., Landecker, T.L., Vaneldik, J.F., Dewdney, P.E.: Radio Sci. **20** (1985) 1118.

85W Weinreb, S.: IEEE Trans. Instrum. Meas. **IM-34** (1985) 670.

86C Cha, A.G.: IEEE Trans. Antennas Propag. **AP-34** (1986) 992.

86G Godwin, M.P., Schoessow, E.P., Grahl, B.H.: Astron. Astrophys. **167** (1986) 390.

86K Keen, N.J., Mischerikow, K.D., Ediss, G.A., Perchtold, E.: Electron. Lett. **21** (1986) 353.

86R Rohlfs, K.: Tools of Radio Astronomy, Springer Verlag (1986).

86T1 Taylor, J.H., Davis, M.M. (Eds.): Proc. Arecibo Upgrading Workshop, NAIC Report, Oct. 1986.

86T2 Thomasson, P.: Q.J.R. Astron. Soc. **27** (1986) 413.

86T3 Thompson, A.R., Moran, J.M., Swenson, G.W.: Interferometry and Synthesis in Radio Astronomy, John Wiley & Sons (1986).

87B Baars, J.W.M., Hooghoudt, B.G., Mezger, P.G., de Jonge, M.J.: Astron. Astrophys. **175** (1987) 319.

87H Harris, A.I., Jaffe, D.T., Stutzki, J., Genzel, R.: Int. J. Infrared and Millimeter Waves **8** (1987) 857.

87I IKI-USSR Acad.Sci., preprint No. 1266 (1987).

87L Leighton, R.B.: A 10-m Telescope for Millimeter and Submillimeter Astronomy, Calif. Inst. Tech. (1987).

87M Mattauch, R.J., Crowe, T.W.: Int. J. Infrared and Millimeter Waves **8** (1987) 1235.

87P Peters, H., Owings, B., Oakley, T., Beno, L.: 41st Symp. Frequency Control; IEEE Ultrasonics, Ferroelectrics & Frequency Control Soc. (1987) p.75.

87R Röser, H.P., Schäfer, F., Schmid-Burgk, J., Schultz, G.V., van der Wal, P., Wattenbach, R.: Int. J. Infrared and Millimeter Waves **8** (1987) 1540.

88B1 Baars, J.W.M., Greve, A., Hooghoudt, B.G., Penalver, J.: Astron. Astrophys. **195** (1988) 364.

88B2 Blundell, R., Carter, M., Gundlach, K.H.: Int. J. Infrared and Millimeter Waves **9** (1988) 361.

88G1 Goldsmith, P.F. (ed.): Instrumentation and Techniques for Radio Astronomy, (reprint collection) IEEE Press (1988).

88G2 Grosbol, P., Harten, R.H., Greisen, E.W., Wells, D.C.: Astron. Astrophyys. Suppl. **73** (1988) 359.

88H Hirabayashi, H., in: Proc. IAU Symp. 129 on Impact of VLBI on Astrophysics and Geophysics (1988) 449.

88K Kardachev, N.S., Slysh, V.I., in: Proc. IAU Symp. 129 on Impact of VLBI on Astrophysics and Geophysics (1988) 433.

88N Norris, R.P.: Sky & Telescope **76** (1988) 615.

88P Pospieszalski, M.J., Weinreb, S.: IEEE Trans. Microwave Theory Tech. **MTT-36** (1988) 552.

88S Sukumar, S., Velusamy, T., Rao, A.P., Swarup, G., Bagri, D.S., Joshi, M.N., Ananthakrishnan, S.: Bull. Astron. Soc. India **16** (1988) 93.

89A Alef, W., in: Very Long baseline Interferometry (M. Felli, R.E. Spencer, eds.) Kluwer Acad. Publ. (1989) p.97.

89B1 Bathker, D.A., Cha, A.G., Roshblatt, D.G., Seidel, B.L., Slobin, S., in: Proc. Int. Symp. Antennas and Propagation, Tokyo, Japan (1989) paper 3B4-3.

89B2 Booth, R.S., Delgado, G., Hagström, M., Johansson, L.E.B., Murphy, D.C., Olberg, M., Whyborn, N.D., Greve, A., Hansson, B., Lindström, C.O., Rydberg, A.: Astron. Astrophys. **216** (1989) 315.

89E Ellison, B.N., Schaffer, P.L., Schaal, W., Miller, R.E., Vail, D.: Int. J. Infrared and Millimeter Waves **10** (1989) 937.

89F Felli, M., Spencer, R.E., (eds.): Very Long Baseline Interferometry, Kluwer Acad. Publ. (1989).

89H Herouni, P.M.: The first radio-optical telescope, in: Proc. 6th Int. Conf. Antennas and Propagation (ICAP89), Univ. Warwick, Coventry, UK, (1989) p.540.

89S Schieder, R., Tolls, V., Winnewisser, G.: Exp. Astron. **1** (1989) 101.

90B Baars, J.W.M., Martin, R.N., in: Proc. 29th Liège Int. Astrophys. Colloq., ESA SP-314 (1990) 293.
90E Erickson, N.R., in: Proc. 1st Symp. Space THz Technology, Ann Arbor, Michigan (1990) p.399.
90F1 Finkelstein, A.M., et al., in: Proc. IAU Symp. 141 on Inertial Coordinate System on the Sky, Leningrad (1989), (J.H. Lieske and V.K. Abalakin, eds.) Kluwer Press (1990) p.293.
90F2 Frisk U., in: Proc. 29th Liège Int. Astrophys. Colloq., ESA SP-314 (1990) 37.
90G Gear, W.K., Cunningham, C.R., in: Proc. 29th Liège Int. Astrophys. Colloq., ESA SP-314 (1990) 353.
90I Ishiguro, M. et al., in: Submillimetre Astronomy (G.D. Watt and A.S. Webster, eds.) Kluwer Acad. Publ. (1990) p.89.
90K1 Kerr, A.R., Pan, S-K.: Int. J. Infrared and Millimeter Waves **11** (1990) 1169.
90K2 Kreysa, E., in: Proc. 29th Liège Int. Astrophys. Colloq., ESA SP-314 (1990) 265.
90M1 Martin, R.N., Baars, J.W.M.: Instrum. Astron. VII, Proc. SPIE **1235** (1990) 503.
90M2 Masson, C., in: Proc. 29th Liège Int. Astrophys. Colloq., ESA SP-314 (1990) 161.
90S Sutton, E.C., in: Proc. 29th Liège Int. Astrophys. Colloq., ESA SP-314 (1990) 199.
91B1 Bos, A.: IEEE Trans. Instr. Meas. **IM-40** (1991) 591.
91B2 Brown, R.L., in Radio Interferometry (T.J. Cornwell and R.A. Perley, eds.) Astron. Soc. Pacific Conf. Ser. **19** (1991) 410.
91P1 Padin, S., Scott, S.L., Woody, D.P., Scoville, N.Z., Seling, T.V., Finch, R.P., Giovanine, C.J., Lawrence, R.P.: J. Astron. Soc. Pacific **103** (1991) 461.
91P2 Poglitsch, A., Beeman, J.W., Geis, N., Genzel, R., Haggerty, M., Haller, E.E., Jackson, J., Rumitz, M., Stacey, G.J., Townes, C.H.: Int. J. Infrared and Millimeter Waves **12** (1991) 859.
91P3 Pospieszalski, M.J., in: Radio Interferometry (T.J. Cornwell and R.A. Perley, eds.) Astron. Soc. Pacific Conf. Ser. **19** (1991) 60.
91S Swarup, G., Ananthakrishnan, S., Kapahi, V.K., Rao, A.P., Subrahmanya, C.R., Kulkarni, V.K.: Curr. Sci. **60** (1991) 95.
91V Vanden Bout, P.A.: Adv. Space Res. **11** (1991) 387.
92E Erickson, N.R., Goldsmith, P.F., Novak, G., Grosslein, R.M., Viscuso, R.J., Erickson, R.B., Predmore, C.R.: IEEE Trans. Microwave Theory Tech. **MTT-40** (1992) 1.
92G Guilloteau, S. et al.: Astron. Astrophys. **262** (1992) 624.
92M Masson, C. (ed.): SMA Design Study, Smithonian Institution (1992).
92Z Zimmermann, R., Zimmermann, Ra., Zimmermann, P., in: Proc. 3rd Int. Symp. Space THz Technology, Ann Arbor, Michigan (1992) p.706.

2 Positions and time determination, astronomical constants

2.1 Determination of astronomical latitude and longitude

2.1.1 Introduction

Textbooks:
Mueller, I.I.: Spherical and practical astronomy applied to geodesy [69M].
Ramsayer, K.: Geodätische Astronomie [70R].
Tardi, P., Laclavère, G.: Astronomie géodésique de position [55T].
Woolard, E.W., Clemence, G.M.: Spherical astronomy [66W].
Sigl, R.: Geodätische Astronomie [91S].
Nautical Almanac, Explanatory Supplement [74NA].

Positions on the earth may be designated by the astronomical latitude and longitude. These coordinates express the direction of the local gravity (astronomical or natural vertical) with respect to a terrestrial reference system. Astronomical and geophysical observations are referred to this system; geodetic reference systems are attached to it.

Because of the motion of the rotational axis of the earth within the earth's body the directly observed (instantaneous) astronomical coordinates have to be corrected to get time-invariant values (cf. subsect. 2.1.2).

Due to the existence of gravity anomalies which influence the direction of the astronomical vertical, astronomical latitudes and longitudes cannot be used to determine geometric positions of stations with respect to other ones or with respect to the centre of the earth. The angle between the astronomical vertical at a station and the corresponding normal of the earth ellipsoid may be up to 10 arcsec in regions of smooth elevations and plains [63S]; in the mountains such deflections of the vertical can reach values up to $20''$ to $50''$ [63S, 57K, 76F] with gradients $> 10''$/km [63S].

2.1.2 Polar motion

The earth's axis of rotation R continually changes its direction slightly within the earth. The term polar motion refers to the motion of a point where – near the north pole – R intersects the surface of the earth, with respect to the conventional origin of the coordinates of the pole. Until the end of the seventies R was given by the instantaneous axis of rotation, which is based on a kinematic definition. According to this fact R displayed a daily motion in the order of $0''.02$.

For the Celestial Ephermeris Pole introduced in 1979 at least conceptually no diurnal motion exists relative to the earth crust as well as to the space [88M].

Polar motion is (at least after the new concept) entirely geophysical in origin and must be determined by observations. Suitable measurements were done until the mid-eighties by optical astronomical methods, but have been replaced in the meantime by space geodetic techniques (e.g. laser ranging to orbiting targets in space, radio interferometry with quasars) because of the much better resolution of the estimated parameters [8, 77IAU, 87E, 90N, 85AGU, 89S].

2.1.3 Definition and observation of astronomical latitude and longitude

The instantaneous latitude φ_S of a station S is the (from observations derived) angle between the direction of the astronomical vertical at S and any plane perpendicular to R. Latitudes are positive on the northern hemisphere.

If M_S is a plane which contains the astronomical vertical at S and a parallel to R, and L the corresponding plane at a designated origin of longitudes, then the angle from L to M_S measured eastward $0^\circ \cdots 360^\circ$ or $0^h \cdots 24^h$ is the instantaneous longitude λ_S of S.

(It should be kept in mind that until 1983 longitudes west of L had positive signs! The new convention is due to a recommendation of the International Astronomical Union from 1982.)

Conventional latitudes and longitudes, $\varphi_{O,S}$ and $\lambda_{O,S}$, respectively, refer to that axis and those planes for which the pole is the adopted origin of the polar coordinates. The conventional origin O of longitude is a fictitious point of 0° conventional latitude; it is thus not affected by polar motion. According to the choice of O, the mean instantaneous longitude of the Greenwich transit circle, old site, is nearly zero; its conventional longitude is about $-0^s.02$ [2/1968, 70R, 71O].

Measurements of star directions with regard to the local vertical, together with appropriate timing, provide the instantaneous latitude and/or the local time. The instantaneous longitude is obtained from the local astronomical time by comparing the astronomical time observed nominally at O [69M, 70R, 91S]. The observed instantaneous coordinates φ_S and λ_S are related to $\varphi_{O,S}$ and $\lambda_{O,S}$ through

$$\varphi_{O,S} - \varphi_S = -x \cos\lambda_S + y \sin\lambda_S,$$

$$\lambda_{O,S} - \lambda_S = -(x \sin\lambda_S + y \cos\lambda_S) \tan\varphi_S.$$

x, y are the coordinates of the pole, generally expressed as angular displacements ($0''.001$ equivalent to 3.1 cm in linear displacement), $+x$ directed to 0°, $+y$ to 270° longitude (90°W). Neglecting geophysical effects, observational and other errors, changes of system and constants, the values of $\varphi_{O,S}$ and $\lambda_{O,S}$ should be time independent.

Astronomical observatories dedicated for regular monitoring of the local latitude or longitude (time) apply photographic zenith tubes [69M, 70R], astrolabes from the Danjon and circumzenithal type [91S], or visual zenith telescopes [69M, 57M]. Field observations for geodetic or geophysical purposes can be performed with portable astrolabes [69M, 70R, 91S], circumzenithals [69M, 91S], photographic zenith cameras [75G], visual transit instruments [69M, 70R, 55T, 91S] and geodetic theodolites [69M, 70R, 55T, 91S]. Most of the mentioned instruments yield simultaneously latitude and longitude (time) from an appropriate set of observations.

The standard deviation of the results for one night can be assumed between $0''.06$ and $0''.10$ for latitude and between $0^s.005$ and $0^s.007$ for longitude or time if observed with instruments like photographic zenith tubes, Danjon astrolabes of photoelectric transit instruments. The long-term stability of these results (derived from the standard deviation of annual versus 10-years mean values) is around $0''.03$ for latitude and $0^s.003$ for longitude or time. Field observations yield standard deviations of about $0''.8$ to $0''.2$ both for latitude and longitude (single-night results) [69M, 70R, 91S]. Visual observations of longitude often reflect systematic errors ("personal equation").

Due to the progress in the development of satellite-based navigation systems [89S, 80J] which allow high-precision geodetic positioning – independent of weather conditions, gravity disturbances and which are able to give real-time results – the application of astronomic positioning has been decreased substantially. However, scientific problems which deal with the direction of the local vertical depend still on the use of astronomic techniques. The orientation of the global reference system established by the astronomic methods will be maintained with some modifications also in the future (subsect. 2.1.4).

2.1.4 Definition and realization of the terrestrial reference system

2.1.4.1 Origin of the coordinates of the pole

By international agreement the coordinates of the pole x, y are referred to the Conventional International Origin (CIO). The CIO approximately corresponds to a mean pole position during the period 1900–1905, based on continuous observations of the five observatories of the International Latitude Service (ILS), which was established in 1899 [79M], see Fig. 1. The latitudes of the five ILS stations are adopted ones (conventional system).

Table 1. Station latitudes defining the CIO [4, 5, 68IAU].

Station	Longitude	Latitude
Mizusawa	+ 141°08′	39°08′03″.602
Kitab	+ 66 53	01.850
Carloforte	+ 8 19	08.941
Gaithersburg	- 77 12	13.202
Ukiah	- 123 13	12.096

The regular latitude observations taken at these stations with zenith telescopes were evaluated by a central bureau to obtain instantaneous values for x and y of the pole. From 1962–1987 the bureau also determined the pole position from all adequate astronomical data available under the designation of the International Polar Motion Service (IPMS). The results from IPMS were nominally on CIO through the adoption of the initial coordinates of the collaborating observatories (50 ··· 60 instruments). ILS and IPMS results were published monthly [1] and annually [2, 3, 4].

In 1987 the International Earth Rotation Service (IERS) was established by both the International Astronomical Union and International Union for Geodesy and Geophysics. The new service started operation in 1988 and replaced the former IPMS and ILS [7/1989]. IERS is mainly based on a set of about 60 observatories, which use high-precision techniques of space geodesy (cf. subsect. 2.1.2).

IERS is responsible for defining and maintaining of conventional terrestrial and celestial reference systems and the connection of these systems by the earth orientation parameters (i.e. the terrestrial and celestial coordinates of the pole and universal time).

The terrestrial coordinates of the IERS-observatories are no longer assumed to be time-invariant: a tectonic motion model, the tides of the solid earth, a no-global-net-rotation and -translation condition are applied to stabilize the IERS Terrestrial Reference Frame (ITRF). The space geodetical techniques yield Cartesian coordinates x, y, z of the observatories related to the centre of the earth's mass (as far as satellite orbits are processed from the observations). The axis of the ITRF is conventional, through this fact it could be connected with CIO during a time span (1978–1987) of observations both with astronomical and space geodetical techniques. Thus the IERS Reference Pole (IRP) and the IERS Reference Meridian (IRM) are adjusted to be consistent with the corresponding directions of the former astronomic services [7/1989]. The coordinates of the pole, the variation of universal time (cf. subsect. 2.1.4.2), and empirical corrections of both nutation components are published in monthly circulars [5, 6] and annually [7, 8].

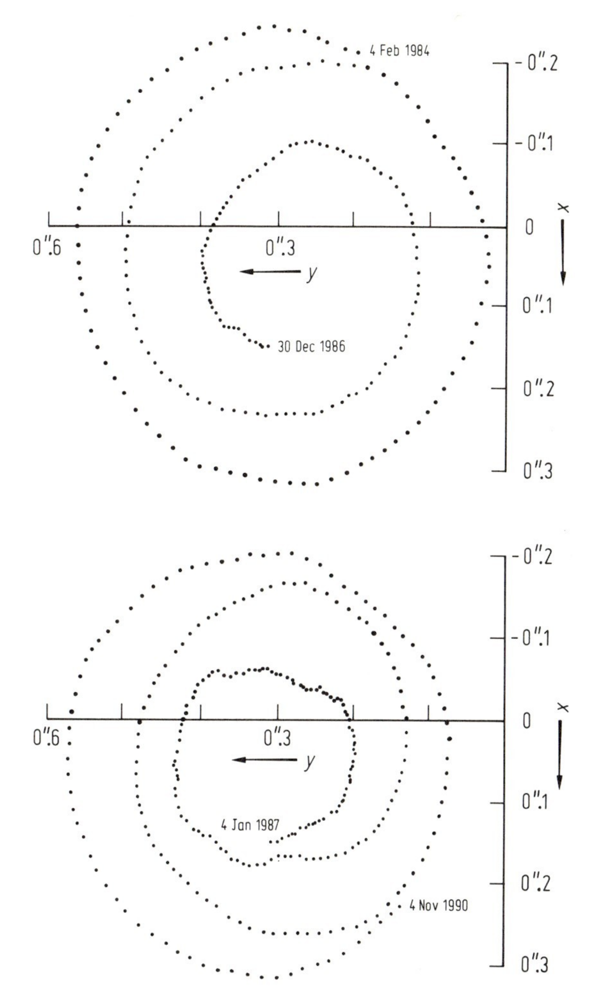

Fig. 1. Polar Motion 1984–1990 (from 7/1990) The origin of the coordinate system was the mean pole position of 1900–1905

2.1.4.2 Origin of longitudes

The origin of longitudes (cf. subsect. 2.1.3) has been defined by the weighted values λ_0 of a number of astronomical observatories, as determined by the Bureau International de l'Heure (BIH) in the "1968 BIH System" [2]. Important contributions came from photographic zenith tubes, Danjon astrolabes, and photoelectric transit instruments. Due to the fact that most of the observatories contributed both to universal time and polar motion by the same data set BIH computed coordinates of the pole too. The BIH reference pole as part of the "1968 BIH System" has been adjusted empirically to the CIO in 1967 and was kept stable independently until 1987 [88M].

The study of earth rotation results in astronomical time on O and the operational coordinates of the pole. In practice this time (denoted by UT1) is referred to UTC which is available worldwide through time signals (see sect. 2.2). Values of UT1-UTC and the length of the day (LOD) are contained in the publications of the IERS mentioned at the end of the last section (2.1.4.1).

2.1.5 See LB VI/2a

2.1.6 Coordinates of observatories

Coordinates of astronomical observatories can be taken from [9]. This reference contains a separate list of optical and radio observatories. Precise coordinates of the space geodetic observatories which represent the global reference frame are published in [10].

2.1.7 See LB VI/2a

2.1.8 Further comments

Polar motion includes two semiperiodic, rotating (counter-clockwise at north) components, its power spectrum is dominated by the Chandler wobble peak and the annual oscillation (see Fig. 1 in subsect. 2.1.4.1). While the former represents a free motion, the second is a forced wobble driven by seasonal redistributions of mass within and between the oceans, atmosphere, ground and surface waters. The exitation mechanism of the Chandler wobble is still unsolved in detail: a combination of seismic, aseismic and meteorological factors could possibly maintain this component of polar motion [88L].

The small secular drift of the mean pole at a rate of $0''.002$ to $0''.003$ per year towards 75° W is attributable to the delayed response of the earth to the melting of the polar ice caps in the late Pleistocene [88L, 80N, 84W, 82Y].

References for 2.1

Circulars and annual publications

1 Bureau International de l'Heure: Circular D (monthly), from 1966 to 1987, Paris.
2 Guinot, B. et al.: Bureau International de l'Heure, Rapport Annuel pour 1967···.
3 International Polar Motion Service: Monthly Notes···, Mizusawa 1962–1987.
4 Yumi, S.: Annual Report of the International Polar Motion Service for the year···, Mizusawa 1962–1987.
5 US Naval Observatory: IERS Bulletin-A and NEOS Earth Orientation Bulletin (weekly issue), Washington 1988···.
6 Bureau Central de l'IERS, Observatoire de Paris: IERS Bulletin B (monthly issue), Paris 1988···.
7 Central Bureau of IERS, Observatoire de Paris, Annual Report for 1989···, Paris 1990···.
8 U.S. Naval Observatory and the National Geodetic Survey: The U.S. National Earth Orientation Service, Annual Report for···, Washington.
9 H.M. Nautical Almanac Office, Royal Greenwich Observatory and Nautical Almanac Office, US Naval Observatory: The Astronomical Almanac for the Year 1981,···. London and Washington. (The publication was entitled "The Astronomical Ephemeris for the Year···" in the UK-version and "The American Ephemeris and Nautical Almanac" in the US-version until 1980).
10 Central Bureau of IERS: IERS Nos. 1, 4, 6, Observatoire de Paris (1989, 1990, 1991).

Special references

55T Tardi, P., Laclavere, G.: Traité de géodésie, Tome II, Astronomie géodésique de position. Paris: Gauthier-Villards (1955).
57K Kaula, M.: Trans. Am. Geophys. Union **38** (1957) 578.
57M Melchior, P.J.: Commun. Obs. Royal Belgique **130** (1957).
63S Straub, G.: Veröff. Deutsche Geod. Komm., Reihe C, Heft 65. München: Bayer. Akad. d. Wiss. (1963).
66W Woolard, E.W., Clemence, G.M.: Spherical astronomy. New York and London: Academic Press (1966).
68IAU Trans. Int. Astron. Union **13B** (1968) 49, 111.
69M Mueller, I.I.: Spherical and practical astronomy as applied to geodesy. New York: Ungar (1969).
70R Ramsayer, K.: Geodätische Astronomie, in: Handbuch der Vermessungskunde, Bd. IIa (Jordan, W., Eggert, O., Kneissl, M., eds.). Stuttgart: Metzler (1970).
71O O'Hora, N.P.J.: Observatory **91** (1971) 155.
74NA Nautical Almanac Offices of the United Kingdom and the United Staates: Explanatory Supplement to the Astronomical Ephemeris and Nautical Almanac. London: H.M. Stationary Office (1974).
75G Gessler, J.: Entwicklung und Erprobung einer transportablen Zenitkamera für astronomisch-geodätische Ortsbestimmungen. Hannover: Veröff. d. Geod. Inst. (1975).
76F Fischer, I.: Int. Hydrographic Rev. **53** (1976) No. 1.
77IAU Trans. Int. Astron. Union **16B** (1977) 60, 155.
79M Moritz, H.: Concepts in Geodetic Reference Frames. Ohio State University (1979) Rep. No. 294.
80IN The Institute of Navigation: The Global Positioning System. Washington (1980).
80J Janiczek, P.M. (ed.): Global Positioning System. Washington (DC): The Institute of Navigation (1980).
80N Nakiboglu, S.M., Lambeck, K.: Deglaciation effects on the rotation of the Earth. Geophys. J. **62** (1980) 49-58.
82Y Yuen, D.A., et al.: The viscosity of the lower mantle as inferred from rotational data. J. Geophys. Res. **87**, No. A 10 (1982) 745.
84W Wu, P., Peltier, W.R.: Pleistocene deglaciation and the Earth's rotation: a new analysis. Geophys. J. **76** (1984) 753-791.
85AGU American Geophysical Union: Lageos Scientific Results. J. Geophys. Res. **90**, No. B11 (1985) 9235-9248.
87E Eubanks, T., Steppe, J.: The Long Term Stability of VLBI Earth Orientation Measurements. Pasadena: Jet Propulsion Lab. (1987) No. 162.
88B Babcock, A., Wilkins, G. (eds.): The Earth's Rotation and Reference Frames for Geodesy and Geodynamics, IAU-Proc. (Symp. No. 128), Dordrecht: Kluwer Academic Publ. (1988).
88L Lambeck, K.: The Earth's Variable Rotation: Some Physical Causes, in [88B], pp. 1-20.
88M Mueller, I.I.: Reference Coordinate Systems: An Update. Ohio State University (1988), Rep. No. 394.
89S Seeber, G.: Satellitengeodäsie. Berlin: de Gruyter (1989).
90N Newhall, X., et. al.: Earth Rotation (UTO-UTC) from Lunar Laser Ranging. Pasadena: Jet Propulsion Lab. (1990) No. 195.
91S Sigl, R.: Geodätische Astronomie. Karlsruhe: Wichmann (1991).

2.2 Time determination

General references

Kovalevsky, J., et al.: Reference frames in astronomy and geophysics [89K]
Vanier, J., Audoin, C.: The quantum physics of atomic frequency standards [89V]
Kartaschoff, P.: Frequency and time [78K]

2.2.0 General trend

During the last ten years a considerable evolution has taken place in time determination:

- atomic time is increasingly accurate and its use is widening,
- Ephemeris Time (ET) is no longer in use for present observations, but is replaced by forms of atomic time (however, ET remains an essential tool for observations prior to 1955),
- although Universal Time is still approximately the basis of civil time, through the Coordinated Universal Time (UTC), it should now be considered only as a parameter of the Earth's orientation in space,
- the discovery of millisecond pulsars, in 1982, has had an important impact on time determination,
- the need for a global relativistic treatment of space-time references has led to new definitions.

A consequence of this evolution is a reorganization of the international services dealing with the rotation of the Earth and with time. On the 31st December 1987, the Bureau International de l'Heure and the International Polar Motion Service ceased to exist. The evaluation of the Earth's orientation parameters is now the responsibility of the International Earth Rotation Service (IERS) [Central Bureau, Observatoire de Paris, 61, avenue de l'Observatoire, 75014 Paris, France]. The responsibility of establishing the International Atomic Time (TAI) was transferred to the Bureau International des Poids et Mesures (BIPM) [Pavillon de Breteuil, 92312 Sèvres Cedex, France].

2.2.1 See LB VI/2a

2.2.2 Sidereal, solar, and universal time

As in LB VI/2a, p. 70,

s_U = second of mean solar time,
s^* = second of mean sidereal time,
s = second of the International System of Units (atomic second).

Since 1984, the Greenwich mean sidereal time (GMST) at 0^h UT1 is given by

$$\text{GMST at } 0^h \text{ UT1} = (24\,110.548\,41 + 8640\,184.812\,866\, T'_U + 0.093\,104\, T'^2_U - 6.2 \cdot 10^{-6} T'^3_U)\, s^*,$$

where

$$T'_U = N/36525,$$

N being the number of days of Univerval Time elapsed since JD 2451 545.0 (2000 January 1, 12^h UT1), taking therefore values ±0.5, ±1.5, ±2.5, ... [82A]. This expression represents the definition of Universal Time, to be used with the fundamental catalogue FK5-based astronomical system.

The relation between the sidereal day, d*, and the day of Universal Time, d_U, becomes

$$1\,d^* = (0.997\,269\,566\,329 - 5.87 \cdot 10^{-11} T' + 6 \cdot 10^{-15} T'^2)\, d_U,$$

where T' is expressed in Julian centuries elapsed since JD 2451 545.0. Approximately

$$1\ d^* = 23\ h_U\ 56\ m_U\ 4.0905\ s_U.$$

The mean angular velocity $\langle\omega\rangle$ of the Earth's rotation, over an interval of duration Δ(TAI) of International Atomic Time, is given by

$$\langle\omega\rangle = (1 + R)\ 7.292\ 115\ 146\ 7064 \cdot 10^{-5}\ \text{rad/s},$$

with $R = \Delta(\text{UT1} - \text{TAI})/\Delta(\text{TAI})$, Δ(UT1 − TAI) being the variation of UT1 − TAI during the same interval. Values of UT1 − TAI: [1, 2].

2.2.3 See LB VI/2a

2.2.4 Atomic time

Relativistic definition of the International Atomic Time (TAI): see subsect. 2.2.9.

Generation of TAI: [89G].

Deviation of the duration of the TAI scale interval from nominal $< 5 \cdot 10^{-14}$ since 1984 (estimates published in [3] and [4]).

Relationship between UTC and TAI: [4].

2.2.5 See LB VI/2a

2.2.6 Fluctuations of the Earth's rotation speed

Continuation of Table 7 in LB VI/2a, p. 76:

Date	ΔT [s]	Date	ΔT [s]
1982.0	52.17	1988.0	55.82
1983.0	52.96	1989.0	56.30
1984.0	53.79	1990.0	56.86
1985.0	54.34	1991.0	57.57
1986.0	54.87	1992.0	58.31
1987.0	55.32	1993.0	59.12

2.2.8 Pulsar time

The first millisecond pulsar (PSR1937+21) was discovered in 1982 [82B]. After corrections required by a period increase, linear with time, and by the motions of the Earth, the frequency stability of this pulsar seems to surpass that of the best atomic time scales. A few other millisecond pulsars have subsequently been observed. Respective role of atomic time and millisecond pulsars: [91G, 91T].

Table 1. Approximate periods and period derivatives of 3 millisecond pulsars, at date t_o [91R].

Pulsar	t_o [JD]	Period [ms]	Period derivative [10^{-20} s s^{-1}]
PSR1855+09	2447526	5.362 100 454 04	1.783
PSR1937+21	2447500	1.557 806 468 82	10.512
PSR1957+20	2447700	1.607 401 684 08	1.685

2.2.9 Relativity in time determination

2.2.9.1 Proper time and coordinate time

Proper time is given by an ideal clock on its world line: it is a measurable quantity. It is assumed that atomic clocks are good realizations of ideal clocks. The SI second (atomic second) must be considered as a proper unit (i.e. realized locally, for instance with a standard fixed in the laboratory where it is used) [91BI]. The use of proper units guarantees the universality of the laws of physics in local experiments. Proper times of identical clocks ususally diverge; for example, two identical clocks, fixed on the ground, at altitudes differing by 2 km, diverge at a rate of 6 µs/a.

Coordinate time is one of the four coordinates in space-time. It is linked to proper time through a relativistic metric form whose choice is not imposed a priori. When the metric form is selected, coordinate time offers a unique and unambiguous way of dating events in an extended space. Thus coordinate times are convenient for worldwide time reference and for astronomy.

In 1991, the International Astronomical Union (UAI) recommended the metric forms to be used for space-time references and specified that the graduation unit (or scale unit) of the space and time coordinate axes should be consistent with the SI metre for proper length and the SI second for proper time [92IAU].

2.2.9.2 Time metrology on the Earth

In 1991, the IAU defined an ideal Geocentric Coordinate Time (TCG). The TCG scale diverges from the proper time of an ideal clock fixed on the Earth, at sea level, by 22 ms/a. In order to avoid the inconvenience of this divergence, the IAU has also defined a Terrestrial Time (TT), which is identical to TCG, but has for scale unit the SI second at sea level. The relation between TCG and TT is, in second,

$$\mathrm{TCG} - \mathrm{TT} = (\mathrm{JD} - 2443\,144.5) \cdot 6.021\,47 \cdot 10^{-5}.$$

Terrestrial Time (TT) is identical to Terrestrial Dynamical Time defined previously (LB VI/2a, subsec. 2.2.4.4, p. 74). The scale TAI must be considered as a realization of TT (but with a time offset so that TT prolongs ET, without time step: TT − TAI = 32.184^{s} on 1977 January 1, 0^{h} TAI). This is in conformity with the relativistic definition of TAI given in 1980 [80BI]:

"TAI is a coordinate time scale defined in a geocentric reference frame with the SI second as realized on the rotating geoid as scale unit."

In addition, for the establishment and dissemination of TAI, it is necessary to adopt a synchronization convention. The concept of "coordinate synchronization" is used [79A, 92IRC]: this leads to formulae which must be applied when comparing distant clocks [90IRC, Rep. 439-5].

When a user needs the SI second (of proper time) in his laboratory, he cannot use the TAI scale interval directly as he obtains it, for example by radio emissions. He has to apply a correction depending on his location and speed. For a user fixed on the Earth, at altitude h (in metres), the SI proper second is obtained by shortening the TAI scale interval (sometimes called improperly the TAI second) by an amount given approximately, in second, by gh/c^2 (g = acceleration of gravity = 9.81 m/s^2, c = velocity of light $\approx 3 \cdot 10^8$ m/s), i.e. $1.09 \cdot 10^{-13}$ s per kilometre of altitude.

2.2.9.3 Solar system dynamics

The time scale TCG can be used to study the motion of satellites of the Earth.

For the motion of planets, a Barycentric Coordinate Time (TCB) has been defined. The relation TCB − TT involves a full 4-dimensional transformation given approximately, in second, by

$$\mathrm{TCB} - \mathrm{TT} = (\mathrm{JD} - 2443\,144.5) \cdot 1.339\,636 \cdot 10^{-3} + c^{-2}\,\mathbf{v}_e \cdot (\mathbf{x} - \mathbf{x}_e) + \mathrm{P},$$

$\mathbf{x}_e$ and $\mathbf{v}_e$ denoting the barycentric position and velocity of the Earth's center of mass and $\mathbf{x}$ the barycentric position of the observer (the corresponding term having a maximum amplitude of 2.1 μs and being diurnal). P denotes a sum of periodic terms, the main term being annual, with an amplitude of 1.7 ms; development of P: [87H, 90F].

In 1976, the IAU defined a Barycentric Dynamical Time (TDB), which is equivalent to TCB but has a rate offset so that the relation TDB − TT involves only periodic terms. Thus, in second,

$$\mathrm{TDB} - \mathrm{TCB} = -(\mathrm{JD} - 2443\,144.5) \cdot 1.339\,636 \cdot 10^{-3}.$$

2.2.10 Dissemination of frequency and time

List of standard frequency and time signal emissions: [90IRC, Rep. 267-7], also [4, part C] (updated yearly). The time signals of these lists transmit UTC.

The "Global Positioning System" (GPS) of the USA, using 24 satellites, disseminates permanenty and in real time UTC (and TAI) with uncertainties of about 1 μs.

The best accuracy in reading of UTC (and TAI) is obtained through [3] and [4] which give the corrections to the readings of laboratory master clocks and to the GPS time in order to obtain UTC. The minimum uncertainty is of order 10 ns (in 1993).

References for 2.2

Circulars and annual publications

1 International Earth Rotation Service: Bulletin B (monthly), No. 1 ···, Paris (1988 ···).
2 International Earth Rotation Service: Annual Report for 1988 ···, Paris (1989 ···).
3 Bureau International des Poids et Mesures: Circular T (monthly), No. 1 ···, Sèvres (1988 ···).
4 Bureau International des Poids et Mesures: Annual report of the BIPM time section, Vol. 1, 1988 ···, Sèvres (1989 ···).

Special references

78K Kartaschoff, P.: Frequency and time. London, New York, San Francisco: Academic Press (1978).
79A Ashby, N., Allan, D.W.: Radio Sci. **14** (1979) 649.
80BI Bureau International des Poids et Mesures: Comité Consultatif pour la Définition de la Seconde, 9e session (1980).

82A Aoki, S., Guinot, B., Kaplan, G.H., Kinoshita, H., McCarthy, D.D., Seidelmann, P.K.: Astron. Astrophys. **105** (1982) 359.
82B Backer, D.C., Kulkarni, S.R., Heiles, C., Davis, M.M., Goss, W.M.: Nature **300** (1982) 615.
87H Hirayama, T., Kinoshita, H., Fujimoto, M.-K., Fukushima, T.: Proc. IAG Symp., UGGI XIX Genenal Assembly (1987).
89G Guinot, B., Thomas, C.: Annual report of the BIPM time section **1** (1989) D-3.
89K Kovalevsky, J., Mueller, I.I., Kolaczek, B. (eds): Reference frames in astronomy and geophysics. Dordrecht, Boston, London: Kluwer (1989).
89V Vanier, J., Audoin, C.: The quantum physics of atomic frequency standards. Bristol: Adam Hilger (1989).
90F Fairhead, L., Bretagnon, P.: Astron. Astrophys. **229** (1990) 240.
90IRC International Radio Consultative Committee (CCIR): Reports of the CCIR, 1990, Annex to Vol. VII, Geneva (1990), pp. 3 and 150.
91BI Bureau International des Poids et Mesures: Le Système International d'Unités (SI), 6e édition (1991) pp. 49 and 109.
91G Guinot, B., Petit, G.: Astron. Astrophys. **248** (1991) 292.
91R Ryba, M.F.: Thesis (1991) Princeton University.
91T Taylor, J.H., jr.: Proc. IEEE **79** (1991) 1054.
92IAU International Astronomical Union: Transactions of the IAU, Vol XXIB. Dordrecht, Boston, London: Kluwer (1992).
92IRC International Radio Consultative Committe (CCIR): Handbook on satellite time and frequency dissemination, Annex on relativity (in preparation).

2.3 The system of astronomical constants

The IAU (1976) System of Astronomical Constants is maintained by the International Astronomical Union, although some changes became necessary in special cases. For a list of the most important constants see LB VI/2a, p. 80. Among these constants, the "constant of nutation" has lost its basic importance. The 1980 IAU Theory of Nutation uses a slightly different value of the constant ($N=9\overset{\prime\prime}{.}2025$ instead of $9\overset{\prime\prime}{.}2055$) which is no longer a scaling factor for the nutation series. The classification of the basic constants as defining, primary, or derived constants was subject to small changes. The constant of nutation has become a derived constant. The velocity of light has become a defining constant (uncertainty $\pm$ 0). The value is that recommended by the fifteenth General Conference on Weights and Measures in 1975. It is understood that this value will remain unchanged even if the meter is redefined in terms of a different wavelength from that now used. Some of the values that are now used in the preparation of the annual ephemeris "The Astronomical Almanac" do not agree exactly with those of the IAU (1976) System of Constants. For these deviations and the other changes see the special supplement to the 1984 edition of this almanac [83A]. Special working groups of the International Astronomical Union collect new determinations of the constants of the system [91F].

References for 2.3

83A The Astronomical Almanac for the year 1984. Washington: U.S. Government Printing Office, and London: H.M. Stationery Office, (1983), Supplement, pp. S6, S8, S11, and S21.
91F Fukushima, T., in: Proc. IAU Colloq. **127** on Reference Systems (J.A. Hughes et al., eds.). Washington: US Naval Observatory (1991) pp. 27–35.

3 The solar system

3.1 The sun

3.1.1 The quiet sun

3.1.1.1 Solar global parameters

Composition: see Grevesse [84G, 89G], Aller [86A].
Colour indices [82T]: $B - V = 0.686 \pm 0.011$, $U - B = 0.183 \pm 0.020$.
Johnson-Mitchell 13-color photometric indices (synthesised) [89M].
For all other global parameters see LB VI/2a, p.82.
For variation of global parameters with solar cycle see subsect. 3.1.2.2.

References for 3.1.1.1

82T Tüg, H., Schmidt-Kaler, T.: Astron. Astrophys. **105** (1982) 400.
84G Grevesse, N.: Phys. Scr. **T8** (1984) 49.
86A Aller, L.H., in: Spectroscopy of Astrophysical Plasmas, (Dalgarno, A., Layzer, D., eds.), Cambridge: Cambridge University Press (1986) 89.
89G Grevesse, N., in: Proc. Sym. Cosmic Abundances of Matter, (Waddington, C.J., ed.), New York: American Institute of Physics (1989) 9.
89M Makarova, E.A., Knyazeva, L.N., Kharitonov, A.V.: Soviet Astron. **33** (1989) 298.

3.1.1.2 Solar interior

General review: [f].

3.1.1.2.1 Standard models

Models 1988 ··· 1990: [a,88B,88T], see Table 1. Accuracy discussed in [90M].

Table 1. Standard solar model with initial X = 0.716, Y = 0.27, Z = 0.014 [88B].

$\mathfrak{M}/\mathfrak{M}_\odot$ = mass fraction
P = pressure
T = temperature
$L/L_\odot$ = scaled luminosity
$R/R_\odot$ = scaled radius
ρ = density

$\mathfrak{M}/\mathfrak{M}_\odot$	P [10^{15} dyn cm^{-2}]	T [10^6 K]	$L/L_\odot$	$R/R_\odot$	ρ [g cm^{-3}]
0.000	229.	15.6	0.000	0.000	148.
0.104	118.	12.5	0.553	0.115	76.4
0.204	76.6	10.9	0.798	0.155	54.0
0.300	51.6	9.74	0.912	0.188	39.9
0.400	33.6	8.70	0.966	0.221	28.8
0.500	21.0	7.76	0.990	0.225	20.1
0.585	13.4	7.01	0.998	0.288	14.2
0.690	6.82	6.08	1.000	0.334	8.34
0.795	2.77	5.09	1.001	0.398	4.06
0.900	0.659	3.88	1.001	0.504	1.27
0.991	0.0144	1.36	1.000	0.802	0.0792

3.1.1.2.2 Non-standard models

Reviews: [a,88B]. For neutrino production and oscillatory modes see Table 2.

3.1.1.2.3 Solar neutrinos

Measured flux of ^{8}B neutrinos [a]:

(2.33 ± 0.25) SNU over 1970 ··· 1988 from chlorine detector.

(0.39±0.15) times value of standard solar model over 1988 ··· 1989 from neutrino-electron scattering.

For possible variation with solar cycle see subsect. 3.1.2.2.

Model neutrino fluxes: see Table 2.

3.1.1.2.4 Global oscillations

Nomenclature of modes [84D]. The modes of a spherically symmetric model are labelled by degree n and order ℓ.

For low-degree modes, the frequencies may be approximated by the quadratic expression [88B]:

$$\nu_{n,\ell} = \nu_{0\ell} + (n - n_0)\Delta\nu_0 + a_\ell (n - n_0)^2 .$$

For very low-degree modes, a linear expression is valid:

$$\nu_{n,\ell} = \nu_{00} - \ell(\ell+1)D_0 + [\Delta\nu_0 + \ell(\ell+1)d_0](n + \ell/2 - n_0) .$$

For low-order modes, the frequencies scale as

$$\nu_{n,\ell} \sim (n + \ell/2 - n_0)\Delta\nu_0 ,$$

Table 2. Calculated neutrino fluxes for various solar models.
1 SNU (solar neutrino unit) = 10^{-36} neutrino absorptions per target atom per second.

Model	Neutrino flux [SNU]	Model properties	Author	Ref.
Standard Solar	7.9 ± 2.6 5.7 ± 1.3		Bahcall and Ulrich (1988) Turck-Chièze et al. (1990)	88B a,p.125
Segregated	> 9		Cox et al. (1989)	89C
Mixed	$3.5 \cdots 9$	Inconsistent with oscillation data	Ulrich and Rhodes (1983)	83U
	< 2	Strong mixing required inconsistent with oscillation data	Lebreton and Maeder (1987)	87L
Strong internal magnetic field	< 6	Inconsistent with oscillation data	Ulrich and Rhodes (1983)	83U
low Z	3.5	Inconsistent with oscillation data	Ulrich and Rhodes (1983)	83U
	1.5	Inconsistent with oscillation data	Bahcall and Ulrich (1988)	88B
high Z	15	Consistent with oscillation data	Ulrich and Rhodes (1983)	83U
high Y	$8 \cdots 9$	Consistent with oscillation data	Bahcall and Ulrich (1988)	88B
Early mass loss	$12 \cdots 15$	$\Delta M > 1.0 M_{\odot}$ $\Delta M > 0.2 M_{\odot}$, inconsistent with oscillation data	Guzik et al. (1987) Turck-Chièze et al. (1988)	87G e,p.629
Modified energy transport	1.0	Uses Weakly-Interacting Massive Particles; consistent with oscillation data	Däppen et al. (1986), Gilliland et al. (1986)	86D 86G
Non-standard physics	1.85	^{8}B branch suppressed, oscillations same as standard model	Bahcall and Ulrich (1988)	88B

and the frequency separations as

$$\nu_{n,\ell} - \nu_{n-1,\ell+2} \sim (4\ell + 6) D_0 .$$

Observations of solar global oscillations give [a]

$$\Delta\nu_0 = 135.15\,\mu\text{Hz}, \quad D_0 = 1.487\,\mu\text{Hz}, \quad \delta_{02} = \nu_{00} - \nu_{02} = 8.8 \cdots 9.9\,\mu\text{Hz}.$$

Coefficients a_ℓ, b_ℓ, c_ℓ of fits to

$$\nu_{n,\ell} = c_\ell + b_\ell(n + \ell/2 - 21.5) + a_\ell(n + \ell/2 - 21.5)^2$$

for $0 \le \ell \le 2$ and $n = 17 \cdots 23$ given in [a].

Observed frequencies: see Table 3.

Table 3. Observed frequencies of solar global oscillations.

Mode	ℓ	n	Frequency range [μ Hz]	Quantity observed	Ref.
low degree *p*	0···3	0···13	260···2000	velocity	90A
	0···3	11···33	1800···4750	velocity	b,p.53
	0···2	17···23	2550···3300	intensity	a,p.279
	2···5	5···34	1000···5000	velocity	d,p.55
intermediate degree *p*	4···100	3···24	1500···4200	velocity, intensity	83D,88D
high degree *p*	30···1320	0···20	1600···5000	velocity	88L
f	18···36		450···650	diameter	88R
g	27···34	14···16	330···360	diameter	90R1
	1,2		10···140	intensity, velocity	b,p.83

The origin of (160.009 ± 0.001) min oscillation is still disputed [90K].

For variation of modes with solar cycle, see subsect. 3.1.2.2.

Discussion of model mode calculations [a, 90D]. Discrepancies between standard solar model results and observation smaller than uncertainties introduced by choice of equation of state and opacities [a]. The standard solar models give [a]

$$\Delta\nu_0 = 135.3 \cdots 136.78\,\mu\text{Hz}, \quad D_0 = 1.500 \cdots 1.557\,\mu\text{Hz}, \quad \delta_{02} = 10.6\,\mu\text{Hz}.$$

3.1.1.2.5 Convection zone

Thermal structure from helioseismic inversion [89S1,91C]. Depth of convection zone $(0.287 \pm 0.003)\,R_\odot$ [91C]. Similar values are found by fitting evolutionary model sequences to observed oscillation frequencies [82S, 83S1].

Dynamical model of upper convection zone [89S2].

3.1.1.2.6 Solar rotation

Oblateness

$$\epsilon = 2 \cdot \frac{R_e - R_p}{R_e + R_p} \quad (R_e \text{ and } R_p \text{ are the equatorial and polar radii})$$

appears to be variable [87D]:

$$\epsilon = (0.6 \cdots 2) \cdot 10^{-5} \quad (1983 \cdots 1985)\,.$$

Surface rotation reviewed [84S1]. Observations:

plasma 1967 ··· 1982 [83H,83S2,84S2,88U],
sunspots 1921 ··· 1982 [84H], 1874 ··· 1976 [c],
photospheric magnetic field pattern by longitude displacement method [83S2], by pattern recurrence [89S3],
coronal magnetic field pattern by extrapolation of photospheric [87H].

There is no significant variation of plasma rotation rate with height in the photosphere/chromosphere [84B, 84P].

Mean differential laws: see Table 4.

Table 4. Mean differential rotation laws, θ = heliographic latitude.

Feature	Period	Angular velocity [degree/d]	Ref.
plasma	1967 ··· 1982	$14.14 - 1.718\sin^2\theta - 2.361\sin^4\theta$	83S2
		$14.05 - 1.492\sin^2\theta - 2.605\sin^4\theta$	84S2
spots	1921 ··· 1982	$14.552 - 2.84\sin^2\theta$	84H
	1974 ··· 1976	$14.551 - 2.87\sin^2\theta$	86B
photospheric fields	1967 ··· 1982	$14.37 - 2.30\sin^2\theta - 1.62\sin^4\theta$	83S2
	1959 ··· 1985	$14.50 - 3.35\sin^2\theta + 2.37\sin^4\theta$	89S3
coronal field	1976 ··· 1985	$13.2 - 0.5\sin^2\theta$	87H

Rotation rates in [μrad s^{-1}] are obtained by multiplying by 0.202, and in [μHz] by multiplying by 0.03215.

Internal rotation found by splitting of frequencies of global oscillation modes of given ℓ, n [c,89B]. The angular velocity in the outer part of the convection zone is almost constant along radii, with a steep radial gradient at the base [89D1]. Close to the surface the values are similar to those of surface magnetic features [90R2]. The radiative core appears to rotate almost solidly, the angular velocity is similar to the mean surface value at the outer edge and about twice that value near the core [b].

Coefficients $\Omega_0(r)$, $\Omega_1(r)$, $\Omega_2(r)$ for the rotation law

$$\Omega(r,\theta) = \Omega_0(r) + \Omega_1(r)\sin^2\theta + \Omega_2(r)\sin^4\theta \quad (\theta \text{ is latitude})$$

for $0.4 \le R/R_\odot \le 1.0$ given in [g] from data in [89L].

The average value of $R_\odot(\partial\Omega/\partial r)$ is given by [90T] to be (0.57 ± 0.29)degree/d at the equator, (-0.57 ± 0.30) at $\theta = \pi/4$ and (-5.8 ± 1.9) at the poles.

Older models of a rotating, spherical convective shell [85G,86G2] are inconsistent with these results [a]. Newer models [89G,90B].

3.1.1.2.6a Large-scale flow pattern

Internal horizontal flow pattern from helioseismology [90H].

Large-scale surface flow patterns reviewed [84S1, 87A, 87B]. Upper limit of 1 ms^{-1} on flows with scales of 400 000 km or more. There is a possible polar pattern of 10 ms^{-1} with scale of 100 000 km near poles [83C]. Latitudinal temperature variation [87K].

Fitting mean differential rotation curves to line shift data leaves residuals (of the order of 10 ms^{-1}) variously interpreted as a double wave (torsional oscillation) in each hemisphere [80H], a double plus single wave [82L1], a more complex torsional shear [85S,87S1], or azimuthal convective rolls [87S2]. The structures drift from close to the pole towards the equator at a steady 2 ms^{-1} [87S1]. Other evidence for azimuthal rolls [g,86R].

Poleward meridional flow also appears in residuals [82L2] having latitude dependence [84S2]:

$$v(\theta) = [(9.1 \pm 1.5) \sin 2\theta + (4.0 \pm 0.6) \sin 4\theta] \, \mathrm{m\,s^{-1}} .$$

Reality of latitude dependence of limb effect discussed [84S1,92C]. Polar zone filaments migrate polewards at 10 ms^{-1} [82T].

Systematic drifts from sunspot proper motions [c].

3.1.1.2.7 General magnetic field

General surface field consists of network clusters and internetwork field [87Z]. Network clusters are unipolar, flux units $2 \cdot 10^{18} \cdots 3 \cdot 10^{19}$ Mx, field strength $1000 \cdots 2000$ G, static, pattern lifetime over 1 day. Intranetwork field has mixed polarities, flux units $10^{16} \cdots 10^{17}$ Mx, field strength ≤ 50 G, moving at 0.3 kms^{-1}, pattern changes over 1 h [90M].

Possible megagauss internal field [a,89D2].

For variation of general field with cycle see subsect. 3.1.2.2.

References for 3.1.1.2

Monographs

a Berthomieu, G., Cribier, M. (eds.): Inside the Sun, Dordrecht: Kluwer Academic Publishers (1990).
b Christensen-Dalsgaard, J., Frandsen, S. (eds.): IAU Symp. 123: Advances in Helio- and Asteroseismology, Dordrecht: D. Reidel Publishing Co. (1988).
c Durney, B.R., Sofia, S. (eds.): The Internal Solar Angular Velocity, Dordrecht: D. Reidel Publishing Co. (1987).
d Gough, D.O. (ed.): Seismology of the Sun and Distant Stars, Dordrecht: D. Reidel Publishing Co. (1986).
e Rolfe (ed.): Seismology of the Sun and Sun-like Stars, ESA SP-286 (1988).
f Sturrock, P.A., Holzer, T.E., Mihalas, D.M., Ulrich, R.K. (eds.): Physics of the Sun. Vol. I. The Solar Interior, Dordrecht: D. Reidel Publishing Co. (1986).
g Tuominen, I., Moss, D., Rüdiger, G. (eds.): The Sun and Cool Stars: activity, magnetism, dynamos, Berlin: Springer-Verlag (1991).

Special references

80H Howard, R., Labonte, B.J.: Astrophys. J. Lett. **239** (1980) L33.
82L1 LaBonte, B.J., Howard, R.: Sol. Phys. **75** (1982) 161.
82L2 LaBonte, B.J., Howard, R.: Sol. Phys. **80** (1982) 361.
82S Scuflaire, R., Gabriel, M., Noels, A.: Astron. Astrophys. **113** (1982) 219.
82T Topka, K., Moore, R., LaBonte, B.J., Howard, R.: Sol. Phys. **79** (1982) 231.
83C Cram, L.E., Durney, B.R., Guenther, D.B.: Astrophys. J. **267** (1983) 442.
83D Duvall, T.L, Harvey, J.W.: Nature (London) **302** (1983) 24.
83H Howard, R., Adkins, J.M., Boyden, J.E., Cragg, T.A., Gregory, T.S., LaBonte, B.J., Padilla, S.P., Webster, L.: Sol. Phys. **83** (1983) 321.
83S1 Shibahashi, H., Noels, A., Gabriel, M.: Astron. Astrophys. **123** (1983) 283.
83S2 Snodgrass, H.B.: Astrophys. J. **270** (1983) 288.
83U Ulrich, R.K., Rhodes, E.J.: Astrophys. J. **265** (1983) 551.
84B Balthasar, H.: Sol. Phys. **93** (1984) 219.
84D Deubner, F.L., Gough, D.O.: Annu. Rev. Astron. Astrophys. **22** (1984) 593
84H Howard, R.F., Gilman, P.A., Gilman, P.I.: Astrophys. J. **283** (1984) 373.
84P Pierce, A.K., Lopresto, J.C.: Sol. Phys. **93** (1984) 155.
84S1 Schröter, E.H.: Sol. Phys. **100** (1984) 141.
84S2 Snodgrass, H.B.: Sol. Phys. **94** (1984) 13.
85G Glatzmeier, G.A.: Astrophys. J. **291** (1985) 300.
85L Livi, S.H.B., Wang, J., Martin, S.F.: Aust. J. Phys. **38** (1985) 855.
85S Snodgrass, H.B.: Astrophys. J. **291** (1985) 339.
86B Balthasar, H., Vázquez, M., Wöhl, H.: Astron. Astrophys. **155** (1986) 87.
86D Däppen, W., Gilliland, R.L., Christensen-Darlsgaard, J.: Nature (London) **321** (1986) 229.
86G1 Gilliland, R.L., Faulkner, J., Press, W.H., Spergel, D.N.: Astrophys. J. **306** (1986) 703.
86G2 Gilman, P.A., Miller, J.: Astrophys. J. Suppl. **61** (1986) 585.
86R Ribes, E.: C. R. Acad. Sci. Paris, t. 302, Serie II **14** (1986) 871.
87A Andersen, B.N.: Sol. Phys. **114** (1987) 207.
87B Bogart, R.S.: Sol. Phys. **110** (1987) 23.
87D Dicke, R.H., Kuhn, J.R., Libbrecht, K.G.: Astrophys. J. **318** (1987) 451.
87G Guzik, J.A., Willson, L.A., Brunish, W.M.: Astrophys. J. **319** (1987) 957.
87H Hoeksema, J.T., Scherrer, P.H.: Astrophys. J. **318** (1987) 428.
87K Kuhn, J.R., Libbrecht, K.G., Dicke, R.H.: Nature (London) **328** (1987) 326.
87L Lebreton, Y., Maeder: Astron. Astrophys. **175** (1987) 99.
87S1 Snodgrass, H.B.: Sol. Phys. **110** (1987) 35.
87S2 Snodgrass, H.B., Wilson, P.R.: Nature (London) **328** (1987) 696.
87Z Zwaan, C.: Annu. Rev. Astron. Astrophys. **25** (1987) 83.
88B Bahcall, J.N., Ulrich, R.K.: Rev. Mod. Phys. **60** (1988) 29.
88D Duvall, T.L., Harvey, J.W., Libbrecht, K.G., Popp, B.D., Pomerantz, M.A.: Astrophys. J. **324** (1988) 1158.
88L Libbrecht, K.G., Kaufman, J.M.: Astrophys. J. **324** (1988) 1172.
88R Rabaey, G.F., Hill, H.A., Barry, C.T.: Astrophys. Space Sci. **143** (1988) 81.
88T Turck-Chièze, S., Cahen, S., Cassé, M., Doom, C.: Astrophys. J. **335** (1988) 415.
88U Ulrich, R.K., Boyden, J.E., Webster, L., Snodgrass, H.B., Padilla, S.P., Gilman, P., Schieber, T.: Sol. Phys. **117** (1988) 291.
89B Brown, T.M., Christensen-Dalsgaard, J., Dziembowski, W.A., Goode, P., Gough, D.O., Morrow, C.A.: Astrophys. J. **343** (1989) 526.
89C Cox, A.N., Guzik, J.A., Kidman, R.B.: Astrophys. J. **342** (1989) 1187.
89D1 Dziembowski, W.A., Goode, P.R., Libbrecht, K.G.: Astrophys. J. **337** (1989) L53.
89D2 Dziembowski, W.A., Goode, P.R.: Astrophys. J. **347** (1989) 540.

89G Gilman, P.A., Morrow, C.A., DeLuca, E.E.: Astrophys. J. **338** (1989) 528.
89L Libbrecht, K.G.: Astrophys. J. **336** (1989) 1092.
89S1 Sekii, T., Shibahashi, H.: Publ. Astron. Soc. Japan **41** (1989) 311.
89S2 Stein, R.F., Nordlund, Å: Astrophys. J. Lett. **342** (1989) L95.
89S3 Stenflo, J.O.: Astron. Astrophys. **210** (1989) 403.
90A Anguera Gubau, M., Pallé, P.L., Pérez Hernández, F., Roca Cortés, T.: Sol. Phys. **128** (1990) 79.
90B Brandenburg, A., Moss, D., Rüdiger, G., Tuominen, I.: Sol. Phys. **128** (1990) 243.
90D Däppen, W., Lebreton, Y., Rogers, F.: Sol. Phys. **128** (1990) 35.
90H Hill, F.: Sol. Phys. **128** (1990) 321.
90K Kotov, V.A., Tsap, T.T.: Sol. Phys. **128** (1990) 269.
90M Morel, P., Provost, J., Berthomieu, G.: Sol. Phys. **128** (1990), 7.
90R1 Rabaey, G.F., Hill, H.A.: Astrophys. J. **362** (1990) 734.
90R2 Rhodes, E.J., Cacciani, A., Korzennik, S., Tomczyk, S., Ulrich, R.K., Woodard, M.F.: Astrophys. J. **351** (1990) 687.
90T Thompson, M.J.: Sol. Phys. **125** (1990) 1.
91C Christensen-Darlsgaard, J., Gough, D.O., Thompson, M.J.: Astrophys. J. **378** (1991) 413.
92C Cavallini, F., Cepatelli, G., Righini, A.: Astron. Astrophys. **254** (1992) 381.

3.1.1.3 Solar energy spectrum

3.1.1.3.1 Absolute energy distribution

Total irradiance varies with activity [87F2]. For 1980 ··· 1987 there was a sinusoidal activity variation of amplitude $(0.039 \pm 0.003)\%$ about the average value $\bar{S} = 1367.72\ \mathrm{W\,m^{-2}}$ [88W]. For details of variation see subsect. 3.1.2.2.

Spectral distribution 0.33 ··· 1.25 μm (flux, central intensity, disk and central brightness temperatures) [84N], 0.2 ··· 0.358 μm (mean irradiance over 1.3 nm passbands, 0.3 nm apart) [87L1].

Solar UV irradiance measurements from satellites (Nimbus 7, 1978 ··· present, 160 ··· 400 nm; Solar Mesosphere Explorer, 1981 ··· present, 115 ··· 305 nm) are not consistent [87L2]. Rocket observations 118 ··· 300 nm [85M].

Review of spectrum 120 ··· 900 nm [89N].

Solar EUV irradiance (Atmospheric Explorer-E, 1976 ··· 1980, 10 ··· 121 nm) [81H,86D1].

Compilation of data 0.255 ··· 1.055 μm giving mean irradiance over 10 nm intervals [89M].

3.1.1.3.2 Relative energy distribution

Table 1. New atlases of the solar spectrum.

Wavelength interval [μm]	Author	Ref.	Data
16 ··· 2	Farmer et al. (1989)	d	central intensity
13 ··· 10	Blatherwick et al. (1982)	a	flux
5.4 ··· 1.0	Delbouille et al. (1981)	c	central intensity
1.3 ··· 0.296	Kurucz et al. (1984)	e	flux
0.195 ··· 0.1175	Cohen (1981)	b	flare intensities
0.1710 ··· 0.1175	Sandlin et al. (1986)	86S1	central and limb intensities, spot, plage

Line wavelengths and identifications: 0.00078 ··· 0.00100 μm (flare) [87F1]; 0.1335 ··· 0.1670 μm (prominence) [90E]; 0.1175 ··· 0.1710 μm [86S1]; 0.2095 ··· 0.3069 μm [f]. Diatomic molecular lines 0.76 ··· 0.81 μm [82B]. Half the lines in the solar spectrum remain unidentified.

Unblended solar lines [84R].

Lines with large Stokes V parameter in active regions outside spots 0.43 ··· 0.67 μm [86S2].

Limb darkening at 30 ··· 200 μm [84L1, 86L], at 800 μm [84L2]. Over 1980 ··· 1982 long-term variation of limb darkening at 0.4451 μm less than 0.1% [84P].

Centre-to-limb variation of various lines [88B], Na D [82P], Al I [82C]. Line asymmetries: Fe I [81D], Fe II [86D2]. Variability of line properties [90N]; latitudinal variation [85C].

Spatial variation in Lα [88F].

Line blocking coefficients: 0.33 ··· 0.686 μm [85N].

3.1.1.3.3 Limb polarization

Linear polarization 10″ from limb: 0.3165 ··· 0.4230 μm [83S1], 0.4200 ··· 0.9950 μm [83S2].

References for 3.1.1.3

Catalogues and monographs

a Blatherwick, R.D., Murcray, F.J., Murcray, F.H., Goldman, A., Murcray, D.G.: Atlas of South Pole IR Solar Spectra 760–960 cm^{-1}, Appl. Opt. **21** (1982) 2658.

b Cohen, L.: An Atlas of Solar Spectra Between 1175 and 1950 Angstroms Recorded on Skylab with the NRL's Apollo Telescope Mount Experiment, NASA Ref. Pub. 1069 (1981).

c Delbouille, L., Roland, G., Brault, J., Testerman, L.: Photometric Atlas of the Solar Spectrum from 1850 to 10000 cm^{-1}, Tucson: Kitt Peak National Observatory (1981).

d Farmer, C.B., Norton, R.H.: A High Resolution Atlas of the Infrared Spectrum of the Sun and Earth Atmosphere from Space, NASA Ref. Pub. 1224 (1989).

e Kurucz, R.L., Furenlid, I., Brault, J., Testerman, L.: Solar Flux Atlas from 296 to 1300 nm, National Solar Observatory Atlas No. 1, Sunspot: NSO (1984).

f Moore, C.E., Tousey, R., Brown, C.M.: The Solar Spectrum 3069 Å to 2095 Å, NRL Report 8653 (1982).

Special references

81D Dravins, D., Lindegren, L., Nordlund, Å.: Astron. Astrophys. **96** (1981) 345.
81H Hinteregger, H.E., Fukai, K., Gilson, B.R.: Geophys. Res. Lett. **8** (1981) 1147.
82B Boyer, R., Sotirovski, P., Harvey, J.W.: Astron. Astrophys. Suppl. Ser. **47** (1982) 145.
82C Cook, J.W., Kjeldseth-Moe, O.: Sol. Phys. **76** (1982) 109.
82P Pierce, A.K., Slaughter, C.: Astrophys. J. Suppl. Ser. **48** (1982) 73.
83S1 Stenflo, J.O., Twerenbold, D., Harvey, J.W.: Astron. Astrophys. Suppl. Ser. **52** (1983) 161.
83S2 Stenflo, J.O., Twerenbold, D., Harvey, J.W., Brault, J.W.: Astron. Astrophys. Suppl. Ser. **54** (1983) 505.
84L1 Lindsey, C., Becklin, E.E., Jefferies, J.T., Orrall, F.Q., Werner, M.W., Gatley, I.: Astrophys. J. **281** (1984) 862.
84L2 Lindsey, C., de Graauw, T., de Vries, C., Lidholm, S.: Astrophys. J. **277** (1984) 424.
84N Neckel, H., Labs, D.: Sol. Phys. **90** (1984) 205.
84P Petro, L.D., Foukal, P.V., Rosen, W.A., Kurucz, R.L., Pierce, A.K.: Astrophys. J. **283** (1984) 426.
84R Rutten, R.J., van der Zalm, E.B.J.: Astron. Astrophys. Suppl. Ser. **55** (1984) 171.
85C Cavallini, F., Cepatelli, G., Righini, A.: Astron. Astrophys. **150** (1985) 256.
85M Mount, G.H., Rottman, G.R.: J. Geophy. Res. **90** (1985) 13031.
85N Neckel, H., Labs, D.; Sol. Phys. **95** (1985) 229.
86D1 Donnelly, R.F., Hinteregger, H.E., Heath, D.F.: J. Geophys. Res. **91** (1986) 5567.
86D2 Dravins, D., Larson, B., Nordlund, Å.: Astron. Astrophys. **158** (1986) 83.
86L Lindsey, C., Becklin, E.E., Orrall, F.Q., Werner, M.W., Jefferies, J.T., Gatley, I.: Astrophys. J. **308** (1986) 448.
86S1 Sandlin, G.D., Bartoe, J.-D.F., Brueckner, G.E., Tousey, R., VanHoosier, M.E.; Astrophys. J. Suppl. Ser. **61** (1986) 801.
86S2 Solanki, S.K., Pantellini, F.G.E., Stenflo, J.O.: Sol. Phys. **107** (1986) 57.
87F1 Fawcett, B.C., Jordan, C., Lemen, J.R., Phillips, K.J.H.: Mon. Not. R. Astron. Soc. **225** (1987) 1013.
87F2 Fröhlich, C.: J. Geophys. Res. **92** (1987) 796.
87L1 Labs, D., Neckel, H., Simon, P.C., Thuillier, G.: Sol. Phys. **107** (1987) 203.
87L2 Lean, J.: J. Geophys. Res. **92** (1987) 839.
88B Balthasar, H.: Astron. Astrophys. Suppl. **72** (1988) 473.
88F Fontenla, J., Reichmann, E.J., Tandberg-Hanssen, E.: Astrophys. J. **329** (1988) 464.
88W Willson, R.C., Hudson, H.S.: Nature (London) **332** (1988) 810.
89M Makarova, E.A., Knyazeva, L.N., Kharitonov, A.V.: Soviet Astron. **33** (1989) 298.
89N Nicolet, M.: Planet. Space Sci. **37** (1989) 1249.
90E Engvold, O., Hansteen, V., Kjeldseth-Moe, O., Brueckner, G.E.: Astrophys. Space Sci. **120** (1990) 179.
90N Neckel, H., Labs, D.: Sol. Phys. **126** (1990) 207.

3.1.1.4 Solar photosphere and chromosphere

General review [d].

3.1.1.4.1 Models

Table 1. Recent semi-empirical solar atmosphere models.

Author	Ref.	Scope	Observational basis
Vernazza et al. (1981)	81V	chromosphere	EUV continua
Maltby et al. (1984)	84M	photosphere-chromosphere	continua 0.3 ··· 2.3 μm
Avrett (1985)	b,p.67	temperature minimum	Ca lines, UV & far IR continua
Fontenla et al. (1991)	91F	low transition region	Lyman lines

Temperature minimum 4400 ··· 4500 K [b]. Evidence for cooler (3700 K) material [81A, 86A]; two-component model [81A].

Temperature plateau in upper chromosphere removed by diffusive effects [90F].

Theoretical models of photosphere: [79K] (LTE with convection); [89A1] (NLTE without convection).

Theoretical model of chromosphere with heating [89A2].

3.1.1.4.2 NLTE studies

Studies of spectral lines of diagnostic importance Ca I [90U], Fe I [83R], Mg I [88M1].

3.1.1.4.3 Morphology of the solar photosphere and chromosphere

3.1.1.4.3.1 Granulation

Reviews [a,c].

Area and size distributions [86R]. Fractal dimension 1.25 for diameters $d > 1000$ km and 2.15 for $d < 1000$ km [86R].

Average lifetime in quiet regions 730 s [d]; 300 s, rising to 600 s when fragments are excluded [89T]. Lifetimes in magnetic regions about twice those in quiet regions [89T].

Mean horizontal flows 370 m s^{-1} in quiet regions and 100 ··· 275 m s^{-1} (dependent on magnetic field concentration) in magnetic regions. RMS horizontal velocities 1.4 km s^{-1} in quiet regions, 0.75 km s^{-1} in magnetic regions [89T].

Mesogranulation: quasi-cellular vertical velocity pattern 60 m s^{-1} amplitude, horizontal scale 5000 ··· 10000 km, lifetime 2 h [81N]. Coherence of pattern disputed [89W1]. Similar scales appear in pattern of horizontal motions of granules [88S] and brightness patterns produced by granule clustering [84O, 90M].

A chromospheric quasi-cellular vertical velocity pattern, 700 m s^{-1} amplitude, is similar too but is not directly correlated with the photospheric mesogranulation [82N]. Chromospheric brightness pattern [87D].

3.1.1.4.3.2 Supergranulation and network

Supergranulation velocity pattern, lifetime $\geq$ 50 h, vertical velocities $<$ 100 $\mathrm{m\,s^{-1}}$ (rms 30 $\mathrm{m\,s^{-1}}$) when magnetic regions excluded [89W2]. Variation of size with latitude [89R].

3.1.1.4.3.3 Oscillations

For global acoustic modes seen in photosphere, see subsect. 3.1.1.2.4. Propagation properties [89D].

In chromosphere, acoustic waves are seen up to 100 μHz. At low frequencies these are global modes but the predicted resonant chromospheric mode at 25 μHz is absent. The propagation properties are difficult to interpret because of the complicated structuring of the chromosphere [84D1,86K,88K,89F].

3.1.1.4.3.4 Network elements

Dynamical properties of elements seen on disc, see subsect. 3.1.1.4.3.3. Morphology and dynamics in C I [84D2].

Spicule physical parameters: $N_e \approx 3\cdot10^{11}$ $\mathrm{cm^{-3}}$, $N_H \approx 2\cdot10^{12}$ $\mathrm{cm^{-3}}$, fraction of hydrogen ionized $\approx$ 0.15 at height of 2100 km [83R]. Kinetic temperatures: 8600 K at 2200 km, 5000 K at 3200 km, 8200 K at 6000 km [88M2].

References for 3.1.1.4

Monographs

a Bray, R.J., Loughhead, R.E., Durrant, C.J.: The Solar Granulation, 2nd ed., Cambridge: Cambridge University Press (1984).

b Lites, B.W. (ed.): Chromospheric Diagnostics and Modelling, Sunspot: Sacramento Peak Obs. (1985).

c Rutten, R.J., Severino, G. (eds.): Solar and Stellar Granulation, Dordrecht: Kluwer (1989)

d Sturrock, P.A., Holzer, T.E., Mihalas, D.M., Ulrich, R.K. (eds.): Physics of the Sun. Vol II. The Solar Atmosphere, Dordrecht: D. Reidel Publishing Co. (1986).

Special references

79K Kurucz, R.L.: Astrophys. J. Suppl. Ser. **40** (1979) 1.

81A Ayres, T.R., Testerman, L.: Astrophys. J. **245** (1981) 1124.

81N November, L..J., Toomre, J., Gebbie, K.B., Simon, G.W.: Astrophys. J. **245** (1981) L123.

81V Vernazza, J.E., Avrett, E.H., Loeser, R.: Astrophys. J. Suppl. Ser. **45** (1981) 635.

82N November, L..J., Toomre, J., Gebbie, K.B., Simon, G.W.: Astrophys. J. **258** (1982) 846.

83R Rutten, R.J., Zwaan, C.: Astron. Astrophys. **117** (1983) 21.

84D1 Damé, L., Gouttebroze, P., Malherbe, J.M.: Astron. Astrophys. **130** (1984) 331.

84D2 Dere, K.P., Bartoe, K.-D.,F., Brueckner, G.E.: Astrophys. J. **281** (1984) 870.

84L Landman, D.A.: Astrophys. J. **284** (1984) 833.

84M Maltby, P., Avrett, E.H., Carlsson, M., Kjeldseth-Moe, O., Kurucz, R.L., Loeser, R.: Astrophys. J. **306** (1984) 284.

84O Oda, N.: Sol. Phys. **93** (1984) 243.

84R Rutten, R.J., van der Zalm, E.B.J.: Astron. Astrophys. Suppl. Ser. **55** (1984) 143.

86A Ayres, T.R., Testerman, L., Brault, J.W.: Astrophys. J. **304** (1986) 542.
86D Dialetis, D., Macris, C., Prokakis, T., Sarris, E.: Astron. Astrophys. **168** (1986) 330.
86K Kneer, F., von Uexküll, M.: Astron. Astrophys. **155** (1986) 178.
86R Roudier, T., Muller, R.,: Sol. Phys. **107** (1986) 11.
87D Damé, L., Martic, M.: Astrophys. J. **314** (1987) L15.
88K Kneer, F., von Uexküll, M.: Astron. Astrophys. **144** (1988) 443.
88M1 Manas, P.J., Avrett, E.H., Loeser, R.: Astrophys. J. **330** (1988) 1008.
88M2 Matsuno, K., Hirayama, T.: Sol. Phys. **117** (1988) 21.
88S Simon, G., Title, A., Topka, K.P., Tarbell, T., Shine, R.A., Ferguson, S.H., Zirin, H., and the SOUP Team: Astrophys. J. **327** (1988) 964.
89A1 Anderson, L.S.: Astrophys. J. **339** (1989) 558.
89A2 Anderson, L.S., Athay, R.G.: Astrophys. J. **336** (1989) 1089.
89D Deubner, F.L., Fleck, B.: Astron. Astrophys. **213** (1989) 423.
89F Fleck, B., Deubner, F.L.: Astron. Astrophys. **224** (1989) 245.
89R Rimmele, T., Schröter, E.H.: Astron. Astrophys. **221** (1989) 137.
89T Title, A.M., Tarbell, T.D., Topka, K.P., Ferguson, S.H., Shine, R.A., and the SOUP Team: Astrophys. J. **336** (1989) 475.
89W1 Wang, H.: Sol. Phys. **123** (1989) 21.
89W2 Wang, H., Zirin, H.: Sol. Phys. **120** (1989) 1.
90F Fontenla, J.M., Avrett, E.H., Loeser, R.: Astrophys. J. **355** (1990) 700.
90M Muller, R., Roudier, T., Vigneau, J.: Sol. Phys. **126** (1990) 53.
90U Uitenbroek, H., in: Proc. 6th Cambridge Workshop on Cool Stars, Stellar Systems and the Sun (Wallerstein, G., ed.), Pacific Conf. Ser. **9** (1990) 103.
91F Fontenla, J.M., Avrett, E.H., Loeser, R.: Astrophys. J. **377** (1991) 712.

3.1.1.5 Solar transition region and quiet corona

General reviews [a, 86M].

3.1.1.5.1 Models

Emission measure $\int N_e^2 \, dh$ as a function of temperature [81R]. Average EUV and X-ray emission relationships [85S].

Lyman continuum absorption [84N].

Static models [81P]; with diffusion [90F]. Dynamic loop models [83M, 84M, 88K]. Ensemble models: hot and cool loops [86A], time-dependent heating [84A], cross-conduction [83B].

3.1.1.5.2 Physical parameters

Low-resolution observations show predominantly red shifts in plasma at $T < 2 \cdot 10^5$ K. At $T = 10^5$ K speeds are $6 \cdots 17$ km s^{-1} (average 12 km s^{-1}) [82R], $3 \cdots 10$ km s^{-1} [90E]. For $T > 2 \cdot 10^5$ K there are no shifts. Evidence for upward and downward propagating waves [81B]. At high resolution, there is no clear relation between downflows and brightness features such as network elements (see subsect. 3.1.1.5.4).

Kinetic temperatures in corona: coronal quiet region, $T = 2.6 \cdot 10^6$ K at $1.5 R_\odot$, $T = 1.2 \cdot 10^6$ K at $4 R_\odot$ [82W]; coronal polar region: $T = 1.2 \cdot 10^6$ K at $1.5 R_\odot$, $T = 0.6 \cdot 10^6$ K at $3.5 R_\odot$ [85W].

3.1.1.5.3 Diagnostics

Review [89D2]. Recent work: T diagnostics [90K1], N_e diagnostics [90K2]. Departures from ionization equilibrium [79J].

3.1.1.5.4 Morphology

Unresolved fine structure in transition region [83F].

In general, the brightness structure in C IV (10^5 K) is correlated with chromospheric network, but downflows, mean 5 $\mathrm{km\,s^{-1}}$, occur in localized features not correlated with fine structure in brightness pattern [84D]. Upflows seen in network regions (expulsion events and spicules) and in cell interiors [86D]. Upflows more frequent in coronal holes [89D3].

Two categories of violent events have been distinguished in cell interiors: explosive (or turbulent) events, transient up and down motions, maximum velocity $\pm 100\ \mathrm{km\,s^{-1}}$, lifetime 60 s, width 1600 km [83B, 89D1]; small-scale jets (or bullets), accelerating upward motions with velocities up to 400 $\mathrm{km\,s^{-1}}$ [83B].

Macrospicules [89D2].

Low and middle latitude coronal holes: cover $0.1 \cdots 3\%$ of visible hemisphere, flux varying from $(2 \cdots 5) \cdot 10^{21}$ Mx at minimum to $(10 \cdots 15) \cdot 10^{21}$ Mx at maximum with little change of area [82H].

Polar coronal holes: cover $\leq 10\%$ of visible hemisphere, flux $3 \cdot 10^{22}$ Mx [78S].

References for 3.1.1.5

Monographs

a Bray, R.J., Cram, L.E., Durrant, C.J., Loughhead, R.E.: Plasma Loops in the Solar Corona, Cambridge: Cambridge Univ. Press (1991).

Special references

78S Svalgaard, L., Duvall, T.L., Scherrer, P.H.: Sol. Phys. **58** (1978) 225.
79J Joselyn, J.A., Munro, R.H., Holzer, T.E.: Astrophys. J. Suppl. Ser. **40** (1979) 793.
81B Bruner, E.C.: Astrophys. J. **247** (1981) 317.
81D Dere, K.P., Mason, H., in: Solar Active Regions (Orrall, F.Q., ed.), Boulder: Colorado Ass. Uni. Press (1981) 129.
81P Pallavicini, R., Peres, G., Serio, S., Vaiana, G.S., Golub, L., Rosner, R.: Astrophys. J. **247** (1981) 692.
81R Raymond, J.C., Doyle, J.G.: Astrophys. J. **247** (1981) 686.
82H Harvey, K.L., Sheeley, N.R., Harvey, J.W.: Sol. Phys. **79** (1982) 149.
82R Roussel-Dupré, D., Shine, R.A.: Sol. Phys. **77** (1982) 329.
82W Withbroe, G.L., Kohl, J.L., Weiser, H., Noci, G., Munro, R.H.: Astrophys. J. **254** (1982) 361.
83B Brueckner, G.E., Bartoe, J.-D.F.: Astrophys. J. **272** (1983) 329.
83F Feldman, U. : Astrophys. J. **275** (1983) 367.
83M Mariska, J.T., Boris, J.P.: Astrophys. J. **267** (1983) 409.
84A Athay, R.G.: Astrophys. J. **287** (1984) 412.
84D Dere, K.P., Bartoe, J.-D.,F., Brueckner, G.E.: Astrophys. J. **281** (1984) 870.
84M Mariska, J.T.: Astrophys. J. **281** (1984) 435.
84N Nishikawa, T.: Sol. Phys. **93** (1984) 361.

85S Schrijver, C.J., Zwaan, C., Maxson, C.W., Noyes, R.W.: Astron. Astrophys. **149** (1985) 123
85W Withbroe, G.L., Kohl, J.L., Weiser, H., Munro, R.H.: Astrophys. J. **297** (1985) 324.
86A Antiochus, S.K., Noci, G.: Astrophys. J. **301** (1986) 440.
86D Dere, K.P., Bartoe, J.-D.,F., Brueckner, G.E.: Astrophys. J. **310** (1986) 456.
86M Mariska, J.T.: Annu. Rev. Astron. Astrophys. **24** (1986) 23.
88K Klimchuk, J.A., Mariska, J.T.: Astrophys. J. **328** (1988) 334.
89D1 Dere, K.P., Bartoe, J.-D.,F., Brueckner, G.E.: Sol. Phys. **123** (1989) 41.
89D2 Dere, K.P., Bartoe, J.-D.F., Brueckner, G.E., Cook, J.W., Socker, D.G., Ewing, J.W.: Sol. Phys. **119** (1989) 55.
89D3 Dere, K.P., Bartoe, J.-D.F., Brueckner, G.E., Recely, F.: Astrophys. J. **345** (1989) L95.
90A Athay, R.G.: Astrophys. J. **362** (1990) 364.
90E Engvold, O., Henze, W.: Astrophys. Space Sci. **170** (1990) 173.
90F Fontenla, J.H., Avrett, E.H., Loeser, R.: Astrophys. J. **355** (1990) 700.
90K1 Keenan, F.P., McCann, S.M., Phillips, K.J.H.: Astrophys. J. **363** (1990) 315.
90K2 Keenan, F.P., McCann, S.M., Widing, K.G.: Astrophys. J. **363** (1990) 310.

3.1.1.6 Radio emission of the quiet sun

Revisions and additions to subsect. 3.1.1.6 in LB VI/2a

3.1.1.6.1. Flux density of the quiet sun

For a recent review of modern techniques used in the measurement of solar flux density and polarization see Nelson et al. [85N].

3.1.1.6.2 The brightness distribution across the solar disk

For a description of modern techniques used in the measurement of solar brightness distribution see Labrum [85L].

Interpretation of the brightness distribution

Recent work on scattering and refraction of radiation in the solar corona has highlighted the importance of these effects for interpreting coronal brightness temperatures and apparent sizes at metric and decametric wavelengths (see comprehensive review by Sheridan and McLean [85S]).

At metre wavelengths and longer, the optical depth τ is small except near the plasma level where $f = f_p$ and $n = 0$. Because the refractive index n is ≤ 1 over most of the ray path, refraction and scattering play a major part in determining the brightness distribution. By contrast, at shorter wavelengths the much larger absorption coefficient κ of the chromosphere makes it impossible for the rays to penetrate down to levels where the refractive index is low, and one may assume that the ray paths are straight lines.

From the ray tracing work of Smerd [50S] (for simplicity, Smerd assumed a spherically symmetrical corona) at frequencies near 100 MHz, the optical depth for the central ray is about 2 or 3. Because the coronal temperature is almost constant the expression for the brightness temperature T_b reduces to $T_b = T_e(1 - e^{-\tau})$ and for $\tau \geq 1$ the brightness temperature $T_b = T_e$ [85S].

Non-central rays are deflected by refraction away from the plasma level, where the absorption is largest. Consequently the optical depth and brightness temperature are less for these rays and there is a progressive decrease of T_b from centre to limb as observed.

At frequencies much above 100 MHz the corona is not optically thick and the optical depth decreases as f^{-2}. As the frequency increases, the height corresponding to $\tau = 1$ decreases, and moves first into the transition region then into the chromosphere. Hence T_b decreases with increasing frequency from a coronal value of about 10^6 K to a chromospheric value about 5000 K.

Below 100 MHz or thereabouts, the observed brightness temperature falls off much more rapidly than predicted by refraction alone, and the size of the observed disc increases considerably. This effect has been partly explained by scattering in an inhomogeneous corona [71A]. More recent models have also invoked ducting of the radiation along under-dense fibres to higher levels of the corona before escaping [77B,79D]. Invoking a filling factor $f \leq 1$ for a bundle of rays and applying Liouville's theorem gives $T_b = T_c/f$ [88M], which may explain the low brightness temperatures observed at frequencies down to 30 MHz.

References for 3.1.1.6

General references

a Solar Radiophysics (McLean, D.J., Labrum, N.R., eds.), Cambridge: Cambridge University Press (1985).

Special references

50S Smerd, S.F.: Aust. J. Sci. Res., Ser. A, **3** (1950) 34.
71A Aubier, M., Leblanc, Y., Boischot, A.: Astron. Astrophys. **12** (1971) 435.
77B Bougeret, J.L., Steinberg, J.L.: Astron. Astrophys. **61** (1977) 777.
79D Duncan, R.A.: Sol. Phys. **63** (1979) 389.
85L Labrum, N.R., in: [a] (1985) ch.7.
85N Nelson, G.J., Sheridan, K.V., Suzuki, S., in: [a] (1985) ch.6.
85S Sheridan, K.V., McLean, D.J., in: [a] (1985) ch.17.
88M Melrose, D.B., Dulk, G.A.: Sol. Phys. **116** (1988) 141.

3.1.2 Solar activity

3.1.2.1 Active regions

Review [a].

3.1.2.1.1 Features of active regions

Gross structure at various wavelengths [82C]. Photospheric/chromospheric fine structure [84K1]. Fine structure of transition region [84K2].

Number and area statistics 1967 ··· 1981 [84T].

Dynamical structure [87K]. Two-dimensional spatial spectra of magnetic and velocity fields [83B].

3.1.2.1.2 Active region development

General evolution of region from photosphere to corona [82G].

Magnetograms 1967 ··· 88 [83G, 89H].

Current systems [85G, 86H, 88A].
Energy balance [84C1, 84C2, 90L].

3.1.2.1.3 Spotgroups

Classification: McIntosh [90M].
Sunspot clustering (nests, complexes of activity) [83G, 86C1, 90B].

3.1.2.1.4 Activity indices, global data (daily values)

Large sunspot groups 1965 ··· 76 [82K].

Variability:
survey [88H1]
irradiance [87F, 88H1]
X-rays (1 ··· 8 Å) [88W]
soft X-rays [89F1]
EUV, UV [86D, 87L2, 88R]
10.7 cm [87T].

Contributions to irradiance variations:
sunspots, Photometric Sunspot Index (PSI) [82H, 86C2, 87L1, 89F2, 90S]
plage, Photometric Plage Index (PPI) [86C2, 87L1].

Periodicities (51 d and 150 ··· 157 d) in irradiance and indices [89L, 90P].

References for 3.1.2.1

General references

a Orrall, F.Q. (ed.): Solar Active Regions, Boulder: Colorado Ass. Uni. Press (1981).

Special references

82C Chiuderi-Drago, F., Bandiera, R., Falciani, R., Antonucci, E., Lang, K.R., Willson, R.F., Shibasaki, K., Slottje, C.: Sol. Phys. **80** (1982) 71.
82G Golub, L., Noci, G., Poletto, G., Vaiana, G.S.: Astrophys. J. **259** (1982) 359.
82H Hudson, H.S., Silva, S., Woodard, M., Willson, R.C.: Sol. Phys. **76** (1982) 211
82K Kopecký, M.: Bull. Astron. Inst. Czech. **33** (1982) 65.
83B Berton, R.: Astron. Astrophys. **127** (1983) 140.
83G Gaizauskas, V., Harvey, K.L., Harvey, J.W., Zwaan, C.: Astrophys. J. **265** (1983) 1056.
84C1 Chapman, G.A.: Nature **308** (1984) 252.
84C2 Chapman, G.A., Herzog, A.D., Lawrence, J.K., Shelton, J.C.: Astrophys. J. **282** (1984) L99.
84K1 Kitai, R., Muller, R.: Sol. Phys. **90** (1984) 303.
84K2 Kjeldseth-Moe, O., Andreasson, Ø., Maltby, P., Bartoe, J.-D.F., Brueckner, G.E., Nicolas, K.R.: Adv. Space Res. **4**(8) (1984) 63.
84T Tang, F., Howard, R., Adkins, J.M.: Sol. Phys. **91** (1984) 75.

85G Gopasyuk, S.I., Kalman, B., Romanov, V.A.: Izv. Krym. Astrofiz. Obs. **72** (1985) 171 (Engl. transl.: Bull. Crimean Astrophys. Obs. **72** (1985) 147).
86C1 Castenmiller, M.J.M., Zwaan, C., van der Zalm, E.B.J. : Sol. Phys. **105** (1986) 237.
86C2 Chapman, G.A., Meyer, A.D.: Sol. Phys. **86** (1986) 103.
86D Donnelly, R.F., Hinteregger, H.E., Heath, D.F.: J. Geophys. Res. **91** (1986) 5567.
86H Haisch, B.M., Bruner, M.E., Hagyard, M.J., Bonnet, R.M.: Astrophys. J. **300** (1986) 428.
87F Fröhlich, C.: J. Geophys. Res. **93** (1987) 796.
87K Klimchuk, J.A.: Astrophys. J. **323** (1987) 368.
87L1 Lawrence, J.K.: J. Geophys. Res. **92** (1987) 813.
87L2 Lean, J.: J. Geophys. Res. **92** (1987) 839.
87T Tapping, K.F.: J. Geophys. Res. **92** (1987) 829.
88A Abramenko, V.I., Gopasyuk, S.I., Ogir',M.B.: Izv. Krym. Astrofiz. Obs. **78** (1988) 151; **79** (1988) 23 (Engl. transl.: Bull. Crimean Astrophys. Obs. **78** (1988) 163; **79** (1988) 21).
88H1 Hickey, J.R., Alton, B.M., Kyle, H.L., Major, E.R.: Adv. Space Res. **8**(7) (1988) 5.
88H2 Hudson, H.S.: Annu. Rev. Astron. Astrophys. **26** (1988) 473.
88R Rottman, G.J.: Adv. Space. Res. **8**(7) (1988) 53.
88W Wagner, W.J.: Adv. Space Res. **8**(7) (1988) 67.
89F1 Feng, W., Ogawa, H.S., Judge, D.L.: J. Geophys. Res. **94** (1989) 9125.
89F2 Fröhlich, C., Pap, J.: Astron. Astrophys. **220** (1989) 272.
89H Howard, R.F.: Sol. Phys. **123** (1989) 271.
89L Lean, J.L., Brueckner, G.E.: Astrophys. J. **337** (1989) 568.
90B Brouwer, M.P., Zwaan, C.: Sol. Phys. **129** (1990) 221.
90L Lawrence, J.K., Chapman, G.A.: Astrophys. J. **361** (1990) 709.
90M McIntosh, P.S.: Sol. Phys. **125** (1990) 251.
90P Pap, J., Tobiska, W.K., Bouwer, S.D.: Sol. Phys. **129** (1990) 165.
90S Steinegger, M., Brandt, P.N., Pap, J., Schmidt, W.: Astrophys. Space Sci. **170** (1990) 127.

3.1.2.2 11-year solar cycle

Variation in total irradiance with activity [88W1]

$$S = (1366.82 + 7.71 \cdot 10^{-3} \cdot R_z)\,\mathrm{W\,m^{-2}}$$
$$= (1366.27 + 8.98 \cdot 10^{-3} \cdot F_{10})\,\mathrm{W\,m^{-2}}$$

and [88H]

$$S = (1366.73 + PSI + 1.53 \cdot 10^{-2} \cdot R_z)\,\mathrm{W\,m^{-2}}$$
$$= (1365.65 + PSI + 1.77 \cdot 10^{-2} \cdot F_{10})\,\mathrm{W\,m^{-2}}$$

where PSI is the photometric sunspot index, R_z the Zürich sunspot number and F_{10} is the 10.7 cm radio flux [see also subsect. 3.1.2.1.4].

Variation of individual properties with cycle:
UV [87L1]
10.7 cm [87T]
21 cm (1981 ··· 1987) [88A]
coronal Fe XIV 530.3 nm (cycles 20, 21) [88A]
coronal Fe X 637.4 nm (cycle 21) [88A]
coronal structures (atlas) [89L]
Ca II K (1974 ··· 1985) [87W] (1969 ··· 1984) [87S1]
He I 1083 nm [91S]
photospheric line strengths (1975 ··· 1985) [82L, 87L2], line asymmetries [87W]

Mg II 280 nm flux variation (1978 ··· 1982) [86H], (1986 ··· 1988) [88D]
solar rotation rate [90H]
sunspot umbral intensity [86M]
total magnetic flux [91L]
polar coronal hole (1940 ··· 1978) [81W].

Variations of global distributions with cycle:
sunspots (synoptic, 1874 ··· 1976) [80Y]
sunspots (harmonic analysis) [90G]
magnetic fields (synoptic, 1967 ··· 1980) [81H]
magnetic fields (harmonic analysis) [86S1, 88S1, 88S2]
torsional oscillations [87S2].

North-South asymmetry in cycle [83M, 86S2].

Polar regions show phenomena of following cycle overlapping the last equatorial phases of the current cycle. Extended cycle length 15 ··· 18 a [81L, 83L, 87S2, 89M, 90B]; 22 a [88W2].

References for 3.1.2.2

80Y Yallop, B.D., Hohenkerk, C.Y.: Sol. Phys. **68** (1980) 303.
81H Howard, R., Labonte, B.J.: Sol. Phys. **74** (1981) 131.
81L Legrand, J.P., Simon, P.A.: Sol. Phys. **70** (1981) 173.
81W Waldmeier, M.: Sol. Phys. **70** (1981) 251.
82L Livingston, W.C., Holweger, H.: Astrophys. J. **252** (1982) 375.
83L Leroy, J.-L., Noens, J.-C.: Astron. Astrophys. **120** (1983) L1.
83M Moussas, X., Papastamatiou, N., Rušin, V., Rybanský, M.: Sol. Phys. **84** (1983) 71.
86H Heath, D.F., Schlesinger, B.M.: J. Geophys. Res. **91** (1986) 8672.
86M Maltby, P., Avrett, E.H., Carlsson, M., Kjeldseth-Moe, O., Kurucz, R.L., Loeser, R.: Astrophys. J. **306** (1986) 284.
86S1 Stenflo, J.O., Vogel, M.: Nature (London) **319** (1986) 285.
86S2 Swinson, D.B., Koyama, H., Saito, T.: Sol. Phys. **106** (1986) 35.
87L1 Lean, J.: J. Geophys. Res. **92** (1987) 839.
87L2 Livingston, W.C., Wallace, L.: Astrophys. J. **314** (1987) 808.
87S1 Sivaraman, K.R., Singh, J., Bagare, S.P., Gupta, S.S.: Astrophys. J. **313** (1987) 456.
87S2 Snodgrass, H.B.: Sol. Phys. **110** (1987) 35.
87T Tapping, K.F.: J. Geophys. Res. **92** (1987) 829.
87W White, O.R., Livingston, W.C., Wallace, L.: J. Geophys. Res. **92** (1987) 823.
88A Altrock, R.C. (ed.): Solar and Stellar Coronal Structure and Dynamics, Sunspot: NSO/Sacramento Peak Obs. (1988).
88D Donnelly, R.F.: Adv. Space Res. **8**(7) (1988) 77.
88H Hudson, H.S.: Annu. Rev. Astron. Astrophys. **26** (1988) 473.
88S1 Stenflo, J.O.: Astrophys. Space Sci. **144** (1988) 321.
88S2 Stenflo, J.O., Güdel, M.: Astron. Astrophys. **191** (1988) 137.
88W1 Willson, R.C., Hudson, H.S.: Nature (London) **332** (1988) 810.
88W2 Wilson, P.R., Altrock, R.C., Harvey, K.L., Martin, S.F., Snodgrass, H.B.: Nature **333** (1988) 748.
89L Loucif, M.L., Koutchmy, S.: Astron. Astrophys. Suppl. Ser. **77** (1989) 45.
89M Makarov, V.I., Sivaraman, K.R.: Sol. Phys. **123** (1989) 367.
90B Bumba, V., Rušin, V., Rybanský, M.: Sol. Phys. **128**(1990) 253.
90G Gokhale, M.H., Javaraiah, J., Hiremath, K.M., in: Solar Photosphere: Structure, Convection and Magnetic Fields (Stenflo, J.O., ed.) Dordrecht: Kluwer Acad. Publ. (1990) 375.

90H Hathaway, D.H., Wilson, R.M.: Astrophys. J. **357** (1990) 271.
91L Livingston, W., in: The Sun and Cool Stars: activity, magnetism, dynamos (Tuominen, I., Moss, D., Rüdiger, G., eds.), Berlin: Springer (1991) 246.
91S Shcherbakov, A.G., Shcherbakova, Z.A. in: The Sun and Cool Stars: activity, magnetism, dynamos (Tuominen, I., Moss, D., Rüdiger, G., eds.), Berlin: Springer (1991) 252.

3.1.2.3 Sunspots

3.1.2.3.1 General characteristics

Reviews [a,85M].

Transition region and corona above spots [81O].
Decay rates proportional to perimeter, independent of maximum size of spot [88M].

3.1.2.3.2 Magnetic field

Vector fields [83K,90A,90L2].
Relative magnetic field strength B as function of radial distance in units of umbral radius ρ [90A]

$$B = 1.0 - 0.601\rho - 0.274\rho^2 .$$

Inclination of field to vertical at edge of penumbra 70° [90A].
Magnetic field strength in transition region 1000 G, gradient of field strength in vertical direction $0.1 \cdots 0.3$ G km^{-1} [83H].

3.1.2.3.3 Spot umbra

Number of spots N having area $\leq A$ is lognormal [88B].
Maps of brightness and inferred thermodynamic parameters [90L2].

Umbral models
Plane parallel: photosphere-chromosphere [a,86M]; chromosphere [82L, 87L].
Multicomponent: photosphere [84V, 90P]; chromosphere [86Y]; high chromosphere [84S].

Fine structure and dynamics
Vertical motions in transition region (C IV): 1.2 ± 5.6 km s^{-1} [83G].
Oscillations:
in chromosphere, vertical phase propagation $10 \cdots 25$ km s^{-1} in $110 \cdots 200$ s band [83V,84L];
in high chromosphere, amplitude $1 \cdots 2$ km s^{-1} in 180 s band, $0.2 \cdots 0.5$ km s^{-1} in 300 s band [86L];
in transition region (C IV), amplitudes up to 3 km s^{-1} in $110 \cdots 200$ s band [84H].
Umbral wave classification [85T, 90H].
Umbral dots [81K].

3.1.2.3.4 Spot penumbra

Intensity, centre-to-limb [88C].
Magnetic field strength 2100 G (inner edge), 900 G (outer edge), no systematic differences between bright and dark regions [90L1].
Evershed effect from photosphere to transition region [88A].

Oscillations: 300 s in photosphere [88L, 90B]; periods increase in chromosphere towards outer edge [88L].

Plane parallel model [89D].

3.1.2.3.5 See LB VI/2a

References for 3.1.2.3

General references

a Cram, L.E., Thomas, J.H.: Physics of Sunspots, Sunspot: NSO/Sacramento Peak Obs. (1981).

Special references

81K Koutchmy, S., Adjabshirzadeh, A.: Astron. Astrophys. **99** (1981) 111.
81O Orrall, F.Q.: Space Sci. Rev. **28** (1981) 423.
82L Lites, B.W., Skumanich, A.: Astrophys. J. Suppl. Ser. **49** (1982) 293.
83G Gurman, J.B., Athay, R.G.: Astrophys. J. **273** (1983) 374.
83H Hagyard, M.J., Teuber, D., West, E.A., Tandberg-Hanssen, E., Henze, W., Beckers, J.M., Bruner, M., Hyder, C.L., Woodgate, B.E.: Sol. Phys. **84** (1983) 13.
83K Kawakami, H.: Pub. Astron. Soc. Japan **35** (1983) 459.
83V von Uexküll, M., Kneer, F., Mattig, W.: Astron. Astrophys. **123** (1983) 263.
84H Henze, W., Tandberg-Hanssen, E., Reichmann, E.J., Athay, R.G.: Sol. Phys. **91** (1984) 33.
84L Lites, B.W.: Astrophys. J. **277** (1984) 874.
84S Staude, J., Fürstenberg, F., Hildebrandt, J., Krüger, A., Jakimiec, J., Obridko, V.N., Siarkowski, M., Sylwester, B., Sylwester, J.: Sov. Astron. (Engl. transl.) **28** (1984) 557.
84V van Ballegooijen, A.A.: Sol. Phys. **91** (1984) 195.
85M Moore, R., Rabin, D.: Annu. Rev. Astron. Astrophys. **23** (1985) 239.
85T Thomas, J.H.: Aust. J. Phys. **38** (1985) 811.
86L Lites, B.W.: Astrophys. J. **301** (1986) 1005.
86M Maltby, P., Avrett, E.H., Carlsson, M., Kjeldseth-Moe, O., Kurucz, R.L., Loeser, R.: Astrophys. J. **306** (1986) 284.
86Y Yun, H.S., Beebe, H.A.: Astrophys. Space Sci. **118** (1986) 173.
87L Lites, B.W., Skumanich, A., Rees, D.E., Murphy, G.A., Carlsson, M.: Astrophys. J. **318** (1987) 930.
88A Alissandrakis, C.E., Dialetis, D., Mein, P., Schmieder, B., Simon, G.: Astron. Astrophys. **201** (1988) 339.
88B Bogdan, T.J., Gilman, P.A., Lerche, I., Howard, R.: Astrophys. J. **327** (1988) 451.
88C Collados, M., del Toro Iniesta, J.C., Vázquez, M.: Astron. Astrophys. **195** (1988) 315.
88L Lites, B.W.: Astrophys. J. **334** (1988) 1054.
88M Moreno-Insertis, F., Vázquez, M.: Astron. Astrophys. **205** (1988) 289.
89D Ding, M.D., Fang, C.: Astron. Astrophys. **225** (1989) 204.
90A Adam, M.G.: Sol. Phys. **125** (1990) 37.
90B Balthasar, H.: Sol. Phys. **125** (1990) 31.
90H Hasan, S.S., Sobouti, Y., in: Solar Photosphere: Structure, Convection and Magnetic Fields (Stenflo, J.O., ed.), Dordrecht: Kluwer Acad. Publ. (1990) 255.

90L1 Lites, B.W., Scharmer, G.B., Skumanich, A.: Astrophys. J. **355** (1990) 329.
90L2 Lites, B.W., Skumanich, A.: Astrophys. J. **348** (1990) 747.
90P Pahlke, K.-D., Wiehr, E.: Astron. Astrophys. **228** (1990) 246.

3.1.2.4 Faculae and plages

3.1.2.4.1 Continuum

Contrast at extreme limb: 525 nm [82C], 525, 800 nm [85L], 626 nm [88L1].
Contrast at disk centre 0.73% at 626 nm [88L2].

3.1.2.4.2 Facula models

Mean fields 500 ··· 1700 G [84S,90D]; weak field component [89Z].
Fraction of area occupied by magnetic fields in facular regions (filling factor) 0.1 ··· 0.6 [90D].
Empirical flux-tube models, minimum temperature in network $T \approx 4700$ K, axial field $B \approx 2000$ G at optical depth unity, $B \approx 300$ G at height of 400 km [90K]. Tubes become cooler with increasing field strength and filling factor [90Z].

3.1.2.4.3 Chromospheric plage

Ca II H and K profiles [83L].
He D_3 profiles [81L].
Densities in transition region 10 times those in quiet Sun [83D].
Oscillations, less power in 300-s band than for quiet Sun, more long-period power [81W]. Oscillations in high chromosphere [81L].
Empirical flux-tube models [87W, 90K], minimum temperature $T \approx 5000$ K [90K].

References for 3.1.2.4

81L Landman, D.A.: Astrophys. J. **244** (1981) 345.
81W Woods, D.T., Cram, L.E.: Sol. Phys. **69** (1981) 233.
82C Chapman, G.A., Klabunde, D.P.: Astrophys. J. **261** (1982) 387.
83D Dumont, S., Mouradian, Z., Pecker, J.-C., Vial, J.-C., Chipman, E.: Sol. Phys. **83** (1983) 27.
83L Lemaire, P.: Sol. Phys. **88** (1983) 31.
84S Stenflo, J.O., Harvey, J.W., Brault, J.W., Solanki, S.: Astron. Astrophys. **131** (1984) 333.
85L Libbrecht, K.G., Kuhn, J.R.: Astrophys. J. **299** (1985) 1047.
87W Walton, S.R.: Astrophys. J. **312** (1987) 909.
88L1 Lawrence, J.K., Chapman, G.A.: Astrophys. J. **335** (1988) 996.
88L2 Lawrence, J.K., Chapman, G.A., Herzog, A.D.: Astrophys. J. **324** (1988) 1184.
89Z Zayer, I., Solanki, S.K., Stenflo, J.O.: Astron. Astrophys. **211** (1989) 463.
90D del Toro Iniesta, J.C., Semel, M., Collados, M., Sánchez Almeida, J.: Astron. Astrophys. **227** (1990) 591.
90K Keller, C.U., Solanki, S.K., Steiner, O., Stenflo, J.O.: Astron. Astrophys. **233** (1990) 583.
90Z Zayer, I., Solanki, S.K., Stenflo, J.O., Keller, C.U.: Astron. Astrophys. **239** (1990) 356.

3.1.2.5 Prominences and ejecta

3.1.2.5.1 See LB VI/2a

3.1.2.5.2 Prominence spectrum

Quiescent prominence 133.5 ⋯ 167 nm [90E].
Erupting prominence 30 ⋯ 63 nm [86W].

3.1.2.5.3 Physical characteristics of quiescent prominences

Reviews [a, b].

Magnetic fields from Stokes polarimetry: $B = 6 \cdots 27$ G, irrespective of type [83A]; prominence map, $B = 6 \cdots (23) \cdots 60$ G [85Q].

Magnetic fields from Hanle effect: high (height > 30000 km) prominences, $B = 5 \cdots 10$ G, low (height < 30000 km) prominences, $B \approx 20$ G [84L2,86B1].

Thermodynamic parameters, temperature T, hydrogen density n_{H}, electron density n_{e}, turbulent velocity v_{t}:

from H_α, Ca II: $T \approx 7500$ K, $n_{\mathrm{H}} \approx 2 \cdot 10^{18}$ cm^{-3}, $v_{\mathrm{t}} \approx 6$ km s^{-1} [87Z],
from Na I, Sr II, Mg I: $T \approx 7000$ K, $n_{\mathrm{e}} \approx 10^{11}$ cm^{-3}, $v_{\mathrm{t}} \approx 5.9$ km s^{-1} [83L,84L1,86L],
from H_β: $n_{\mathrm{e}} < 10^{10}$ cm^{-3} [86B2],
from UV: $T = 30000 \cdots 100000$ K, $n_{\mathrm{e}} \leq 3 \cdot 10^{11}$ cm^{-3} [83P].

Internal motions reviewed [a, b]; activated prominence [88S].

3.1.2.5.4 Ejections

Reviews, see subsect. 3.1.2.7.1.

Classification into filament eruption and filament shedding [86T].

Active region filament motions [85S, 90V].

Eruptive prominences: at about $2R_\odot$, $T = 20000 \cdots 25000$ K, $n_{\mathrm{e}} = (0.5 \cdots 5) \cdot 10^8$ cm^{-3}, mass $m = (1.5 \cdots 15) \cdot 10^{15}$ g [86A,86I].

Spray: $n_{\mathrm{e}} \approx (1 \cdots 3) \cdot 10^{11}$ cm^{-3} [86W].

Surges: relation to X-ray bursts [90H]; trajectories, maximum velocities $20 \cdots 300$ km s^{-1} upwards, $30 \cdots 230$ km s^{-1} downwards [83B].

Coronal mass ejections, survey 1979 ⋯ 1981 [85H].

3.1.2.5.5 Interface prominence-corona

Reviews [a,b].

Temperature structure [88E].

References for 3.1.2.5

General references

a Priest, E.R. (ed.): Dynamics and Structure of Quiescent Solar Prominences, Dordrecht: Kluwer Academic Publ. (1989).
b Roždjak, V., Tandberg-Hanssen, E. (eds.): Dynamics of Quiescent Prominences, Berlin: Springer (1990).

Special references

83A Athay, R.G., Querfeld, C.W., Smartt, R.N., Landi Degl'Innocenti, E., Bommier, V.: Sol. Phys. **89** (1983) 3.
83B Banos, G., Dara-Papamargariti, H.: Astron. Astrophys. **120** (1983) 181.
83L Landman, D.A.: Astrophys. J. **270** (1983) 265.
83P Poland, A.I., Tandberg-Hanssen, E.: Sol. Phys. **84** (1983) 63.
84L1 Landman, D.A.: Astrophys. J. **279** (1984) 438.
84L2 Leroy, J.L., Bommier, V., Sahal-Bréchot, S.: Astron. Astrophys. **131** (1984) 33.
85H Howard, R.A., Sheeley, N.R., Koomen, M.J., Michels, D.J.: J. Geophys. Res. **90** (1985) 8173.
85Q Querfeld, C.W., Smartt, R.N., Bommier, V., Landi Degl'Innocenti, E., House, L.L.: Sol. Phys. **96** (1985) 277.
85S Schmieder, B., Raadu, M.A., Malherbe, J.M.: Astron. Astrophys. **142** (1985) 249.
86A Athay, R.G., Illing, R.M.E.: J. Geophys. Res. **91** (1986) 10961.
86B1 Bommier, V., Leroy, J.L., Sahal-Bréchot, S.: Astron. Astrophys. **156** (1986) 79.
86B2 Bommier, V., Leroy, J.L., Sahal-Bréchot, S.: Astron. Astrophys. **156** (1986) 90.
86I Illing, R.M.C., Athay, R.G.: Sol. Phys. **105** (1986) 173.
86L Landman, D.A.: Astrophys. J. **305** (1986) 546.
86T Tang, F.: Sol. Phys. **105** (1986) 399.
86W Widing, K.G., Feldman, U., Bhatia, A.K.: Astrophys. J. **308** (1986) 982.
87Z Zhang, Q.Z., Livingston, W.C., Hu, J., Fang, C.: Sol. Phys. **114** (1987) 245.
88E Engvold, O., in: Solar and Stellar Coronal Structure and Dynamics, (Altrock, R.C., ed.) Sunspot: NSO/Sacramento Peak Obs. (1988) 151.
88S Schmieder, B., Poland, A., Thompson, B., Démoulin, P.: Astron. Astrophys. **197** (1988) 281.
90E Engvold, O., Hansteen, V., Kjeldseth-Moe, O., Brueckner, G.E.: Astrophys. Space Sci. **170** (1990) 179.
90H Harrison, R.A., Sime, D.G., Pearce, G.: Astron. Astrophys. **238** (1990) 347.
90V Vršnak, B.: Sol. Phys. **127** (1990) 129.

3.1.2.6 Coronal active region

General review [a].

Differential emission measure for whole active region [81T, 82D2].
Loop structure in active region [81W, 82D1, 85H]. Evolution of loops [80S, 85H, 88H].

Thermodynamic parameters (temperature T, electron density n_e, basal pressure p) in loops:
loops with $T > 10^5$ K, $p \approx 5$ dyn cm^{-2} for length $3 \cdot 10^9$ cm to $p \approx 0.2$ dyn cm^{-2} for length $2 \cdot 10^{10}$ cm [81P],

loops with $T \approx 2 \cdot 10^6$ K, $n_e = (2.2 \cdots 3.7) \cdot 10^9$ cm^{-3}, $p = 1.3 \cdots 2.2$ dyn cm^{-2} [80C],
loops with $T = 2.5 \cdot 10^4 \cdots 2 \cdot 10^6$ K, $n_e = 5 \cdot 10^9 \cdots 10^{10}$ cm^{-3} [85D],
loops with $T = (2.5 \cdots 4) \cdot 10^5$ K, $p \approx 5$ dyn cm^{-2} [86T], $n_e = 2 \cdot 10^8 \cdots 10^{10}$ cm^{-3} [82R].

3.1.2.6.1 Visible

H_α loops: $T = 15000 \cdots 25000$ K, $n_e = (3 \cdots 7.5) \cdot 10^{10}$ cm^{-3}, $p = 0.2 \cdots 0.5$ dyn cm^{-2} [85L]. Flow velocities along the loop $40 \cdots 120$ km s^{-1} [84L].

3.1.2.6.2 – 3.1.2.6.4 see LB VI/2a.

References for 3.1.2.6

General references

a Bray, R.J., Cram, L.E., Durrant, C.J., Loughhead, R.E.: Plasma Loops in the Solar Corona, Cambridge: Cambridge Uni. Press (1991).

Special references

80C Cheng, C.-C.: Astrophys. J. **238** (1980) 743.
80S Sheeley, N.R.: Sol. Phys. **66** (1980) 79.
81P Pallavicini, R., Peres, G., Serio, S., Vaiana, G.S., Golub, L., Rosner, R.: Astrophys. J. **247** (1981) 692.
81T Teske, R.G., Mayfield, E.B.: Astron. Astrophys. **93** (1981) 228.
81W Webb, D.F., Zirin, H.: Sol. Phys. **69** (1981) 99.
82D1 Dere, K.P.: Sol. Phys. **75** (1982) 189.
82D2 Dere, K.P.: Sol. Phys. **77** (1982) 77.
82R Raymond, J.C., Foukal, P.: Astrophys. J. **253** (1982) 323.
84L Loughhead, R.E., Bray, R.J.: Astrophys. J. **283** (1984) 392.
85D Doyle, J.G., Mason, H.E., Vernazza, J.E.: Astron. Astrophys. **150** (1985) 69.
85H Habbal, S.R., Ronan, R., Withbroe, G.L.: Sol. Phys. **98** (1985) 323.
85L Loughhead, R.E., Bray, R.J., Wang, J.-L.: Astrophys. J. **294** (1985) 697.
86T Tsiropoula, G., Alissandrakis, C., Bonnet, R.M., Gouttebroze, P.: Astron. Astrophys. **167** (1986) 351.
88H Haisch, B.M., Strong, K.T., Harrison, R.A., Gary, G.A.: Astrophys. J. Suppl. Ser. **68** (1988) 371.

3.1.2.7 Flares

3.1.2.7.1 General

Reviews [b,c,d,f,g]. Flare loops reviewed [a].

Classifications:
according to hard X-rays [86T],

according to physical processes [89B].

Well-observed (multiwavelength) flares:
21 May 1980 (2-ribbon) [83H, 85D1]
25 June 1980 (pre-flare) [83S, 85B2]
1 November 1980 (2-ribbon) [84T]
12 November 1980 [85M]
16 May 1981 (2-ribbon) [83F].

Periodicities $152 \cdots 154$ d (hard X-rays and microwaves) [84R, 85B2, 87B].

Bursts in UV (smaller-scale flares?), $n_e \approx 10^{11} \cdots 10^{12}$ cm^{-3}, timescale $1 \cdots 10$ min, 50% show upward or downward motions $6 \cdots 10$ km s^{-1} [87H].

3.1.2.7.2 Flare spectrum

Visible spectrum $360 \cdots 590$ nm (white-light flare) [84D1, 84D2].

Hard X-ray review [85D2]. X-ray spectrum $10 \cdots 100$ Å[85A].

Gamma ray reviews [84C, 85B1]. Continuum > 300 keV [88V]. Gamma-ray lines, intensities, central wavelengths, widths [90M2].

High-resolution mapping: radio [90W2], hard and soft X-rays [89N], microwave and soft X-ray [88A].

3.1.2.7.3 Flare physics

See general reviews above and [e,h].

Impulsive flare analysis, (XUV) [88W], (X-ray) [88L2]. Compact flare X-ray analysis $T \approx 16 \cdot 10^6$ K, $n_e \approx 3 \cdot 10^{12}$ cm^{-3} [88L1].

Chromospheric heating at temperature minimum [90M1].

Flare loop arcade, inner loops $T = (1.5 \cdots 2) \cdot 10^6$ K, $n_e \approx 2 \cdot 10^{10}$ cm^{-3}, outer loops $T = (2 \cdots 5) \cdot 10^6$ K, $n_e \approx 7 \cdot 10^{10}$ cm^{-3}, apex densities 50% greater than footpoints [90D, 90W1].

Flare model classification scheme [80S].

Gamma-ray physics [84C, 85B1, 87R].

3.1.2.7.4 Flare particle emission

Electron spectra $0.1 \cdots 100$ MeV [89M].

References for 3.1.2.7

General references

a Bray, R.J., Cram, L.E., Durrant, C.J., Loughhead, R.E.: Plasma Loops in the Solar Corona, Cambridge: Cambridge Uni. Press (1991).
b Haisch, B.M., Rodonò, M. (eds.): Solar and Stellar Flares, Sol. Phys. **121** (1989).
c Kundu, M.R., Woodgate, B., Schmahl, E.J. (eds.): Energetic Phenomena on the Sun, Dordrecht: Kluwer Acad. Publ. (1989).
d Priest (ed.): Solar Flare Magnetohydrodynamics, New York: Gordon and Breach Sci. Publ. (1981).

e Somov, B.V.: Physical Processes in Solar Flares, Dordrecht: Kluwer Acad. Publ. (1992).
f Sturrock, P.A.: Solar Flares, Boulder: Colorado Ass. Uni. Press (1980).
g Tandberg-Hanssen, E., Emslie, A.G.: The Physics of Solar Flares, Cambridge: Cambridge Uni. Press (1988).
h Uchida, Y., Canfield, R.C., Watanabe, T., Hiei, E. (eds.): Flare Physics in Solar Activity Maximum 22, Berlin: Springer (1991).

Special references

80S Spicer, D.S., Brown, J.C.: Sol. Phys. **67** (1980) 385.
83F Fárník, F., Kaastra, J.,, Kálmán, B., Karlický, M., Slottje, C., Valníček, B.: Sol. Phys. **89** (1983) 355.
83H Harvey, J.W.: Adv. Space Res. **2**(11) (1983) 31.
83S Schmahl, E.J.: Adv. Space Res. **2**(11) (1983) 73.
84C Chupp, E.L.: Annu. Rev. Astron. Astrophys. **22** (1984) 359.
84D1 Donati-Falchi, A., Falciani, R., Sambuco, A.M., Smaldone, L.A.: Astron. Astrophys. Suppl. Ser. **55** (1984) 425.
84D2 Donati-Falchi, A., Falciani, R., Smaldone, L.A.: Astron. Astrophys. **131** (1984) 256.
84R Rieger, E., Share, G.H., Forrest, D.J., Kaubach, G., Reppin, C., Chupp, E.L.: Nature **312** (1984) 623.
84T Tandberg-Hanssen, E., Kaufmann, P., Reichmann, E.J., Teuber, D.L., Moore, R.L., Orwig, L.E., Zirin, H.: Sol. Phys. **90** (1984) 41.
85A Acton, L.W., Bruner, M.E., Brown, W.A., Fawcett, B.C., Schweizer, W., Speer, R.J.: Astrophys. J. **291** (1985) 865.
85B1 Bai, T., Dennis, B.R.: Astrophys. J. **292** (1985) 699.
85B2 Bogart, R.S., Bai, T.: Astrophys. J. **299** (1985) L51.
85D1 de Jager, C., Švestka, Z.: Sol. Phys. **100** (1985) 435.
85D2 Dennis, B.R.: Sol. Phys. **100** (1985) 465.
85K Kundu, M.R., Gaizauskas, V., Woodgate, B.E., Schmahl, E.J., Shine, R., Jones, H.P.: Astrophys. J. Suppl. Ser. **57** (1985) 621.
85M MacNeice, P., Pallavicini, R., Mason, H.E., Simnett, G.M., Antonucci, E., Shine, R.A., Rust, D.M., Jordan, C., Dennis, B.R.: Sol. Phys. **99** (1985) 167.
86T Tanaka, K.: Astrophys. Space Sci. **118** (1986) 101.
87B Bai, T., Sturrock, P.A.: Nature **327** (1987) 601.
87H Hayes, M., Shine, R.A.: Astrophys. J. **312** (1987) 943.
87R Ramaty, R., Murphy, R.J.: Space Sci. Rev. **45** (1987) 213.
88A Alissandrakis, C.E., Schadee, A., Kundu, M.R.: Astron. Astrophys. **195** (1988) 290.
88L1 Linford, G.A., Lemen, J.R., Strong, K.T.: Adv. Space Sci. **8**(11) (1988) 178.
88L2 Linford, G.A., Wolfson, C.J.: Astrophys. J. **331** (1988) 1036.
88V Vestrand, W.T.: Sol. Phys. **118** (1988) 95.
88W Widing, K.G., Cook, J.W.: Astrophys. J. **320** (1988) 913.
89B Bai, T., Sturrock, P.A.: Annu. Rev. Astron. Astrophys. **27** (1989) 421.
89M Moses, D., Dröge, W., Meyer, P., Evenson, P.: Astrophys. J. **346** (1989) 523.
89N Nitta, N., Kiplinger, A.L., Kai, K.: Astrophys. J. **337** (1989) 1003.
90D Doyle, J.G., Widing, K.G.: Astrophys. J. **352** (1990) 754.
90M1 Metcalf, T.R., Canfield, R.C., Saba, J.L.R.: Astrophys. J. **365** (1990) 391.
90M2 Murphy, R.J., Share, G.H., Letaw, J.R., Forrest, D.J.: Astrophys. J. **358** (1990) 298.
90W1 Widing, K.G., Doyle, J.G.: Astrophys. J. **352** (1990) 760.
90W2 Willson, R.F., Klein, K.L., Kerdraon, A., Lang, K.R., Trottet, G.: Astrophys. J. **357** (1990) 662.

3.1.2.8 Radio emission of the disturbed sun

Revisions and additions to subsect. 3.1.2.8 in LB VI/2a

3.1.2.8.1 – 3.1.2.8.3 see LB VI/2a

3.1.2.8.4. Noise storms

Storm bursts: Theoretical interpretation
A recent review of observational evidence and theory show that fundamental plasma emission is the preferred mechanism for both type I bursts and storm continuum [85K].

3.1.2.8.6 Solar radio bursts

3.1.2.8.6.1 Microwave bursts

Fine structure in centimetric and decimetric bursts
Fine structure occurs in about 30% of bursts at these high frequencies and can be sub-divided into at least four different classes of events. For typical examples, see the atlas of Allaart et al. [90A]. Similarly, in the decimetric range there are also at least four major classes; see catalogue of Gudel and Benz [88G].

Spike bursts
Much has been written about these interesting bursts since the first descriptions by [77D] and [78S]. They are characterised by rise times ≤ 10 ms, durations ≤ 40 ms, bandwidths $\leq 1\%$, flux densities up to 10^{-18} W m^{-2} Hz^{-1}, and degrees of circular polarization up to 100%. From light travel-time arguments, the occasional occurrence of rise times as short as 1 ms imply source sizes ≤ 300 km and brightness temperatures $\geq 10^{13}$ K; such high temperatures indicate coherent emission (for more details see [85D]). The preferred mechanism is the electron-cyclotron maser [80H] at the second harmonic of the gyro-frequency [82M] although debate still continues on how the radiation manages to escape from the source region.

3.1.2.8.6.2 Fast-drift bursts (type III bursts)

Type V bursts
It is assumed that the type V burst is generated by the same fast electrons responsible for the prior type III burst, but that a part of the electron beam is trapped and stored in a large magnetic loop (Fig.8 in LB VI/2a, p.299) or escapes along rapidly diverging field lines [85S1].

Polarization of type III and type V bursts
The Culgoora spectropolarimeter has been used to study the circular polarization of fundamental (F) and second harmonic (H) type III bursts and type V bursts (for more details see [85S1]). The degree of circular polarization of type V bursts is always low ≤ 0.1 with an average value of 0.07 and varies from centre to limb as the cosine of the viewing angle. The sense is usually opposite to that of the preceding type III (which is rather surprising if both are O-mode plasma radiation, as is usually assumed). For type III F-H pairs the fundamental is more highly polarized than the harmonic with average values of $\langle P_H \rangle = 0.12$ and $\langle P_F \rangle = 0.35$. There is a tendency for P_F/P_H to decrease from the centre to limb. Generally the degree of polarization in a given burst does not vary with frequency (down to 25 MHz).

A theory of polarization of H radiation has been developed (see [80M] for details) which shows that O-mode emission for H emission only occurs when the wave vectors of the Langmuir waves are confined within about 20 degrees of the field direction. However, the theory gives much higher values for the magnetic field than expected.

The elliptical polarization found in earlier observations could not be confirmed by later observations and has been now discredited [73G].

Theoretical discussion

For a recent review of the application of weak turbulence and quasi-linear theory to Langmuir wave generation by type III electron streams see [85G]. Also, for a review on plasma emission mechanisms see [85M1].

3.1.2.8.6.3 Slow-drift bursts (type II bursts)

The spectral structure

The spectrum at a fixed time is generally narrow-band but complex. One conspicuous characteristic is the appearance of the second harmonic, which occurs in about 50 ··· 60% of the bursts. The intensity of the second harmonics is comparable with that of the fundamental emission. Higher harmonics are not detected. Frequency splitting can also occasionally be seen in the spectra. This effect is the appearance of two distinct maxima separated from each other by an amount equal to about 10% of the mean frequency, i.e. 10 MHz for a fundamental at 100 MHz and a corresponding split of 20 MHz for the second harmonic. The split persists during the frequency drift over a large part of the frequency range. Possible causes for this phenomenon, such as the magnetic field, a Doppler shift or a geometrical effect, have been investigated [LB VI/2a]. Smerd et al. suggest that the two bands correspond to emission in front of and behind the exciting shock front [74S].

Theoretical model

For a more recent review see [85N].

Polarization

Ordinary type II bursts are only weakly polarized [58K] but fast-drift herringbone structure can be highly polarized [66S, 80S]. These results are consistent with the hypothesis that the herringbone structure is due to type III-like emission from electrons escaping ahead of and behind the shock front.

Shock-accellerated events

As well as the herringbone structure seen in some type II events there is a form of interplanetary event called SA [81C] which bears a resemblance to kilometric type III bursts but is always associated with a metre wavelength type II burst with herringbone structure. These SA events are thought to be excited in the interplanetary medium by electrons streaming far ahead of the shock front. Both the herringbones and SA events are the most direct evidence for particle accelleration by shock waves in the solar corona.

Interplanetary type II bursts

Observations of slow drift type II bursts in the interplanetary medium have been made with instruments aboard the Voyager spacecraft [80B] and in ISEE-3 [82C].

Blast wave or driven shock wave

McLean [59M] proposed that type II bursts are driven by an ejected mass of gas. The observation that many type II bursts are closely associated with coronal transients with speeds $\leq 400\ \mathrm{km\,s^{-1}}$ supports this view with the excitation coming from a stand-off shock ahead of the coronal ejection (piston) which is travelling at the Alfvén speed.

Uchida [74U] proposed an alternative interpretation in terms of a blast wave. He suggested that refraction and focusing of an MHD blast wave in regions of low Alfvén velocity could account for the multiple-ray paths sometimes observed in type II bursts.

This question is still undecided as there appears to be good observational evidence supporting both points of view. It is possible that both mechanisms play roles at different times during solar flares [85N].

3.1.2.8.6.4 Continuum bursts (type IV bursts)

Type IVmF bursts
A further sub-division of flare continuum bursts into sub-classes FCE and FCII has been suggested by Robinson and Smerd [75R] based on Culgoora radioheliograph observations. In this scheme the impulsive phase event is called an FCE and usually lasts for about 15 minutes; this is sometimes immediately followed by an FCII event which is closely associated with the type II burst. The FCII finally blends with a stationary type IV burst (also called storm continuum) discussed below. The emission mechanism for both types of flare continuum is thought to be the scattering of Langmuir waves into fundamental plasma radiation [85R].

Type IVmA bursts (moving type IV)
The emission mechanism was formerly assumed to be a pure gyrosynchrotron process, but the high brightness temperatures measured by the Culgoora radioheliograph argue strongly for coherent plasma emission during the early stages of the moving type IV burst whilst high polarization in the later stages favours gyro-synchrotron emission. Simultaneous observation by the SMM coronagraph show that the electron density in the moving source region is very high, consistent with either plasma or gyro-synchrotron radiation [85S2].

For a review of gyro-synchrotron theory see [85M2].

References for 3.1.2.8

General references

a Solar Radiophysics (McLean, D.J., Labrum, N.R., eds.), Cambridge University Press (1985).
b Radio Physics of the Sun: Int. Astron. Union Symp. **86** (Kundo, M.R., Gergely, T.E., eds.), Dordrecht: Reidel (1980).

Special references

58K Komesaroff, M.: Aust. J. Phys. **11** (1958) 201.
59M McLean, D.J.: Aust. J. Phys. **12** (1959) 404.
66S Stewart, R.T.: Aust. J. Phys. **19** (1966) 209.
73G Grognard, R.J.-M., McLean, D.J.: Sol. Phys **29** (1973) 149.
74S Smerd, S.F., Sheridan, K.V., Stewart, R.T., in: Coronal Disturbances, Int. Astron. Union Symp. **57** (Newkirk, G., ed.) Dordrecht: Reidel (1974) p.389.
74U Uchida, Y.: Sol. Phys. **28** (1974) 495.
75R Robinson, R.D., Smerd, S.F.: Proc. Astron. Soc. Aust. **2** (1975) 374.
77D Dröge, F.: Astron. Astrophys. **57** (1977) 285.
78S Slottje, C.: Nature **275** (1978) 520.
80B Boischot, A., Riddle, A.C., Pearce, J.B., Warwick, J.W.: Sol. Phys. **65** (1980) 397.
80H Holman, G.D., Eichler, D., Kundu, M.R. in: [b] (1980) p.457.
80M Melrose, D.B., Dulk, G.A., Gary, D.E.: Proc. Astron. Soc. Aust. **4** (1980) 50.
80S Suzuki, S., Stewart, R.T., Magun, A., in: [b] (1980) p.241.

81C Cane, H.V., Stone, R.G., Fainberg, G., Stewart, R.T., Steinberg, J.L., Hoang, S.: Geophys. Res. Lett. **8** (1981) 1285.
82C Cane, H., Stone, R.G., Fainberg, G., Steinberg, J.L., Hoang, S.: Sol. Phys. **78** (1982) 187.
82M Melrose, D.B., Dulk, G.A.: Astrophys. J. **259** (1982) 844.
85D Dulk, G.A. in: [a] (1985) ch.4.
85G Grognard, R.J.-M. in: [a] (1985) ch.11.
85K Kai, K., Melrose, D.B., Suzuki, S., in: [a] (1985) ch.16.
85M1 Melrose, D.B., in: [a] (1985) ch.8.
85M2 Melrose, D.B., in: [a] (1985) ch.9.
85N Nelson, G.J., Melrose, D.B., in: [a] (1985) ch.13.
85R Robinson, R.D., in: [a] (1985) ch.15.
85S1 Suzuki, S., Dulk, G.A., in: [a] (1985) ch.12.
85S2 Stewart, R.T., in: [a] (1986) ch.14.
88G Gudel, M., Benz, A.O.: Astron. Astrophys. Suppl. **75** (1988) 243.
90A Allaart, M.A.F., van Nieuwkoop, J., Slottje, C., Sondaar, L.H.: Atlas of Fine Structure in Solar Microwave Bursts, Sterrekundig Institut Utrecht (1990).

3.2 The planets and their satellites

3.2.1 Mechanical data of the planets and satellites

3.2.1.1 Orbital elements of the planets

Table 1 shows osculating orbital elements for the nine major planets. Subsequent epochs of osculation differ by 1000 days. The epochs refer to 0^h ephemeris time (ET), or more precisely, to 0^h barycentric dynamical time (cf. sect. 2.2). Similar tabulations at small intervals appear in the annual issues of "The Astronomical Almanac". J. Schubart has computed the data of Table 1 in analogy to tables given on page E3 of these issues. He has used the same basic ephemeris "DE200/LE200" that was derived at the Jet Propulsion Laboratory in U.S.A.. For additional information that is related to the elements of the planets see LB VI/2a, p.129.

Table 1. Heliocentric osculating orbital elements of the major planets referred to the mean ecliptic and equinox J2000.0.

Epoch	Inclina-	Longitude of		Mean	Daily	Eccen-	Mean
	tion	asc. node	perihelion	distance	motion	tricity	longitude
0^h ET	i	Ω	ϖ	$a\,[AU]$	n	e	L
Mercury							
1993 Jan. 13	7°00539	48°3392	77°4442	0.387099	4°092336	0.205620	279°2965
1995 Oct. 10	7.00522	48.3366	77.4505	0.387098	4.092347	0.205641	51.6323
1998 July 6	7.00504	48.3323	77.4542	0.387099	4.092334	0.205620	183.9740
2001 Apr. 1	7.00490	48.3295	77.4591	0.387098	4.092353	0.205641	316.3096
2003 Dec. 27	7.00469	48.3255	77.4632	0.387099	4.092340	0.205620	88.6517
2006 Sep. 22	7.00460	48.3226	77.4669	0.387098	4.092351	0.205640	220.9872
Venus							
1993 Jan. 13	3°39460	76°6998	131°6960	0.723325	1°602155	0.006749	65°3609
1995 Oct. 10	3.39481	76.6916	131.4539	0.723330	1.602137	0.006741	227.4892

Table 1. (continued)

Epoch	Inclina-tion	Longitude of asc. node	Longitude of perihelion	Mean distance	Daily motion	Eccen-tricity	Mean longitude
0^h ET	i	Ω	ϖ	$a\,[AU]$	n	e	L
1998 July 6	3.39459	76.6843	131.3656	0.723328	1.602144	0.006767	29.6190
2001 Apr. 1	3.39458	76.6753	131.6765	0.723342	1.602099	0.006772	191.7510
2003 Dec. 27	3.39465	76.6687	131.4699	0.723328	1.602145	0.006734	353.8804
2006 Sep. 22	3.39460	76.6621	131.1912	0.723329	1.602142	0.006759	156.0122
Earth (the elements refer to the Earth/Moon barycenter)							
1993 Jan. 13	0°.00096	0°.6418	102°.8470	1.000003	0°.985605	0.016705	112°.5811
1995 Oct. 10	0.00052	344.9392	102.9857	0.999989	0.985625	0.016750	18.1928
1998 July 6	0.00026	4.2625	102.9884	0.999990	0.985624	0.016723	283.8019
2001 Apr. 1	0.00025	175.1505	102.8488	0.999990	0.985624	0.016678	189.4086
2003 Dec. 27	0.00054	179.8911	103.0083	0.999996	0.985614	0.016705	95.0187
2006 Sep. 22	0.00089	175.7108	103.0223	0.999988	0.985627	0.016717	0.6291
Mars							
1993 Jan. 13	1°.85033	49°.5753	336°.0177	1.523626	0°.524067	0.093481	102°.0481
1995 Oct. 10	1.85005	49.5744	336.0049	1.523761	0.523997	0.093376	266.0824
1998 July 6	1.84993	49.5622	335.9907	1.523594	0.524083	0.093384	70.1173
2001 Apr. 1	1.84961	49.5617	335.9689	1.523724	0.524016	0.093396	234.1519
2003 Dec. 27	1.84935	49.5409	336.0579	1.523678	0.524040	0.093541	38.1757
2006 Sep. 22	1.84929	49.5382	336.1009	1.523674	0.524042	0.093437	202.2117
Jupiter							
1993 Jan. 13	1°.30464	100°.4696	15°.6871	5.202988	0°.083087	0.048300	182°.9502
1995 Oct. 10	1.30463	100.4706	15.7139	5.202465	0.083099	0.048403	266.0470
1998 July 6	1.30465	100.4726	15.6635	5.203010	0.083086	0.048542	349.1522
2001 Apr. 1	1.30415	100.5104	15.2987	5.204003	0.083062	0.048877	72.1839
2003 Dec. 27	1.30375	100.5102	14.8088	5.201910	0.083113	0.048968	155.2687
2006 Sep. 22	1.30375	100.5098	14.6933	5.202040	0.083110	0.048933	238.3917
Saturn							
1993 Jan. 13	2°.48682	113°.6735	93°.4970	9.525356	0°.033531	0.053811	324°.8002
1995 Oct. 10	2.48542	113.6447	90.9596	9.551792	0.033392	0.052336	358.2147
1998 July 6	2.48525	113.6398	88.5241	9.578816	0.033251	0.054071	31.6824
2001 Apr. 1	2.48560	113.6290	91.4251	9.583858	0.033224	0.057066	65.3842
2003 Dec. 27	2.48579	113.6258	94.0289	9.578433	0.033253	0.057153	98.9125
2006 Sep. 22	2.48720	113.6296	93.6858	9.551249	0.033395	0.054509	132.3202
Uranus							
1993 Jan. 13	0°.77226	74°.0219	172°.8397	19.216801	0°.011700	0.047762	283°.2434
1995 Oct. 10	0.77333	74.0880	176.6428	19.293470	0.011630	0.044837	295.0478
1998 July 6	0.77317	74.0676	173.9420	19.284657	0.011638	0.042773	307.0252
2001 Apr. 1	0.77207	73.9143	168.8747	19.178952	0.011735	0.046796	318.8050
2003 Dec. 27	0.77180	73.8594	169.7591	19.129435	0.011780	0.050085	330.3233
2006 Sep. 22	0.77194	73.9631	172.4829	19.176268	0.011737	0.048095	341.8685
Neptune							
1993 Jan. 13	1°.77202	131°.7531	43°.8105	30.104114	0°.005967	0.006637	289°.4598
1995 Oct. 10	1.77009	131.7750	3.0630	30.258135	0.005922	0.007985	295.5641
1998 July 6	1.76752	131.7966	16.2751	30.224221	0.005932	0.010860	301.8815
2001 Apr. 1	1.76918	131.7901	54.2035	30.006703	0.005996	0.011208	307.8670
2003 Dec. 27	1.77157	131.7866	69.1811	29.955205	0.006012	0.009089	313.5324
2006 Sep. 22	1.77093	131.7842	36.0944	30.098634	0.005969	0.006895	319.3433

Table 1. (continued)

Epoch	Inclina-	Longitude of		Mean	Daily	Eccen-	Mean
	tion	asc. node	perihelion	distance	motion	tricity	longitude
0^h ET	i	Ω	ϖ	$a\,[AU]$	n	e	L
Pluto							
1993 Jan. 13	17°14012	110°2901	223°9482	39.771574	0°003930	0.254284	228°6992
1995 Oct. 10	17.12035	110.3863	224.6786	39.798998	0.003926	0.254418	232.9355
1998 July 6	17.12962	110.3546	224.6234	39.442077	0.003979	0.247584	237.0330
2001 Apr. 1	17.16656	110.2436	223.5942	39.240448	0.004010	0.244750	240.7181
2003 Dec. 27	17.16378	110.2477	223.5746	39.497144	0.003971	0.250063	244.4650
2006 Sep. 22	17.12748	110.3143	224.5301	39.784761	0.003928	0.254308	248.5497

3.2.1.2 Dimensions and mechanical properties, rotation of the planets

Tables 2 and 3 update the respective tables of the preceding edition [81G] in the cases of the three outermost planets. The successful Voyager II mission to Uranus [90M, 86D] and Neptune [89S] has given new information on these bodies, especially on the rotation. New data for Pluto [89B, 89T, 93N] have resulted from observations of its satellite and of related phenomena.

Table 2. Dimensions and mechanical properties of the outermost planets.

D_{equ} equatorial diameter
D_{pol} polar diameter
D diameter; for Uranus, Neptune: $D = (2\ D_{equ} + D_{pol})/3$
$1/m$ reciprocal mass (including satellites), solar mass is unity
$\bar{\varrho}$ mean density
J_2 ellipticity coefficient of gravity field

Planet	D_{equ} [km] D_{pol} [km]	D [km]	$1/m$ [$1/m_\odot$]	$\bar{\varrho}$ [gcm^{-3}]	J_2	Ref.
Uranus	equ. 51120 pol. 49950	50730	22902.9	1.27	0.0033	90M
Neptune	equ. 49530 pol. 48680	49247	19412.2	1.64	0.0034	89S
Pluto		2300	135 000 000	2.0		89T, 89B

Table 3. Rotation of the outermost planets.

Planet	Sidereal rotation period [d]	[h]	Ref.	Inclination of equator to orbit	Ref.
Uranus		17.24	90M	98°	81G
Neptune		16.11	89S	29°	81G
Pluto	6.3872		89T	120°	93S

References for 3.2.1.2

81G Gondolatsch, F., in: Landolt-Börnstein, NS, Vol. VI/2a (1981) 132 ··· 136.
86D Desch, M.D., Connerney, J.E.P., Kaiser, M.L.: Nature **322** (1986) 42.
89B Beletic, J.W., Goody, R.M., Tholen, D.J.: Icarus **79** (1989) 38.
89S Stone, E.C., Miner, E.D.: Science **246** (1989) 1417; see the associated papers by numerous authors on p.1466 and p.1498.
89T Tholen, D.J., Buie, M.W.: Bull. Am. Astron. Soc. **21** (1989) 981.
90M Miner, E.D.: Uranus – The Planet, Rings and Satellites. New York, London, etc.: Ellis Horwood (1990) 211.
93N Null, G.W., et al.: Sky and Telescope **85** (1993) 9, see Astron. J. **105** (1993) 2319.
93S Schubart, J.: Estimation from data on orbital plane of satellite.

3.2.1.3 Satellites and ring systems of the planets

3.2.1.3.1 Mechanical data of the satellites

Table 4 replaces and extends the analogous former table [81G] with respect to the most important information. Most lunar data are retained. Values on orbital dimension and period correspond to a table in the 1992 edition of "The Astronomical Almanac" [91AA], but Saturn's eighteenth satellite Pan appears in addition, see [91S] for data on Pan. The mean diameter, D, equals two times the mean radius given in the 1988 report of an IAU/IAG/COSPAR working group [89D] in most cases. For values on the system of Neptune see [89S]. A value of D for Charon is chosen according to [89T1] and [91E]. Values of satellite mass, in some cases published together with an estimate of mean density, are available for Phobos [89A], Deimos [80V], the Galilean satellites of Jupiter [85C], some satellites of Saturn [89C] and of Uranus [90M], and for Triton [89T2]. In case of Saturn, the masses given in [81G] for satellites I–IV are retained. Additional values of mean density result from the quotient of the mass and the volume of a sphere with diameter D.

Table 4. Mechanical data of the satellites.

a semi-major axis of the orbit
P sidereal period around the planet, retrograde motion indicated by letter r
D mean diameter
m mass
$\bar{\varrho}$ mean density

Satellite*)		a [10^3km]	P [d]	D [km]	m [g]	$\bar{\varrho}$ [gcm^{-3}]	Discovery
Earth							
	Moon	384	27.322	3475	$7.35 \cdot 10^{25}$	3.34	
Mars							
I	Phobos	9.4	0.319	22	$10.8 \cdot 10^{18}$	2.0	1877
II	Deimos	23.5	1.262	12	$2.0 \cdot 10^{18}$	2	1877
Jupiter							
XVI	Metis	128	0.295	40			1980
XV	Adrastea	129	0.298	20			1979
V	Amalthea	181	0.498	172			1892
XIV	Thebe	222	0.674	100			1980
I	Io	422	1.769	3643	$8.93 \cdot 10^{25}$	3.53	1610
II	Europa	671	3.551	3130	$4.80 \cdot 10^{25}$	2.99	1610

Table 4. (continued)

Satellite*)		a [10^3km]	P [d]		D [km]	m [g]	$\bar{\varrho}$ [gcm^{-3}]	Discovery
III	Ganymede	1070	7.155		5268	$14.82 \cdot 10^{25}$	1.94	1610
IV	Callisto	1883	16.689		4806	$10.76 \cdot 10^{25}$	1.85	1610
XIII	Leda	11094	239		10			1974
VI	Himalia	11480	251		170			1904
X	Lysithea	11720	259		24			1938
VII	Elara	11737	260		80			1905
XII	Ananke	21200	631	r	20			1951
XI	Carme	22600	692	r	30			1938
VIII	Pasiphae	23500	735	r	36			1908
IX	Sinope	23700	758	r	28			1914
Saturn								
XVIII	Pan	133.6	0.575		20			1990
XV	Atlas	137.7	0.602		32			1980
XVI	Prometheus	139.4	0.613		100			1980
XVII	Pandora	141.7	0.628		84			1980
XI	Epimetheus	151.4	0.694		119			1980
X	Janus	151.5	0.694		178			1976
I	Mimas	185.5	0.942		398	$3.7 \cdot 10^{22}$	1.12	1789
II	Enceladus	238	1.370		498	$8.4 \cdot 10^{22}$	1.3	1789
III	Tethys	295	1.888		1046	$6.3 \cdot 10^{23}$	1.05	1684
XIII	Telesto	295	1.888		22			1980
XIV	Calypso	295	1.888		19			1980
IV	Dione	377	2.737		1120	$11.6 \cdot 10^{23}$	1.58	1684
XII	Helene	377	2.737		32			1980
V	Rhea	527	4.518		1528	$23.1 \cdot 10^{23}$	1.24	1672
VI	Titan	1222	15.95		5150	$13.5 \cdot 10^{25}$	1.89	1655
VII	Hyperion	1481	21.28		283			1848
VIII	Iapetus	3561	79.33		1436	$1.6 \cdot 10^{24}$	1.0	1671
IX	Phoebe	12952	550	r	220			1898
Uranus								
VI	Cordelia	49.8	0.335		26			1986
VII	Ophelia	53.8	0.376		30			1986
VIII	Bianca	59.2	0.435		42			1986
IX	Cressida	61.8	0.464		62			1986
X	Desdemona	62.7	0.474		54			1986
XI	Juliet	64.4	0.493		84			1986
XII	Portia	66.1	0.513		108			1986
XIII	Rosalind	69.9	0.558		54			1986
XIV	Belinda	75.3	0.624		66			1986
XV	Puck	86.0	0.762		154			1985
V	Miranda	129.4	1.413		472	$6.3 \cdot 10^{22}$	1.2	1948
I	Ariel	191	2.520		1158	$12.7 \cdot 10^{23}$	1.6	1851
II	Umbriel	266	4.144		1169	$12.7 \cdot 10^{23}$	1.5	1851
III	Titania	436	8.706		1578	$34.9 \cdot 10^{23}$	1.7	1787
IV	Oberon	584	13.463		1523	$30.3 \cdot 10^{23}$	1.6	1787

Table 4. (continued)

Satellite*)		a [10^3km]	P [d]		D [km]	m [g]	$\overline{\varrho}$ [gcm^{-3}]	Discovery
Neptune								
III	Naiad	48.2	0.294		54			1989
IV	Thalassa	50.1	0.311		80			1989
V	Despina	52.5	0.335		180			1989
VI	Galatea	62.0	0.429		150			1989
VII	Larissa	73.6	0.555		190			1989
VIII	Proteus	117.6	1.122		400			1989
I	Triton	355	5.877	r	2705	$21.4 \cdot 10^{24}$	2.05	1846
II	Nereid	5513	360		340			1949
Pluto								
I	Charon	19.6	6.387		1200			1978

*) The final Roman numbers and names of the IAU are given here. In the literature sometimes provisional numbers given at the discovery or arbitrarily chosen by some authors are still used.

3.2.1.3.2 Ring systems of the planets Saturn, Uranus, Neptune

Table 5 shows in its three parts data about the ring systems of Saturn, Uranus, and Neptune. For the faint ring of Jupiter see [81G]. Each part of Table 5 starts with a value for the radius of the planetary equator. The values for Saturn's rings correspond to a table in [84G]. For the system of Uranus see [90M]. Special names refer to the most prominent features in Neptune's system [91IAU], see [89S].

References for 3.2.1.3

80V Veverka, J., Burns, J.A.: Annu. Rev. Earth Planet. Sci. **8** (1980) 527.

81G Gondolatsch, F., in: Landolt-Börnstein, NS, Vol. VI/2a (1981) 136···141.

84G Gehrels, T., Matthews, M.S., eds.: Saturn. Tucson, Ariz.: The University of Arizona Press (1984) 473.

85C Campbell, J.K., Synnott, S.P.: Astron. J. **90** (1985) 364.

89A Avanesov, G.A., et al.: Nature **341** (1989) 585.

89C Campbell, J.K., Anderson, J.D.: Astron. J. **97** (1989) 1485.

89D Davies, M.E., chairman: Report of the IAU/IAG/COSPAR working group on cartographic coordinates and rotational elements of the planets and satellites: 1988, Celestial Mechanics Dyn. Astron. **46** (1989) 187···204.

89S Stone, E.C., Miner, E.D.: Science **246** (1989) 1417.

89T1 Tholen, D.J., Buie, M.W.: Bull. Am. Astron. Soc. **21** (1989) 981.

89T2 Tyler, G.L., et al.: Science **246** (1989) 1466.

90M Miner, E.D.: Uranus – The Planet, Rings and Satellites. New York, London, etc.: Ellis Horwood (1990) 262 and 287.

91AA The Astronomical Almanac for the year 1992. Washington: U.S. Government Printing Office, and London: H.M. Stationery Office (1991) F2.

91E Elliot, J.L., Young, L.A.: Icarus **89** (1991) 244.

91IAU IAU working group for planetary system nomenclature: Trans. Int. Astron. Union **XXIA** (1991) 619.

91S Showalter, M.R.: Nature **351** (1991) 709.

Table 5. Planetary ring data.

R_{equ} radius of planetary equator
r radius; in case of extended features or eccentric rings, r refers to the center or semi-major axis, respectively
w radial width of gaps, or of narrow rings (variation indicated)
IE inner edge of a ring
OE outer edge of a ring

Saturn			Uranus			Neptune	
Feature	r [10^3km]	w [km]	Ring	r [10^3km]	w [km]	Ring	r [10^3km]
R_{equ}	60.3		R_{equ}	25.6		R_{equ}	24.8
D Ring, IE	67.0		1986U2R, IE	37		Galle = 1989N3R	41.9
C Ring, IE	74.5		1986U2R, OE	39.5			
Maxwell Gap	87.5	253	6	41.84	1···3	Leverrier = 1989N2R	53.2
B Ring, IE	92.0		5	42.23	2···3		
B Ring, OE	117.5		4	42.57	2···3	Adams = 1989N1R	62.9
Huygens Gap	117.7	430	Alpha	44.72	4···13	contains arcs	
Cassini Division	119.8	4540	Beta	45.66	7···12		
A Ring, IE	122.2		Eta	47.18	1···2		
Encke Gap	133.6	328	Gamma	47.63	1···4		
Keeler Gap	136.5	31	Delta	48.30	3···7		
A Ring, OE	136.8		1986U1R	50.02	2···3		
F Ring	140.4	50	Epsilon	51.15	20···95		
G Ring	170						
E Ring, IE	180						
E Ring, OE	480						

3.2.1.4/5 Special data for earth and moon

The careful compilation by Gondolatsch of mechanical data for earth [81G1] and moon [81G2] is not repeated in this volume. Beginning with the issue for 1984, the annual ephemeris "The Astronomical Almanac" is based on fundamental ephemerides of the planets and the moon prepared at the Jet Propulsion Laboratory, California [83AA1]. These ephemerides are consistent with the IAU (1976) system of astronomical constants [81F] apart from some modifications [83AA2]. The respective changes of some values given in [81G1] or [81G2] are small or negligible at the used accuracy. For a more recent discussion of numerical standards see [89M].

References for 3.2.1.4/5

81F Fricke, W., in: Landolt-Börnstein, NS, VI/2a (1981), subsect. 2.3.3; see the respective remarks in this volume.

81G1 Gondolatsch, F., in: Landolt-Börnstein, NS, VI/2a (1981), subsect. 3.2.1.4.

81G2 Gondolatsch, F., in: Landolt-Börnstein, NS, VI/2a (1981), subsect. 3.2.1.5.

83AA1 The Astronomical Almanac for the year 1984, Washington and London (1983), preface.

83AA2 The Astronomical Almanac for the year 1984, Washington and London (1983), supplement, pp. S11/12.

89M McCarthy, D.D., chairman of working group, International Earth Rotation Service, Observatoire de Paris, IERS Technical Note **3** (1989).

3.2.1.6 See LB VI/2a

3.2.2 Planetary physical data

3.2.2.1 Introduction

Much had happened in planetary sciences in the decade since the publication of LB VI/2a. The Voyager spacecraft have completed their grand tour of the outer solar system and are now on their ways to the interstellar space. The spaceprobes to comets Giacobini-Zinner and Halley in 1985 and 1986 have raised a lot of excitements and have indeed become one of the landmarks in modern scientific achievements. The end of the eighties saw the Phobos mission to Mars which unfortunately did not fully attain the original mission goals because of technical failures. The early part of the nineties is marked by the completion of the radar mapping of the Venus' surface by the Magellan Venus Orbiter and then the successful closeup observations of the asteroids, Graspa and Ida, by the Galileo spacecraft. Of all these many new results in planetary physics we have opted to focus on the Voyager results in the present volume. This is because the progress made by this Outer Planets mission has been truely breathtaking and hence is worth special mentioning. Following the previous approach, we provide a brief description in each section to highlight the physical properties of major importance. We do not attempt a comprehensive listing of references except for a few which are considered to be basic to the subject matter. Also all the tables – except for Table 3 – are mostly culles from K.R. Lang's Astrophysical Data: Planets and Stars [92L] to avoid duplication in efforts. There are, however, slight additions to some of the tables.

Table 1. Physical data of planets. From [92L].

R_{equ} = equatorial radius
f = oblateness
= $(R_{equ} - R_{pol})/R_{equ}$
M = mass
P = sidereal rotational period
i = inclination of equator to orbit
mg = magnetic moment
α = tilt angle of magnetic axis to rotational axis
J_2 = second zonal harmonic of gravitational potential

	R_{equ} [km]	f	M [g]	P	i	mg/R_{equ}^3 [G]	α	J_2
Mercury	2439	0.0	$3.3022 \cdot 10^{26}$	58.6462 d	7.0°	0.0035	< 10°	$(8 \pm 6) \cdot 10^{-5}$
Venus	6051	0.0	$4.8690 \cdot 10^{27}$	243.01 d	177.4°			$(6 \pm 3) \cdot 10^{-6}$
Earth	6378	$3.3529 \cdot 10^{-3}$	$5.9742 \cdot 10^{27}$	23h 56m 4.009s	23.45°	0.31	11.5°	$1.083 \cdot 10^{-3}$
Mars	3397	$7.4 \cdot 10^{-3}$	$6.4191 \cdot 10^{26}$	24h 37m 22.66s	23.98°	≤ 0.0004		$1.959 \cdot 10^{-3}$
Jupiter	71492	$6.487 \cdot 10^{-2}$	$1.8992 \cdot 10^{30}$	9h 55m 29.7s	3.08°	4.3	9.6°	$1.474 \cdot 10^{-2}$
Saturn	60268	$9.796 \cdot 10^{-2}$	$5.6865 \cdot 10^{29}$	10h 39m 22.4s	26.73°	0.21	0.8°	$1.648 \cdot 10^{-2}$
Uranus	25559	$2.293 \cdot 10^{-2}$	$8.6849 \cdot 10^{28}$	17.24 ± 0.01 h	97.92°	0.228	58.6°	$3.343 \cdot 10^{-3}$
Neptune	24764	$1.71 \cdot 10^{-2}$	$1.0235 \cdot 10^{29}$	16.11 ± 0.05 h	28.8°	0.133	47°	$3.411 \cdot 10^{-3}$
Pluto	1123		$1.36 \cdot 10^{25}$	6.387 d				

3.2.2.2 Planetary interiors

The interior structures of the outer gaseous planets are generally considered to be composed of three layers:

(a) a rocky core (composed of 38% SiO_2, 25% MgO, 25% FeS and 12% FeO by mass);
(b) an icy mantle (composed of 56.5% H_2O, 32.5% CH_4 and 11% NH_3 by mass); and
(c) an envelope of solar composition with hydrogen and helium as the major species [89H].

According to Gautier and Owen [89G], the renormalized helium mass fraction Y of the atmospheres is 0.18 ± 0.04 (Jupiter), 0.06 ± 0.05 (Saturn), and 0.26 ± 0.05 (Uranus and Neptune), respectively [89C]. The depletion of the helium in Saturn may be the result of helium-hydrogen phase separation in Saturn's metallic-hydrogen zone [75S]. There are also enrichments of heavy elements in the atmospheres: the atmospheric mixing ratio of CH_4 is about twice the solar mixing ratio in Jupiter, two to six times in Saturn, and may be as much as 25 times in Uranus and Neptune [89G]. As for the central cores, they all appear to be of similar mass (10 ··· 15 Earth masses) according to model calculations. This feature is consistent with the theoretical scenario that all Jovian planets formed first with the formation of a solid core of 10 ··· 15 Earth masses in the solar nebula which then triggered the subsequent gravitational collapse of the gaseous envelope [80M]. Such a two-stage process was challenged by Stevenson [85S] who suggested that there should be a continuous capture of large-sized solid objects even after the formation of the central cores. Fig. 1 shows the updated models for the density distributions in Jupiter, Saturn and Uranus according to the published work of Hubbard and Marley [89H].

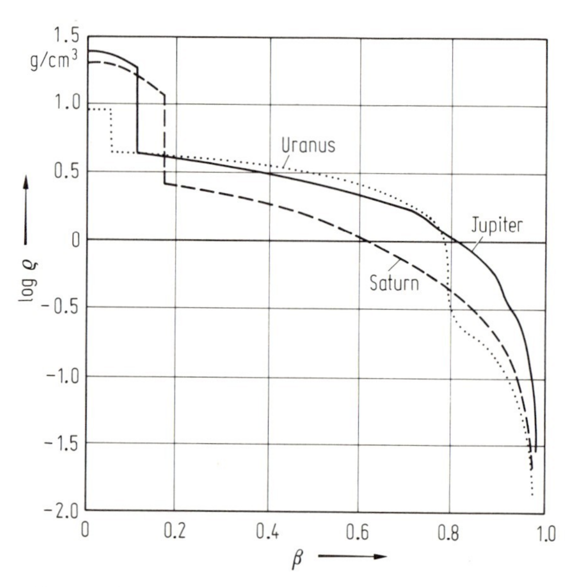

Fig. 1. Distribution of density (ρ) in Jupiter, Saturn and Uranus as a function of normalized radius (β). From Hubbard and Marley [89H].

3.2.2.3 and 3.2.2.4 see LB VI/2a

3.2.2.5 Planetary and satellite atmospheres

From ground-based obsevations, the dynamics of the Jovian atmosphere are long known to be characterized by latitudinal bands and zonal jet streams as well as the Giant Red Spot (GRD). Whether similar behaviour might be found in other Jovian planets had always been a central issue in planetary metereology. The Voyager flyby observations of the outer planetary systems thus gave us the first opportunity of comparing the atmospheric dynamics of Jupiter, Saturn, Uranus and Neptune. The upper atmospheres of these gaseous planets are all mixtures of molecular hydrogen (H_2) and helium. The visible cloud features are ammonia (NH_3) with colouration by sulfur, phosphorus, and carbon compunds. At Uranus and Neptune, cirrus clouds of methane (CH_4) could form at high altitudes. At low altitudes, ammonium hydrosulfide (NH_4SH) and water will condense into cloud layers. The cloud-top pressure is estimated to be at the level between 0.3 and 3 bar and that of the cloud-base between 5 and 15 bar [90I].

Table 2. Planetary and satellite atmospheres. From [92L].

p_0 = surface pressure
T_0 = surface temperature
a_0 = surface acceleration

	p_0 [bar]	T_0 [K]	a_0 [cm s^{-2}]	Major composition
Mercury		450	395	
Venus	90	730	888	CO_2(0.96), N_2(0.035)
Earth	1	288	978	N_2(0.77), O_2(0.21)
Mars	$7 \cdot 10^{-3}$	218	373	CO_2(0.95), N_2(0.027)
Jupiter			2320	H_2(0.86), He(0.14)
Saturn			877	H_2(0.92), He(0.074)
Uranus			946	H_2($\approx$ 0.89), He($\approx$ 0.11)
Neptune			1370	H_2($\approx$ 0.89), He($\approx$ 0.11)
Pluto	$(5 \cdots 20) \cdot 10^{-6}$	$35 \cdots 37$	72	N_2 or CO_2
Io	$(5 \cdots 35) \cdot 10^{-6}$	130	181	SO_2 and S_2(?)
Titan	1.6	95	135	N_2($\approx$ 0.98), CH_4($\approx$ 0.01)
Triton	$1.4 \cdot 10^{-5}$	38	77.7	N_2, CH_4 (mixing ratio $\approx 2.5 \cdot 10^{-6}$ at 40 km altitude)

One of the major discoveries by Voyager concerns the strength of the zonal winds at Saturn. The equatorial value of 500 m/s is almost three times faster than the wind speed at Jupiter. Moreover, the latitudinal wind profile is always eastward. The record of largest wind speed in the solar system belongs to Neptune, however. As summarized in Fig. 2, the retrograte equatorial jet reaches a wind speed of about 600 m/s. It is noteworthy that Jupiter and Saturn have prograde equatorial jets and Uranus and Neptune retrograde equatorial jets. To a large extent, Neptune's zonal wind profile is quite similar to that of Uranus. One major difference is that a large discrete cloud feature called the Giant Dark Spot (GDS) was found at 20°S latitude. Similar to the GRS of Jupiter, the circulation of the GDS is anticyclonic. Besides the GDS, several other discrete features of smaller sizes have also been identified at Neptune.

A hexagonal feature co-rotating with the planet was found at the North Pole of Saturn by using the filtered mosaics from Voyager [88G]. The outermost region of this hexagonal belt-zone structure reaches 65°N. This interesting pattern (see Fig. 3) is most likely the result of large-scale planetary waves. Another co-rotating feature at Saturn has to do with the source region of strong radio waves near the equator as detected by the Planetary Radio Astronomy experiment on Voyager 1 [81W]. The generation of such Saturn Electrostatic Discharge (SED) might be related to the occurrence of lightning activities in a few long-lasting thunderstorm zones near the equator [83Z]. Both the hexagonal feature in the polar region and the SED emission mechanism at the equator will be investigated by the Cassini mission.

Pluto was recently inferred to have an atmosphere of gaseous N_2 or CO_2 [93S]. The surface temperature of Pluto at $35 \cdots 37$ K is enough to maintain a N_2 or CO_2 atmosphere via sublimation. This new

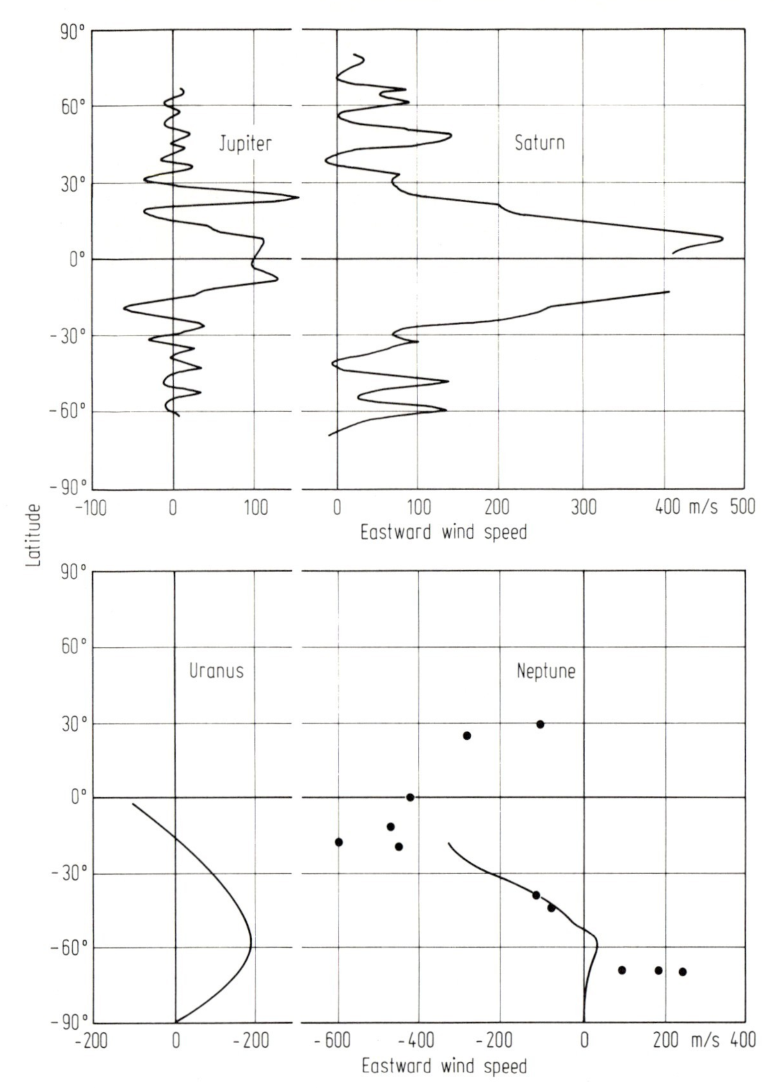

Fig. 2. Zonal velocity versus latitude for the four outer planets. Velocities are measured relative to the planetary interiors. From Ingersoll [90I].

finding is at variance with previous work [80F, 89Y] which suggested that the atmosphere is mainly CH_4 and that a second component of heavier mass might also exist. Because of Pluto's large obliquity and very eccentric orbit, very strong seasonal and orbital variations in its volatile atmosphere are expected.

The near-infrared reflectance spectra of the largest minor planet, Ceres, was consistent with the possible presence of water frost on its surface [81L]. Recent IUE (International Ultraviolet Explorer) observations showed a clear signature of OH emission in the vicinity of Ceres [92A] and hence proved that there must be a source of water ice, either near the polar regions or buried under the surface. The gas production rate was estimated to be on the order of a few times 10^{25} molecule/s.

From ground-based microwave observations of Io at 222 GHz using the 30-m IRAM radiotelescope, an atmosphere of SO_2 with a surface pressure of $4 \cdots 35$ nbar was detected [90L]. Simultaneous measurements at 168.8 GHz put the surface pressure of H_2S below 10^{-10} bar.

Fig. 3. A composite filtered mosaic of the north pole of Saturn from the Voyager imaging observations. From Godfrey [88G].

Table 3. Abundances of Titan's minor constituents at the equator.

Gas		Mole fraction		
		Updated values [91C]	Previous estimates*)	Model predictions
Acethylene	C_2H_2	$2.2^{+0.7}_{-0.9} \cdot 10^{-6}$	$3.0 \cdot 10^{-6}$ $2.0 \cdot 10^{-6}$	$6.0 \cdot 10^{-5}$
Ethylene	C_2H_4	$9.0^{+3}_{-5} \cdot 10^{-8}$	$1.0 \cdot 10^{-6}$ $4.0 \cdot 10^{-7}$	$3.5 \cdot 10^{-7}$
Ethane	C_2H_6	$1.3^{+0.5}_{-0.7} \cdot 10^{-5}$	$2.0 \cdot 10^{-4}$	
Methyl acethylene	C_3H_4	$4.4^{+1.7}_{-2.1} \cdot 10^{-9}$	$3.0 \cdot 10^{-8}$	$6.0 \cdot 10^{-7}$
Propane	C_3H_8	$7.0^{+4}_{-4} \cdot 10^{-7}$	$2.0 \cdot 10^{-5}$ $1.2 \cdot 10^{-5}$	$9.0 \cdot 10^{-6}$
Diacethylene	C_4H_2	$1.4^{+0.6}_{-0.7} \cdot 10^{-9}$		$8.0 \cdot 10^{-9}$
Hydrogen cyranide	HCN	$1.6^{+0.4}_{-0.6} \cdot 10^{-7}$		$9.0 \cdot 10^{-6}$
Cyanoacethylene	HC_3N	$\leq 1.5 \cdot 10^{-9}$		
Cyanogen	C_2N_2	$\leq 1.5 \cdot 10^{-9}$		
Carbon dioxide	CO_2	$1.4^{+0.3}_{-0.5} \cdot 10^{-8}$	$1.5 \cdot 10^{-9}$	$4.5 \cdot 10^{-9}$

*) See Coustenis et al. [89C3] for detailed references.

The flyby observations by Voyager 1 showed that Saturn's largest satellite, Titan, has a very thick atmosphere with a surface pressure of 1.6 bar. The main atmospheric component is nitrogen plus a number of complex hydrocarbon molecules as minor species (see Table 3). The mixing ratio of methane (CH_4) is on the order of 2% near the homopause. Due to photolytic processes, a variety of hydrocarbon molecules are produced in the upper atmosphere leading to the continuous sedimentation of organic polymers on the surface. The presence of such aerosol particles produces the orange colour of Titan's atmosphere and its large optical opaqueness. The accumulation of ethane (C_6H_6) and organic polymers over the eons might eventually lead to the formation of liquid ($CH_4 - C_2H_6$) surface at Titan. Both the surface pressure of 1.6 bar and the temperature of 96 K are consistent with this scenario [83L]. The possible surface coverage (i.e., global oceans vs. localized puddles) is still under debate.

Triton, the largest satellite of Neptune, was found to possess a nitrogen atmosphere as well. The surface pressure ($\approx$ 14 μbar) is extremely small, however. The surface temperature of 34 K is sufficient to maintain the measured surface pressure under the condition of vapour pressure equilibrium. The plume-like structures in the lower atmosphere are probably indicative of the presence of sublimation-driven wind systems [89B]. The mixing ratio of CH_4 was found to be about $2.5 \cdot 10^{-6}$ at 40 km altitude [86S]. Such low abundance of the methane gas might in fact be the reason why Triton has a robust ionosphere with a peak electron density of $4 \cdot 10^4$ electrons/cm^3 while that of Titan should be much smaller with $n_e < 3 \cdot 10^3$ electrons/cm^3 [81T, 89T].

3.2.2.6 Planetary magnetic fields

The encounter of the Ulysses spacecraft with the Jovian system in February 1992 broke the hiatus since the last flyby of the Voyager 2 spacecraft in 1979. One of the major results is that the size of the Jovian magnetosphere was significantly larger during the Ulysses encounter. This large change may be caused by a time-variation in the solar wind dynamic pressure or by the internal effect of the Io plasma torus. Clarification of the actual controlling mechanism must await the long-term measurements of the Galileo Orbiter starting in 1995.

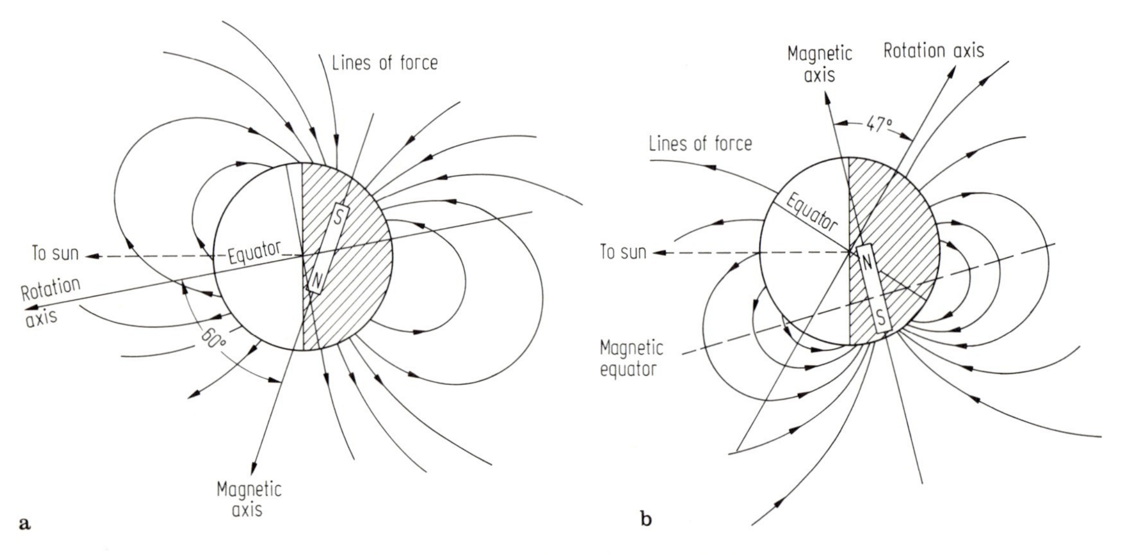

Fig. 4. Diagrams of the OTD field lines of Uranus (a) and Neptune (b). From Ness et al. [86N, 89N].

While the magnetic fields of Jupiter, Uranus and Neptune all have very large quadrupolar components and appreciable displacements from the planetary centres, Saturn has the most symmetrical magnetic field of all planets which have been visited by spaceprobes. The dipole off-set of Jupiter according to the OTD (Offset Tilted Dipole) model [76A] is 0.068 R_J (Jovian radius) with a tilt angle of 9.5°, and much discussions have been given to the possible magnetic anomaly effect on the Jovian magnetosphere [83H]. However, this is dwarfed by the much larger off-sets and tilt angles of Uranus (0.3 R_U and 60° and Neptune (0.55 R_N and 47°). The magnetic-field-line structures of Uranus and Neptune are shown in Fig. 4. Such large quadrupole moments in these two planets might be related to their internal structures possibly characterized by a dynamo region close to the planetary surface [89N].

3.2.2.7 see LB VI/2a

3.2.2.8 Planetary satellites

A fundamental contribution of the Voyager mission to planetary science lies in the close-up looks of various planetary satellites in the outer solar system. This new information has completely revolutionized our concept and understanding of these objects. For examples, the crater frequency distributions on different parts of a satellite provide us with insights into the solar-system-wide impact history and its resurfacing process.

The high-resolution images of different satellites testify to the view that almost each one of them has its own individual characteristics. Take Miranda, which was an object of little scientific interest before the Voyager flyby observations of the Uranian system, the close-up pictures now show that it has actually one of the most spectacular geomorphic structures among planetary satellites. As shown in Fig. 5, half of Miranda's surface is approximately divided into three rectangular sections of young terrain; and an enormous system of fault scarps can be traced across the whole globe [86S]. Such features might be associated with the repeated process of catastrophic impact collision and reaccretion of this satellite during his early history.

In the Saturnian system, the inner satellite, Mimas, also shows a record of heavy surface bombardment by small stray bodies [81S]. Its density of small craters is even higher than that on the lunar highland. Furthermore, it has a crater which size is about one-third of the satellite diameter and hence the corresponding impact event must have been near the limit of catastrophic breakup of the whole object. Moving away from the planet, we find Enceladus which has the highest albedo (≈ 0.8) in the Saturnian system. If this peculiar surface property is the result of recent resurfacing events, no completely satisfactory explanations have been put forward. The tidal interaction with Dione with the 2 : 1 orbital resonant relation remains one possibility. Another interesting feature deals with the close correlation of Enceladus' orbit with the peak brightness distribution of the E ring. The small dust particles, which might be of water ice composition, have been suggested to be ejected from Enceladus itself [91S]. The Enceladus-E-ring relation remains a major unsolved issue in the study of the Saturnian system. Dione and Rhea are two other icy satellites which display evidence of extensive resurfacing events. In addition, both satellites are covered by bright wispy streaks on the darker trailing hemispheres. Such hemispherical dichotomy with an albedo difference of about 20% between the leading side and the trailing side – which is also found in Tethys – might be caused by surface impact effects by interplanetary meteorites and comets, interaction with the magnetospheric plasma and the E-ring particles [90B]. Interesting enough, the second largest satellite of Saturn, Iapetus, shows the opposite behaviour of hemispherical brightness variation. Its trailing side is much brighter than the leading side. The pattern of dark albedo contour on the leading side centres at the apex of orbital motion with the corresponding albedo being as low as $0.02 \cdots 0.03$. In comparison, the brightest region on the trailing side has a surface reflectance as high as 0.6 [84S]. The reflectance map of Iapetus in Fig. 6 underlines this brightness asymmetry which might be unique in the solar system.

Hemispherical brightness variations of the Jovian satellites have been well studied by ground-based measurements [77J]. The leading sides of both Europa and Ganymede are considerably brighter than the trailing sides while Callisto shows the opposite effect. The Voyager observations have provided further details to such global albedo and colour variations. Fig. 7 illustrates a global mosaic of the albedo pattern of Europa which might be the result of exogenic agents such as impact meteoritic gardening, magnetospheric sputtering and/or surface implantation of the SO_2 ions from Io [86M].

Table 4. Orbital elements of planetary satellites. From [92L]. See also Table 4 in subsect. 3.2.1.3.1 and Table 4 in LB VI/2a, p.136.

a = distance from planetary center (= semi-major axis)
P = orbital period
e = eccentricity = $(a^2 - b^2)^{1/2}/a$ (b = semi-minor axis)
i_E = inclination of the satellite's orbit to the planet's equator
i_O = inclination to the satellite's orbit to the planet's orbital plane
R = radius
ρ = mean density
A = geometrical visual albedo

Satellite*)		$a[10^3$ km]	$a[R_p]$	P[d]	e	i_E	i_O	R [km]	ρ [g/cm³]	A
Earth										
	Moon	384.4	60.2	27.3217	0.05490	18.2° ··· 28.6°	5.1°	1738	3.34	0.12
Mars										
I	Phobos	9.37	2.76	0.3189	0.0150	1.1°		14 × 10	≤ 2	0.06
II	Deimos	23.52	6.90	1.262	0.0008	0.9° ··· 2.7°		8 × 6	≤ 2	0.07
Jupiter										
XV	Adrastea	128	1.80	0.295	≈ 0.0	≈ 0.0°				
XVI	Metis	128	1.80	0.295	≈ 0.0	≈ 0.0°				
V	Amalthea	181	2.55	0.489	0.003	0.4°				
XVI	Thebe	221	3.11	0.675	≈ 0.0	≈ 0.0°				
I	Io	422	5.95	1.769	0.004	0.0°		1815	3.55	0.63
II	Europa	671	9.47	3.551	0.000	0.5°		1569	3.04	0.64
III	Ganymede	1070	15.1	7.155	0.0001	0.2°		2631	1.93	0.43
IV	Callisto	1880	26.6	16.69	0.010	0.2°		2400	1.83	0.17
XIII	Leda	11110	156	240	0.146		26.7°	≈ 5		
VI	Himalia	11470	161	251	0.158		27.6°	90 ± 10		0.03
X	Lysithea	11710	164	260	0.130		29.0°	≈ 10		
VII	Elara	11740	165	260	0.207		24.8°	40 ± 5		0.03
XII	Ananke	20700	291	617	0.17		147°	≈ 10		
XI	Carme	22350	314	692	0.21		164°	≈ 15		
VIII	Pasiphae	23300	327	735	0.38		145°	≈ 20		
IX	Sinope	23700	333	758	0.28		153°	≈ 15		

(continued)

Table 4. (continued)

Satellite*)		$a[10^3$ km]	$a[R_p]$	P[d]	e	i_E	i_O	R [km]	ρ [g/cm^3]	A
Saturn										
XV	Atlas	137.7	2.276	0.602	0.002	0.3°		20 × 10		0.4
XVI	Prometheus	139.4	2.310	0.613	0.004	0.0°		70 × 50 × 40		0.6
XVII	Pandora	141.7	2.349	0.629	0.004	0.1°		55 × 45 × 35		0.6
X	Janus	151.4	2.510	0.694	0.009	0.3°		110 × 100 × 80		0.5
XI	Epimetheus	151.5	2.511	0.695	0.007	0.1°		70 × 60 × 50		0.5
I	Mimas	186	3.08	0.942	0.020	1.5°		196	1.44	0.7
II	Enceladus	238	3.95	1.370	0.004	0.0°		250	1.13	1.0
III	Tethys	295	4.88	1.888	0.000	1.1°		530	1.20	0.8
XIII	Telesto	295	4.88	1.888				17 × 14 × 13		0.6
XIV	Calypso	295	4.88	1.888				17 × 11 × 11		0.8
IV	Dione	377	6.26	2.737	0.002	0.0°		560	1.41	0.60
XII	Helene	377	6.26	2.737	0.005	0.2°				
V	Rhea	527	8.73	4.518	0.001	0.4°		765	1.33	0.60
VI	Titan	1222	20.3	15.95	0.029	0.3°		2575	1.88	0.21
VII	Hyperion	1481	24.6	21.28	0.104	0.4°		205×130×110		
VIII	Iapetus	3561	59	79.33	0.028		14.7°	730	1.15	0.12
IX	Phoebe	12954	215	550	0.163		150°	110 ± 10		0.06
Uranus										
VI	Cordelia	49.7	1.94	0.333				≈ 20		< 0.1
VII	Ophelia	53.8	2.10	0.375				≈ 25		< 0.1
VIII	Bianca	59.2	2.32	0.433				≈ 25		< 0.1
IX	Cressida	61.8	2.42	0.463				≈ 30		< 0.1
X	Desdemona	62.7	2.45	0.475				≈ 40		< 0.1
XI	Juliet	64.6	2.53	0.492				≈ 40		< 0.1
XII	Portia	66.1	2.58	0.513				≈ 40		< 0.1
XIII	Rosalind	69.9	2.73	0.558				≈ 30		< 0.1
XIV	Belinda	75.3	2.94	0.621				≈ 30		< 0.1
XV	Puck	86.0	3.36	0.763				85 ± 5		< 0.1

Table 4. (continued)

Satellite*)		$a[10^3$ km]	$a[R_p]$	P[d]	e	i_E	i_O	R [km]	ρ [g/cm^3]	A
Uranus (continued)										
V	Miranda	129.9	5.08	1.413	0.017	3.4°		242 ± 5	1.26 ± 0.39	0.07 ± 0.02
I	Ariel	190.9	7.47	2.521	0.0028			580 ± 5	1.65 ± 0.30	0.34 ± 0.02
II	Umbriel	266.0	10.41	4.146	0.0035			595 ± 10	1.44 ± 0.28	0.19 ± 0.01
III	Titania	436.3	17.07	8.704	0.0024			805 ± 5	1.59 ± 0.09	0.28 ± 0.02
IV	Oberon	583.4	22.82	13.463	0.0007			775 ± 10	1.50 ± 0.10	0.24 ± 0.01
Neptune										
III	Naiad	48.0	1.94	0.296				29 ± 6		0.06
IV	Thalassa	50.0	2.01	0.313				40 ± 8		0.06
V	Despina	52.5	2.12	0.333				79 ± 12		0.06
VI	Galatea	62.0	2.50	0.429				74 ± 10		0.06
VII	Larissa	73.6	2.97	0.554				96 ± 7		0.06
VIII	Proteus	117.6	4.75	1.121				208 ± 8		0.06
I	Triton	354.8	14.33	5.877	0.00	160.0°		1350	2.05	0.06
II	Nereid	5513.4	222.65	365.2	0.75		27.6°	170 ± 25		0.16

*) The final Roman numbers and names of the IAU are given here. In the literature sometimes provisional numbers given at the discovery or arbitrarily chosen by some authors are still used.

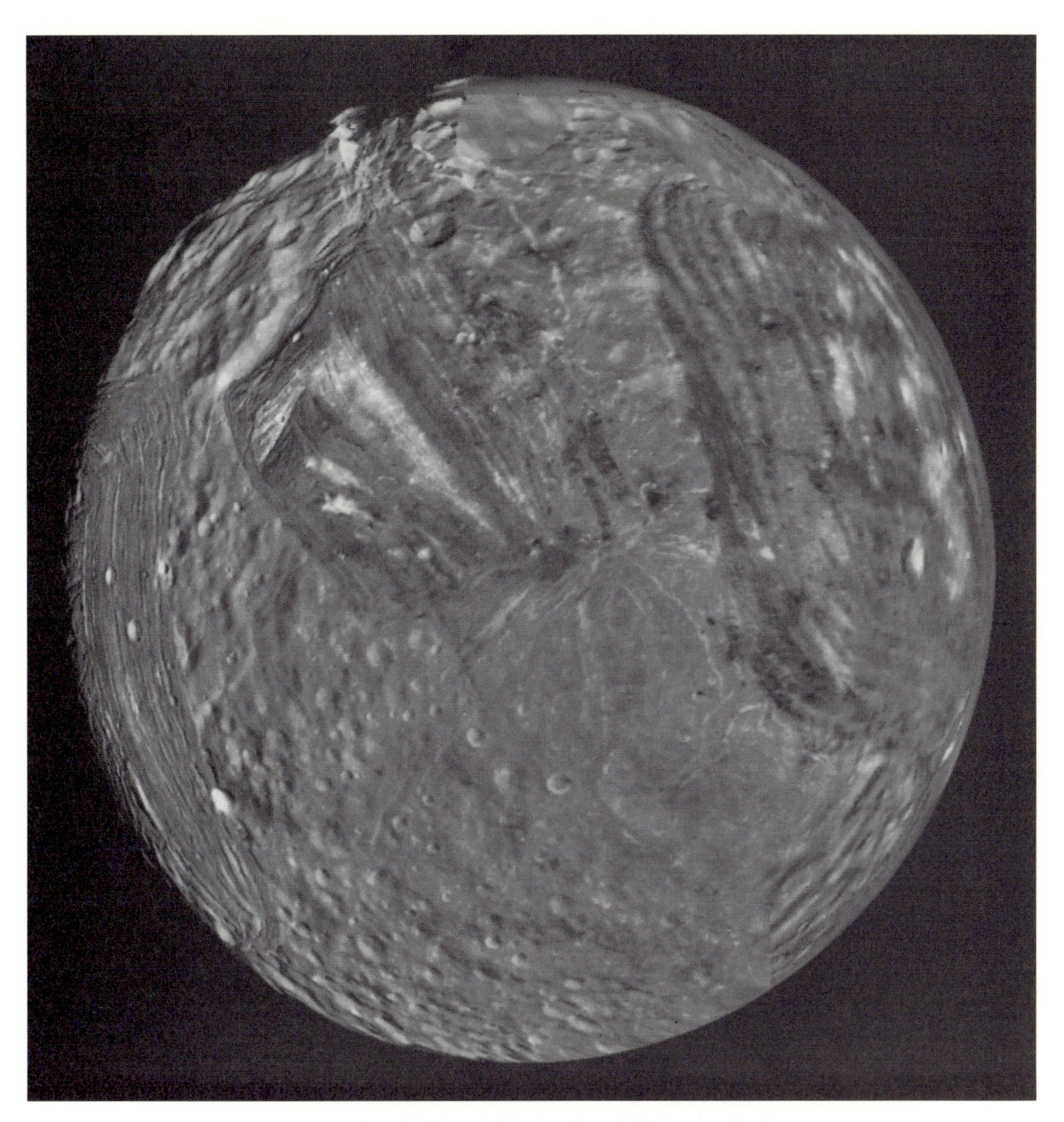

Fig. 5. Mosaic of high-resolution images of Miranda from Voyager 2. From Smith et al. [86S].

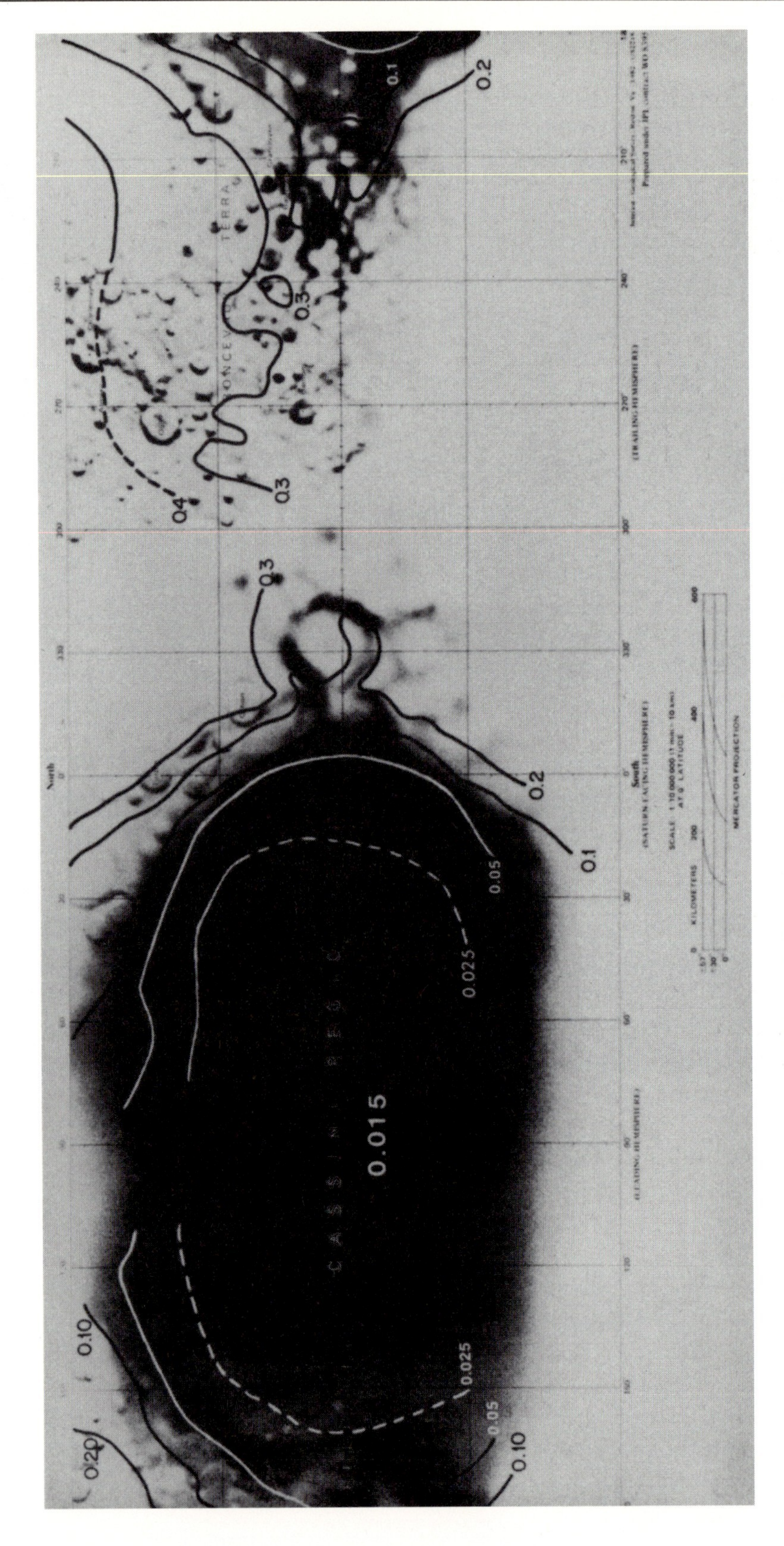

Fig. 6. Reflectance map of Iapetus. From Squires et al. [84S].

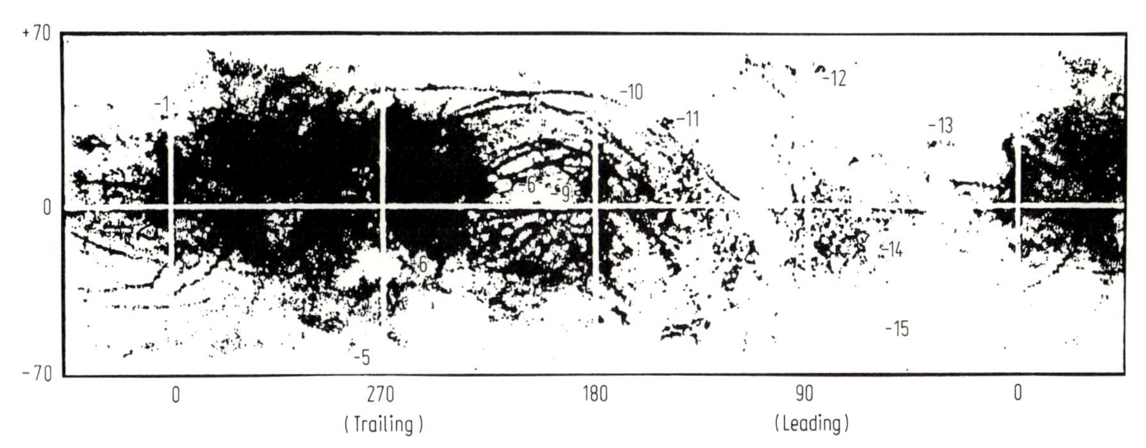

Fig. 7. A global mosaic of Europa. From McEwen [86M].

3.2.2.9 Planetary rings

The wealth of new data on the planetary ring systems gleaned from the Voyager mission is spectacular indeed. One could be almost sure that no similar quantum jump in our knowledge in planetary astronomy would happen again except perhaps for the close-up views of extra-solar systems. For example, the optical detection of the Jovian ring by Voyager 1 came as a full surprise [79S1,79S2] even though the near-ring plane plasma observations by the Pioneer 11 spacecraft had already shown signatures of the presence of a tenuouos dust ring; see Fig. 8. The thin ring system with a width of 6500 km and a normal optical depth of about $5 \cdot 10^{-5}$ is enveloped in a halo of submicron dust particles. While the vertical thickness of the "main" ring is quite small ($\approx$ 30 km), the halo has a vertical size of about $2 \cdot 10^4$ km extending between the planetary surface and the inner edge of the optical ring. The two inner satellites, J XV Adrastea and J XVI Metis, are believed to be the parent bodies of the Jovian ring particles [82J].

Fig. 8. A view of the Jupiter ring system from Voyager 2. From Smith et al. [79S1].

Table 5. Physical data of planetary rings. From [92L]. See also Table 5 in subsect. 3.2.1.3.2.

d = distance from planetary center w = radial width
R_p = radius of the planet τ = optical depth (at 3.6 cm)

	Designation	d [km]	d [R_p]	w [km]	τ	Remark
Jupiter		122800 ··· 129200	1.72 ··· 1.81	4500	$3 \cdot 10^{-5}$	Source satellites are likely to be XIV Adrastea (R = 10 km and a = 1.8064 R_J) and XVI Metis (R = 20 km and a = 1.7992 R_J).
		71398 ··· 122800	1.00 ··· 1.72	325	$\leq 7 \cdot 10^{-6}$	
Saturn	D ring	69970 ··· 74510	1.110 ··· 1.235	30 ··· 500	$< 10^{-2}$	The sharp division of the B and C rings is probably the result of electrodynamic interaction with Saturn's magnetic field. The source satellite of the E ring is likely to be Enceladus.
	C ring	74510 ··· 92000	1.235 ··· 1.525	10000	$\approx 10^{-2}$	
	B ring	92000 ··· 117580	1.525 ··· 1.949		1.0	
	Cassini Division	119000	1.972		$\approx$ 0.06	
	A ring	122170 ··· 136780	2.025 ··· 2.267		$\approx$ 0.5	
	Encke Gap	135706	2.214			
	F ring	140300	2.326		0.01 ··· 1.0	
	G ring	170000	2.818		$10^{-4} \cdots 10^{-5}$	
	E ring	180000 ··· 480000	2.984 ··· 7.956		$10^{-6} \cdots 10^{-7}$	
Uranus	1986 U2R	37000 ··· 39500	1.448 ··· 1.545	1 ··· 3	$10^{-4} \cdots 10^{3}$	The ϵ ring is confined by the shepherding satellites pair: 1986U7 and 1986U8.
	6	41850	1.637	2 ··· 3	0.2 ··· 0.3	
	5	42240	1.653	2 ··· 3	0.5	
	4	42580	1.666	7 ··· 12	0.3	
	α	44730	1.750	7 ··· 12	0.3 ··· 0.4	
	β	45670	1.787		0.2	
	η	47180	1.856	$\approx$ 2	0.1 ··· 0.4	
	γ	47630	1.864	1 ··· 4	1.3 ··· 3.3	
	δ	48310	1.890	1 ··· 9	0.3 ··· 0.4	
	1986 U1R	50040	1.956	1 ··· 2	0.1	
	ϵ	51160	2.002	22 ··· 93	0.5 ··· 2.1	
Neptune	1989 N3R	41900	$\approx$ 1.69	$\approx$ 1700	10^{-4}	
	1989 N2R	53200	2.15	< 15	0.01 ··· 0.02	
	1989 N4R	53200 ··· 59000	2.15 ··· 2.4	5800	10^{-4}	
	1989 N1R	62930	$\approx$ 2.53	$\approx$ 20	0.01 ··· 0.1	

A rich variety of dynamical phenomena related to ring-satellite interactions were discovered in the Saturnian ring system. These range from the shepherding satellite effect in the gravitational confinement of the F ring to the numerous density wave structures in the A ring and the C ring. The reader is referred to the review chapters on planetary rings in the book edited by Greenberg and Brahic [84G] for detailed descriptions. Focus here will be on the gross structures of the main ring system (i.e., the A, B, and C rings) and the physical properties of the ring particles. In Fig. 9 is shown the optical depth profile across the ring system obtained from the radio science occultation experiment on Voyager at 3.6 cm wavelength [83T]. There are several sharp boundaries delineating the A, B and C rings. The outer edge of the A ring is determined by the gravitational (7 : 6) resonant effect of S X Janus, and the Cassini Division by the 3 : 2 resonance with Mimas. The boundary between the B and C rings is presumably unrelated to graviatational effect but rather to the electrodynamical interaction between the charged dust particles in the ring system and the planetary magnetic field. Another indication of electrostatic effect concerns the occurrence of the so-called spokes in the B ring which are radial features consisting of small dust particles [81S]. One surprising result from theoretical studies of the ring system is that the dynamical lifetime of the Saturnian rings might be as short as just a few hundred million years. Its origin is therefore still a major mystery.

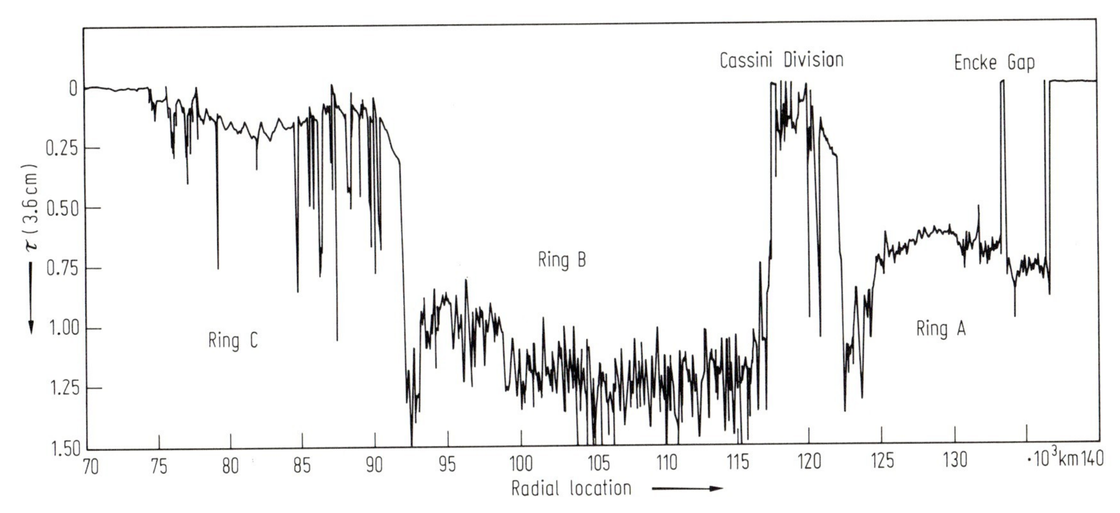

Fig. 9. An overview of the normal optical depth of Saturn's rings at 3.6 cm wavelength from the Voyager radio wave experiment. From Tyler et al. [83T].

Most of the particles in the Saturnian ring system are of centimeter-to-several-meter size. Their size distribution tends to follow a differential power law of the form $n(r) = n_0 r^{-q}$ with q between 2.8 and 3.4 [84C]. The meter-sized objects in fact contain most of the mass. The ring particles must be predominantly of water ice composition even though contamination by silicate materials of meteoritic or cometary origin is suggested by infrared observations. The total ring mass has been estimated to be on the order of $3 \cdot 10^{22}$ g. If this mass were to be reassembled in a precursor satellite, it would have a radius of about 130 km.

As for the Uranian rings, there are two major points of special interest. Firstly, this ring system is composed mainly of narrow rings with the ϵ ring being the broadest one (width $20 \cdots 96$ km). After their first discovery from stellar occultation measurements [77E] many details had been filled in by subsequent ground-based observations. For example, the orbits of the ring particles in the eccentric ϵ ring were known to precess in such a way that the individual streamlines will remain separate from each other. The average optical depth of the ϵ ring is about 1.2 at its widest part. The optical depth also varies nearly inversely to the width. It has been suggested that the self-gravitational force of the ring itself might be effective in countering the differential precession effect caused by the oblateness of the central planet [82G]. The sharp edges of the ϵ ring are presumably the result of the gravitational perturbation by a pair of shepherding satellites XI and XII, Juliet and Portia, provisional names 1986 U8 and U7, which were detected by the Voyager 2 spacecraft [86S]. The other peculiar feature associated with the Uranian ring system has to do with the near-absence of small dust particles in its vicinity. The basic reason appears to be the existence of a very extended hydrogen exosphere which can sweep the small dust particles off the ring system via gaseous drag. Electrostatic charging effect of the dust particles might also play a role in the instability of their orbital motion. That the ring particles are extremely dark (albedo ≈ 0.05) might be indicative of the surface irradiation effect by energetic charged particles in case the ring particles are made up of hydrocarbon-rich material [89C1].

Fig. 10. A Voyager observation of the Uranian rings and the shepherding satellites 1986 U7 and 1986 U8 of the ϵ ring. From Smith et al. [86S].

The discovery of ring arcs around Neptune from stellar occultation measurements from the Earth [89C3,86H] has led to the idea that the azimuthal confinement might be caused by the resonant effect of one single shepherding satellite [86G]. Applying the Voyager observations of new small satellites in the vicinity of the ring system, Porco [91P] has shown that the ring arc, 1989N1R, with a width of 15 km might indeed be confined by the 42 : 43 outer Linblad resonance of VI Galatea. As with other planetary rings, the ring arcs of Neptune might have originated from the catastrophic disruption of small satellites near the planet.

References for 3.2.2

75S Stevenson, D.: Phys Rev. B **12** (1975) 3999.
76A Acuna, M.H., Ness, N.F.: J. Geophys. Res. **81** (1976) 2917.
77E Elliot, J.L., Dunham, E.W., Mink, D.J.: Nature **267** (1977) 328.

77J Johnson, T.V., Pilcher, C.B., in: Planetary Satellites (J.A. Burns, ed.), Univ. Arizona Press (1977) 232-268.

79S1 Smith, B.A., and the Voyager Imaging Team: Science **204** (1979) 951.

79S2 Smith, B.A., and the Voyager Imaging Team: Science **206** (1979) 927.

80F Fink, U., Smith, B.A., Benner, D.C., et al.: Icarus **44** (1980) 62.

80M Mizuno, H.: Prog. Theoret. Phys. **64** (1980) 544.

81L Lebofsky, L.A., Feierberg, M.A., Tokunaga, A.T. et al.: Icarus **48** (1981) 453.

81S Smith, B.A., and the Voyager Imaging Team: Science **212** (1981) 163.

81T Tyler, G.L., Eshleman, V.R., Anderson, J.D., et al.: Science **212** (1981) 201.

81W Warwick, J.W., et al.: Science **212** (1981) 239.

82G Goldreich, P., Tremaine, S: Annu. Rev. Astron. Astrophys. **20** (1982) 249.

82J Jewitt, D.C., in: Satellites of Jupiter (D. Morrison, ed.), Univ. Arizona Press (1982) 44-64.

83H Hill, T.W., Dessler, A.J., Goertz, C.K., in: Physics of the Jovian Magnetosphere (A.J. Dessler, ed.), Cambridge Univ. Press (1983) 353-394.

83L Lunine, J.I., Stevenson, D., Yung, Y.: Science **222** (1983) 1229.

83T Tyler, G.L., Marouf, E.A., Simpson, R.A., et al.: Icarus **54** (1983) 160.

83Z Zarka, P., Pedersen, B.M.: J. Geophys. Res. **88** (1983) 9007.

84C Cuzzi, J.N., Lissauer, J.J., Esposito, L.W., et al., in: Planetary Rings, (R. Greenberg and A. Brahic, eds.), Univ. Arizona Press (1984) 73.

84G Greenberg, R., Brahic, A.: Planetary Rings, Univ. Arizona Press (1984).

84S Squires, S.W., Buratti, B., Veverka, J., Sagan, C.: Icarus **59** (1984) 426.

85S Stevenson, D.: Icarus **62** (1985) 4.

86C Covault, C.E., et al.: Icarus **67** (1986) 126.

86G Goldreich, P., Tremaine, S., Borderies, N.: Astron. J. **92** (1986) 490.

86H Hubbard, W.B., et al.: Nature **319** (1986) 636.

86M McEwen, A.S.: J. Geophys. Res. **91** (1986) 8077.

86N Ness, N.F., Acuna, M.H., Behannon, K.W., et al.: Science **233** (1986) 85.

86S Smith, B.A., and the Voyager Imaging Team: Science **233** (1986) 43.

88G Godfrey, D.A.: Icarus **76** (1988) 335.

89B Broadfoot, A.L., Atreya, S.K., Bertaux, J.L., et al.: Science **246** (1989) 1459.

89C1 Cheng, A.F., Johnson, R.E., in: The Origin and Evolution of Planetary and Satellite Atmospheres (S.K. Atreya and J.B. Pollack, eds.), Univ. Arizona Press (1989) 682-722.

89C2 Conrath, B., et al.: Science **246** (1989) 1454.

89C3 Coustenis, A. Bezard, B., Gautier, D.: Icarus **80** (1989) 54.

89G Gautier, D., Owen, T., in: The Origin and Evolution of Planetary and Satellite Atmospheres, (S.K. Atreya, J.B. Pollack and M.S. Matthews, eds.), Univ. of Arizona Press (1989) 487-512.

89H Hubbard, W.B., Marley, M.S.: Icarus **78** (1989) 102.

89N Ness, N.F., Acuna, M.H., Burlaga, L.F., et al.: Science **246** (1989) 1473.

89S Smith, B.A., and the Voyager Imaging Team, Science **246** (1989) 1422.

89T Tyler, G.L., Sweetnam, D.N., Anderson, J.D., et al.: Science **246** (1989) 1466.

89Y Yelle, R., Lunine, J.I.: Nature **339** (1989) 288.

90B Buratti, B.J., Mosher, J.A., Johnson, T.V.: Icarus **87** (1990) 339.

90I Ingersoll, A.P.: Science **248** (1990) 308.

90L Lellouch, E., Belton, M., de Pater, I., et al.: Nature **346** (1990) 638.

91C Coustenis, A., Bezard, B., Gautier, D., Marten, A.: Icarus **89** (1991) 152.

91P Porco, C.C.: Science **253** (1991) 995.

91S Showalter, M.R., Cuzzi, J.N., Larson, S.M.: Icarus **94** (1991) 451.

92A A'Hearn, M.F., Feldman, P.D.: Icarus (1992).

92L Lang, K.R.: Astrophysical Data: Planets and Stars. New York: Springer-Verlag (1992).

93S Stern, A.S., Weintraub, D.A., Festou, M.C.: Science **261** (1993) 1713.

3.3 Small bodies in the solar system

3.3.1 The asteroids (minor planets)

3.3.1.1 Characterization of asteroid orbits

In March 1992 [92MPC2] the count of permanently numbered asteroids had risen to 5151 objects, among which only one object considered lost had remained (719 Albert). From the most recent census of asteroids the total number is estimated to 300000 for objects with a diameter of more than 1 km [91T]. New discoveries did not change the general distribution of asteroid orbits but added some rare types of objects, mostly in orbits that seem to be interrelated with other small bodies of the solar system like comets. These objects include an analogue to the Trojan asteroids that are moving in 1/1 resonance with Jupiter, for Mars [90MPC], and a trans-Saturnian asteroid in a highly eccentric orbit [92MPC1].

Table 1. Selected orbital elements for members of outstanding new asteroid classes.

Object	Epoch	Semimajor axis [AU]	Inclination	Eccentricity
5145 1992AD	1992 Jun. 27.0	20.4801	24.68°(2000.0)	0.5759
5261 1990MB	1990 Nov. 5.0	1.5235	20.28°(1950.0)	0.0648

Dynamical classification of near-Earth asteroids

A dynamical classification for near-Earth asteroids (actually or at least secularly with a perihelion distance q smaller than 1.3 AU) has been proposed in [89M1] according to the criteria given in Table 2. The restrictive criteria for the "comet" class do not imply that members of other classes must not be of cometary origin.

Table 2. Near-Earth asteroid dynamical classification (q = perihelion distance).

Characteristics	Class designation
Resonant orbit with Jupiter, no close approaches to it	Alinda
Other:	
Not Earth crossing, but secularly q smaller than 1 AU	Kozai
Not Earth crossing, and secularly q greater than 1 AU	Eros
Earth crossing:	
Resonant orbit with Earth	Toro
Non-resonant orbit with Earth:	
Close approaches to Earth dominant	Geographos
Close approaches to Earth not dominant:	
Jupiter crossing	Comet
Non-Jupiter crossing	Oljato

3.3.1.2 Orbit distribution

Visualizations of the radial and vertical orbit distribution of asteroids are given in [91H1]. The distribution of orbits of Earth-approaching asteroids leads to the encounter parameters in Table 3.

Table 3. Very close encounters of asteroids with Earth (1900-2100). Δ = geocentric distance.

Year	Object	Δ [AU]	Ref.
1921	2340 Hathor	0.0099	91H2
1937	1937UB (Hermes)	0.0049	91H2
1938	1989UP	0.0052	91H2
1976	2340 Hathor	0.0078	91H2
1988	1988TA	0.0099	91H2
1989	4581 Asclepius	0.0046	91H2
1991	1991BA	0.0011	91IC1
1991	1991VG	0.0031	91IC2
2069	2340 Hathor	0.0066	91H2
2086	2340 Hathor	0.0056	91H2

3.3.1.3 Physical properties

Masses

Recently only two new masses of asteroids could be determined: (10) Hygiea, $(4.7 \pm 2.3) \cdot 10^{-11}$ solar masses [87S], (704) Interamnia, $(3.7 \pm 1.7) \cdot 10^{-11}$ solar masses [92L].

Sizes and rotation properties

A large collection of asteroid diameter data is found in [89T2]. A summary on asteroid rotation rates was published in [89B1]. Pole position determinations are listed in [89M2].

Spectral data

Results of broad band (UBV) photometric data are compiled in the section of magnitudes, colours, albedos and diameters of the Asteroids II data base [89T2]. Detailed light curve information in various colours

Table 4. Passband characteristics of the eight colour asteroid survey [82T]. λ_{eff} = effective wavelength, FWHM = full width at half maximum.

Filter	λ_{eff} [μm]	FWHM [μm]
s	0.337	0.047
u	0.359	0.060
b	0.437	0.090
v	0.550	0.057
w	0.701	0.058
x	0.853	0.081
p	0.948	0.080
z	1.041	0.067

Table 5. Asteroid taxonomic classification A···V [89B2, 89G, 89T1].

	Example asteroid	Albedo	Spectrum properties	Inferred mineralogy
A	246	moderately high	strong absorption feature longward of 0.7 μm, extremely reddish shortward of 0.7 μm	olivine
B	2	moderately low	UV absorption feature, flat to slightly reddish	hydrated silicates + opaques
C	10	low	UV absorption feature, flat to slightly reddish	hydrated silicates, carbon, organics
D	279	low	featureless, flat to slightly reddish shortward of 0.55 μm, very red longward of 0.7 μm	organic rich
E	44	high	featureless, flat to slightly reddish	Fe-free silicates
F	704	low	weak UV absorption feature, flat to slightly bluish	hydrated silicates + opaques
G	1	low	very strong UV absorption feature, flat longward of 0.4 μm	hydrated silicates + opaques
K	221	moderate	no formal description, intermediate between S and C	olivine, pyroxene, carbon
M	16	moderate	featureless, flat to slightly reddish	NiFe
P	46	low	featureless, flat to slightly reddish	organic rich
Q	1862	moderately high	strong absorption features shortward and longward of 0.7 μm	pyroxene, olivine, NiFe
R	349	moderately high	strong absorption features shortward and longward of 0.7 μm	pyroxene, olivine
S	3	moderate	moderate to strong absorption feature shortward of 0.7 μm, moderate to absent absorption feature longward of 0.7 μm	NiFe, olivine, pyroxene
T	114	low	moderate absorption feature shortward of 0.85 μm, flat longward of 0.85 μm	similar to P, D?
V	4	moderately high	strong absorption features shortward and longward of 0.7 μm	pyroxene, plagioclase feldspar, olivine

is stored in the Asteroid Photometric Catalogue and its updates [87L, 88L, 89L]. More recent extensive light curve collections are found in [92H]. A special compilation of outer (Cybele, Hilda, Trojan) asteroid light curves was published in [92B]. Based on the Asteroid Photometric Catalogue an ample set of phase curves was derived in [90L], a large number of models in [90M1]. Intermediate band spectrophotometric data of 589 asteroids based on a specially designed eight colour system ([82T], Table 4) are listed in [85Z]. Based on this eight colour survey and the albedo the asteroid taxonomy has been further developed [89T1]. The basic characteristics of this classification are summarized in Table 5.

Data on infrared emission for about two thousand objects were derived from the IRAS satellite [86M]. Information on the surface roughness and materials is also available for a growing number of asteroids by radar techniques [89O].

3.3.1.5 Interrelations of asteroids with other solar system objects

Several asteroids have properties that place them close to comets and interplanetary matter in an evolutionary context. Table 6 summarizes the evidence for the strongest candidates of being dormant or extinct cometary nuclei. In Table 7 the relationship between some asteroid groups formed by collisions, and concentrations of interplanetary dust is shown.

Table 6. Objects with possible relationship to comets.

Object	Evidence for cometary origin	Ref.
944	Jupiter crosser	79K
2060	observed coma	90M2
2101	anomalous radar reflection	85O
2201	interactions with interplanetary magnetic field	84R
3200	association with Geminid meteor stream	83W
4015	identified with comet P/Wilson-Harrington	92IC
5145	planet crossing orbit in outer solar system	90MPC

Table 7. IRAS dust bands [89S].

Band name	Asteroid family association
α	Themis family
β	Koronis family
γ	Eos family
J	Io family
K	Io family

References for 3.3.1

General references

a Binzel, R. P., Gehrels, T., Matthews, M. S., eds.: Asteroids II. Tucson: University of Arizona Press (1989).

b Lagerkvist, C.-I., Rickman, H., eds.: Asteroids Comets Meteors. Uppsala: Uppsala Universitet Reprocentralen (1983).

c Lagerkvist, C.-I., Lindblad, B. A., Lundstedt, M., Rickman, H., eds.: Asteroids Comets Meteors II. Uppsala: Uppsala Universitet Reprocentralen (1986).

d Lagerkvist, C.-I., Rickman, H., Lindblad, B. A., Lindgren, M., eds.: Asteroids Comets Meteors III. Uppsala: Uppsala Universitet Reprocentralen (1990).

e Schmadel, L. D.: Dictionary of Minor Planet Names. Berlin: Springer (1992).

Special references

79K Kresak, L., in: Asteroids (T. Gehrels, ed.). Tucson: University of Arizona Press (1979) p.289.

82T Tedesco, E.F., Tholen, D.J., Zellner, B.: Astron. J. **87** (1982) 1585.

83W Whipple, F.L.: IAU Circular 3881 (1983).

84R Russell, C.T., Aroian, R., Arghavani, M., Nock, K.: Science **226** (1984) 43.

85O Ostro, S.J.: Publ. Astron. Soc. Pac. **97** (1985) 877.

85Z Zellner, B., Tholen, D.J., Tedesco, E.F.: Icarus **61** (1985) 355.

86M L. Matson, ed.: Infrared Astronomical Satellite Asteroid and Comet Survey: Preprint Version No. 1, JPL Document D-3698 (1986).

87L Lagerkvist, C.-I., Barucci, M.A., Capria, M.T., Fulchignoni, M., Guerriero, L., Perozzi, E., Zappala, V., eds.: Asteroid Photometric Catalogue. Rome: Consiglio Nazionale Delle Ricerche (1987).

87S Scholl, H., Schmadel, L.D., Röser, S.: Astron. Astrophys. **179** (1987) 311.

88L Lagerkvist, C.-I., Barucci, M.A., Capria, M.T., Fulchignoni, M., Magnusson, P., Zappala, V., eds: Asteroid Photometric Catalogue. First Update. Rome: Consiglio Nazionale Delle Ricerche (1988).

89B1 Binzel, R.P., Farinella, P., Zappala, V., Cellino, A.: published in [a], p. 416.

89B2 Bell, J.F., Davis, D.R., Hartmann, W.K., Gaffey, M.J.: published in [a], p. 921.

89G Gaffey, M.J., Bell, J.F., Cruikshank, D.P.: published in [a], p. 98.

89L Lagerkvist, C.-I., Barucci, M.A., Capria, M.T., Dahlgren, M., Erikson, A., Fulchignoni, M., Magnusson, P., eds.: Asteroid Photometric Catalogue. Second Update. Uppsala: Uppsala Universitet Reprocentralen (1992).

89M1 Milani, A., Carpino, M., Hahn, G., Nobili, A.M.: Icarus **78** (1989) 212.

89M2 Magnusson, P.: published in [a], p. 1180.

89O Ostro, S.J.: published in [a], p. 192.

89S Sykes, M.V., Greenberg, R., Dermott, S.F., Nicholson, P.D., Burns, J.A., Gautier, T.N., III: published in [a], p. 336.

89T1 Tholen, D.J., Barucci, M.A.: published in [a], p. 298.

89T2 Tedesco, E.F.: published in [a], p. 1090.

90L Lagerkvist, C.-I., Magnusson, P.: Astron. Astrophys. Suppl. Ser. **86** (1990) 119.

90M1 Magnusson, P.: Uppsala Astronomical Observatory Report No. 54 (1990).

90M2 Meech, K.J., Belton, M.J.S.: Astron. J. **100** (1990) 1323.

90MPC Minor Planet Circular 16880 (1990).

91H1 Hoffmann, M.: Rev. Mod. Astron. **4** (1991) 165.

91H2 Hahn, G.: Adv. Space Res. **11** (1991) 29.
91IC1 IAU Circular 5127 (1991).
91IC2 IAU Circular 5387 (1991).
91T Tedesco, E.F., Veeder, G.J.: paper presented at the International Conference on Asteroids, Comets, Meteors, Flagstaff (1991).
92B Binzel, R.B., Sauter, L.M.: Icarus **95** (1992) 222.
92H Harris, A.W., Young, J.W., Dockweiler, T., Gibson, J., Poutanen, M., Bowell, E.: Icarus **95** (1992) 115.
92IC IAU Circular 5585 (1992).
92L Landgraf, W., in: Chaos, Resonance and Collective Dynamical Phenomena in the Solar System (S. Ferraz-Mello, ed.). Dordrecht: Reidel (1992), p. 179.
92MPC1 Minor Planet Circular 19850 (1992).
92MPC2 Minor Planet Circular 19853 (1992).

3.3.2 Meteors and meteorites

No really fundamental new facts have appeared since the last edition LB VI/2a p. 187-202. Therefore only a few additions are given.

3.3.2.0 and 3.3.2.1 see LB VI/2a

3.3.2.2 Meteorites

3.3.2.2.1 Definition

A meteorite is an extraterrestrial solid object which survived passage through the earth's atmosphere and reached the earth's surface as a recoverable object [74W]. This includes interplanetary dust particles collected in the stratosphere [85B], micrometeorites found in polar ice [91M] and spherules from the deep sea [85B].

3.3.2.2.2 Significance of meteorite study

The most recent and complete account of meteorite research is found in [88K].

3.3.2.2.3 – 3.3.2.2.6 see LB VI/2a

3.3.2.2.7 Rare gases

A new discussion of noble gases in meteorites is found in [83O].

3.3.2.2.8 Isotopic anomalies

Isotopic anomalies also have been encountered in interplanetary dust particles as well as in interstellar grains extracted from primitive meteorites [91Z].

There appears to be a general agreement that isotopic anomalies reflect nucleosynthesis in and condensation around certain stars. But no widely accepted scenario has yet been developed for the relationship in time and space between measured anomalies and the possible production processes.

3.3.2.2.9 Origin of meteorites

In addition to meteorites with almost certain asteroidal origin, a few meteorites are rocks expelled from the Earth's moon [83GRL]. Another subclass of (differentiated) meteorites, the Shergotty-Nakhla-Chassigny (SNC) class, most probably stems from Mars [83B, 82S]. Their study yields unique elemental and isotopic information relevant to the geologic and atmospheric evolution of Mars [85O, 86W, 87H].

3.3.2.2.10 See LB VI/2a

References for 3.3.2.2

74W Wasson, J.: Meteorites. Berlin: Springer (1974).

82S Shih, C.-Y., Nyquist, L.E., Bogard, D.D., McKay, G.A., Wooden, J.L., Bansal, B.M., Wiesman, H.: Geochim. Cosmochim. Acta **46** (1982) 2323.

83B Bogard, D.D., Johnson, P.: Science **221** (1983) 651.

83GRL Special issue Geophys. Res. Lett. **10** (1983) 773-840.

83O Ozima, M., Podosek, F.A.: Noble Gas Geochemistry. Cambridge: Cambridge Univ. Press (1983).

85B Brownlee, D.E.: Annu. Rev. Earth Planet. Sci. **13** (1985) 134.

85O Ott, U., Begemann, F.: Nature **317** (1985) 509.

86W Wänke, H., Dreibus, G., Jagoutz, E., Palme, H., Spettel, B., Weckwerth, G.: Lunar Planet. Sci. **XVII** (1986) 919.

87H Hunten, D.M., Pepin, R.O., Walker, J.C.G.: Icarus **69** (1987) 532.

88K Kerridge, J.F., Mathews, M.S. (eds.): Meteorites and the Early Solar System. Tucson: Univ. Arizona Press (1988).

91M Maurette, M.: Nature **351** (1991) 44.

91Z Zinner, E.: Space Sci. Rev. (Jessberger, E.K. ed.) **56/1+2** (1991) 147.

3.3.3 Comets

Concerning the mechanical data of comets the situation has not changed very much during the past decade. But our knowledge about the physics of comets has improved considerably mainly due to the missions to comet Halley in 1986. The general aspects and implication of comets research are reviewed for instance in the monograph [i] (see reviews [90H1, 90H2, 93F]) and in the proceeding [a] (see review [91W2]). The number of observed comets has grown and thus some tables are updated. For definition of orbital elements and orbital characteristics see LB VI/2a p.202 ff.

3.3.3.1 Mechanical data

Number of comets, Oort Cloud, Kuiper Belt

Comets are divided according to their periodicity in two populations: short-period comets and long-period comets. For definition see Table 2. More that 500 comets are recorded before 1600 AD (mostly in Chinese records). The large cloud of comets surrounding the planetary system was first suggested by the Dutch astronomer J. Oort in 1950. The existence of this Oort cloud is supported by the unusual distribution of orbital energies of the observed long-period comets (plotted in Fig. 2 as function of the original semi-major axis a (which is equivalent to the inverse orbital energy). The dimension of this cloud is about 10^4 AU. Cometary orbits in the cloud evolve under gravitational interaction with nearby stars, giant molecular interstellar clouds, as well as planetary perturbation. There are evidences for the existence of a more dense inner Oort cloud of comets which acts as a reservoir of comets in outer clouds which are stripped away. However, these clouds are reservoirs also for periodical comets which have a life-time of less than 10^6 years. The total number of comets in Oort clouds is estimated in the range of 10^{11} to 10^{12} with total mass equivalent to the mass of one or more terrestrial planets. As a third component to the Oort cloud, the "Kuiper Belt" is suggested. The comets in this (somewhat speculative) belt have semi-major axes of 50 to 500 AU. Reviews about the Oort cloud, comet origin and evolution see in [90O2, 90R, 91W1, 91F1].

The Tables 1 and 2 show the number of comets with known orbits and distribution of the orbital forms (eccentricities); they are updated tables of the corresponding Tables 1 and 2 in LB VI/2a, pp. 203 and 204. The Tables 1–4 are partly compiled from [C89M].

Interstellar comets

Although interstellar comets may exist, no comet on a clearly interstellar heliocentric trajectory has been observed.

Nomenclature of comets

Preliminary: name or names (up to three) of discoverer(s), year of the discovery and lowercase letter in the order of discovery: a, b,..., z; a_1, b_1,..., z_1; a_2, b_2, ..., etc.

Final: Year of perihelion passage followed by roman numeral in the order of perihelion passage: comet Arend-Roland 1956 h = 1957 III.

If the discovery is provided by some space-borne instrumentation, then the acronym of the spacecraft is used instead of the discoverer's name: 1983 XVI = IRAS, 1984 XII = SOLWIND 5.

Short-period comets ($P < 200$ years) are marked by letter P/ followed by the discoverer's name or by the name of the astronomer who first determined the periodicity and orbital elements of the given comet: P/Halley = comet 1982 i = 1986 III.

Short-period comets observed all around the orbit (Encke, Gunn, Machholz, Schwassmann-Wachmann 1) are denoted only by name.

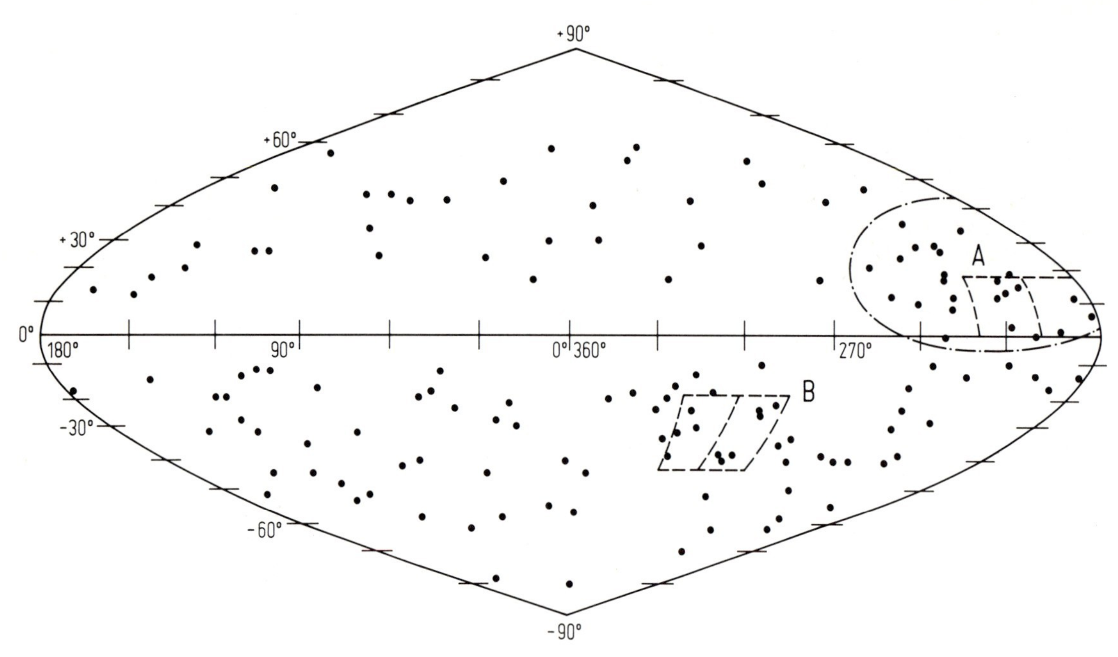

Fig. 1. Distribution of aphelion points of 142 "new" comets with original semi-major axis a > 500 AU on the celestial sphere in the galactic coordinates. A and B denote two regions of excess found by Rhea Lüst (dash-dotted curve) [84L] and by Delsemme (dashed frames) [86D]. (According to [91F1]).

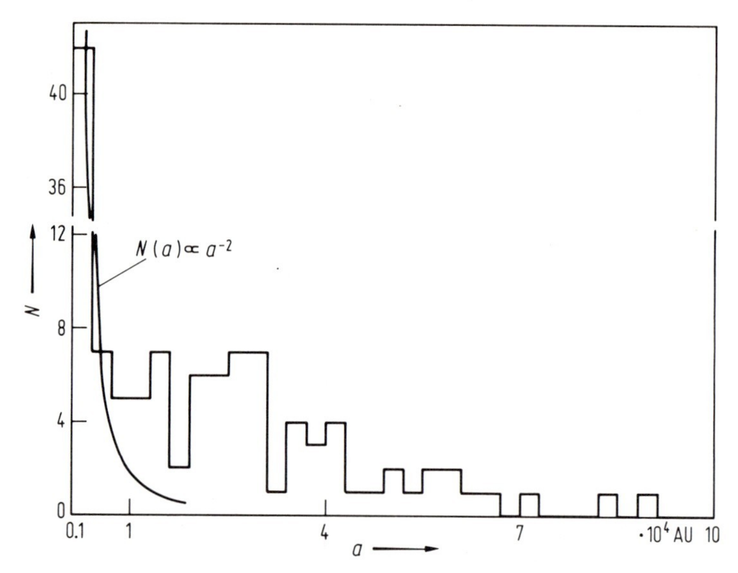

Fig. 2. Frequency distribution of original semi-major axes a of observed long-period comets [83E, 87E]. (According to [91F1]).

Table 1. Number of appearance of comets with known orbits. (Compiled from [C89M]).
Interval: compiled according to year of perihelion passage
N: total number of appearances
N_0: total number of discoveries
N_1: total number of newly discovered short-period comets ($P < 200\,a$; $e < 0.97$)
$N(e = 1.0)$: number of comets with parabolic orbits
$N(e > 1.0)$: number of comets with hyperbolic orbits

Interval	N	N_0	N_1	$N(e = 1.0)$	$N(e > 1.0)$
0 ··· 500	10				
500 ··· 1000	15				
1000 ··· 1500	44				
1500 ··· 1600	17				
1600 ··· 1700	23				
1700 ··· 1800	63				
1800 ··· 1820.0	24	22	5	13	0
1820 ··· 1840	31	22	1	17	0
1840 ··· 1860	74	61	9	27	5
1860 ··· 1880	73	52	5	29	2
1880 ··· 1900.0	106	78	15	29	14
1900 ··· 1910.0	41	29	6	8	10
1910 ··· 1920	47	28	6	5	7
1920 ··· 1930	50	29	9	6	6
1930 ··· 1940	54	30	4	5	5
1940 ··· 1950	75	47	11	12	12
1950 ··· 1960	83	41	7	7	15
1960 ··· 1970	94	42	8	12	4
1970 ··· 1980	144	63	18	10	16
1980 ··· 1989	214	111	33	27	18

Table 2. Distribution N of orbital forms for 1292 apparitions of 810 individual comets observed between 240 BC and end of 1989. (Compiled from [C89M]).

Orbital form	e	N	%
Elliptical orbits	<1	347	42.8
short-periodic ($P < 200\,a$)	<0.97	155[1])	19.1
long-periodic ($P > 200\,a$)	$0.97 \cdots < 1$	192	23.7
Parabolic orbits	1	341	42.1
Hyperbolic orbits	>1	121	14.9
Strongly hyperbolic orbits	≥ 1.0057	1	

[1]) Of these comets, 93 have been observed at two or more apparitions and 62 at one apparition only.

Catalogues of cometary orbits

A catalogue of cometary orbits [C89M] (sixth edition 1989) lists 1292 apparitions of 810 individual comets observed between 240 BC and the end of 1989. Since 1992 the updated version of this catalogue is available also on floppy disk. Another important catalogue is the Catalogue of Short-Period Comets [C86B] (first edition 1986), where the results of new long-term integrations of the equations of motion of all known short-period comets are presented together with their observing conditions at each perihelion passage, and various data characterizing the orbital evolution are compiled.

In recent years the number of observed comets has increased by over 30 per year. Many of the newly discovered or rediscovered comets are periodic. Therefore, the list of periodic comets in Tables 3 and 4 (updated tables of the corresponding tables in LB VI/2a, pp. 205 and 206) is intended to be complete for comets observed through August 1991 only. Information concerning the discoveries of new comets, newly computed orbits and other relevant data is distributed by the Central Bureau for Astronomical Telegrams and Minor Planet Center, Smithsonian Astrophysical Observatory, Cambridge, MA 02138 USA, by telegrams, electromail, and in IAU Circulars and/or in Minor Planet Circulars. The circulars were partly used for compilation of Tables 3, 4, and 4a. The orbital elements are for equinox J 2000.0 for orbits after 1986, otherwise 1950.0.

Orbital characteristics

Concerning the behaviour of orbital elements see LB VI/2a, p. 209. Further information about orbital characteristics of comets is shown in Figs. 1–3. For review of orbital characteristics see for instance [83E, 87E, C86B, C86M, C89M, 91Y1].

Seven new comets with great values of the perihelion distance ($q \geq 4.0$ AU) can be added to Table 5 in LB VI/2a (e = eccentricity):

Comet	q [AU]	e
1985 XIV Hartley	4.000	1.000
1991 r Helin-Alu	4.850	1.000
1979 VI Torres	4.496	1.001
1981 XV Elias	4.743	1.001
1987 IV Shoemaker	5.031	1.003
1986 XIV Shoemaker	5.458	1.003
1984 XV Shoemaker	5.489	0.995

Sungrazing comet group

The comets with very small perihelion distance ($q < 0.02$ AU) are called "sun-grazers" (or "sungrazers") see Table 5. Most of them are member of a sungrazing group of comets first studied by H. Kreutz in 1891. The Kreutz group is characterized by orbital parameters such that the revolution period is a few centuries, and the perihelion ecliptical longitude and latitude (for equinox 1950.0) are confined to $L = 287^\circ.0 \pm 0^\circ.4$ and $B = +35^\circ.2 \pm 0^\circ.2$ respectively. The most spectacular member of this group, Great September Comet 1882 II, is a well known case of a tidal-induced breakup into individual fragments having periods 6.7, 7.7, 8.8 and 9.6 centuries. It is assumed that all Kreutz sungrazers are results of successive fragmentation of earlier members as they passed the perihelion. The comets 1882 II and 1965 VIII must have separated from each other around the early part of the 12th century and probably have been observed as one comet in 1106. The number of sungrazers was considerably enlarged by detection of 16 comets by the space-borne SOLWIND and Solar Maximum Mission (SMM) coronographs. None of them survived perihelion passage. These discoveries would more than double the number of the Kreutz group comets. Discussion see [89M].

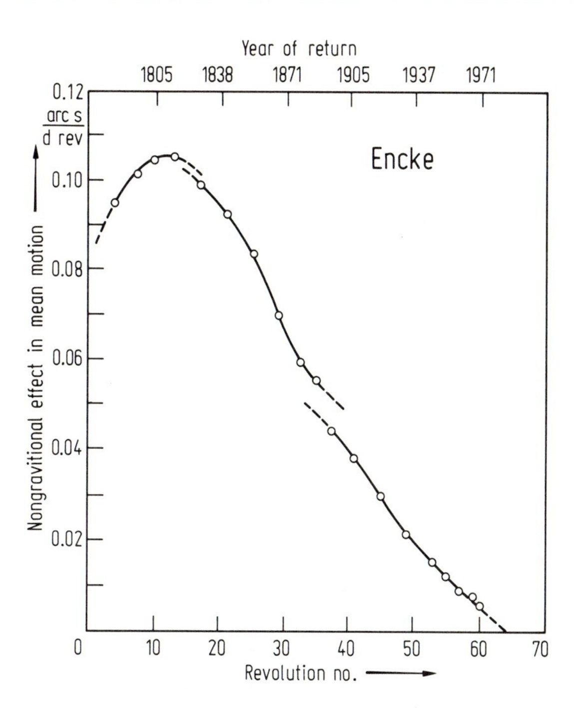

Fig. 3. Nongravitational perturbation of the mean orbital motion of P/Encke between 1786 and 1987 expressed in arcsec day^{-1} $revolution^{-1}$. The decrease of nongravitational acceleration since the early 19th century can be explained by a gradual reduction of the outgasing active area located at the north hemisphere, but, since 1890, also by gradual increase of activity on the south hemisphere of the nucleus. In comparison with the nongravitational acceleration of other short-period comets, that of P/Encke is known to have been changing fairly smoothly. (Adapted from [87N, 91S2]).

Table 3. Periodic comets of more than one appearance before 1992 (references see text).

T = time of perihelion passage
P = period
q = perihelion distance
e = eccentricity
ω = argument of perihelion in the orbital plane
Ω = longitude of the ascending node
Q = aphelion distance
i = inclination of orbit to ecliptic
For details of definition see LB VI/2a, p. 202.

No.	Name	Appearance			T	P	q	e	ω	Ω	i	Q
		first	number	last		[a]	[AU]		[deg]	[deg]	[deg]	[AU]
1	Encke[1])	1786 I	55	1990 XXI	1990.83	3.28	0.331	0.850	186.2	334.0	11.9	4.09
2	Grigg-Skjellerup	1902 II	16	1987 X	1987.46	5.10	0.995	0.664	359.3	212.6	21.1	4.93
3	Du Toit-Hartley	1945 II	3	1987 IX	1987.45	5.21	1.199	0.601	251.7	308.6	2.9	4.81
4	Machholz[1])	1986 VIII	2	1991	1991.55	5.25	0.127	0.958	14.5	93.8	60.0	5.91
5	Tempel 2	1873 II	18	1988 XIV	1988.71	5.29	1.383	0.544	191.0	119.1	12.4	4.69
6	Honda-Mrkos-Pajdusakova	1948 XII	8	1990 f	1990.70	5.30	0.541	0.822	325.7	88.7	4.2	5.54
7	Schwassmann-Wachmann 3	1930 VI	3	1989 d_1	1990.38	5.35	0.936	0.694	198.8	69.3	11.4	5.19
8	Neujmin 2[2])	1916 II	2	1927 I	1927.04	5.43	1.338	0.567	193.7	328.0	10.6	4.84
9	Brorsen[2])	1846 III	5	1879 I	1879.24	5.46	0.590	0.810	14.9	102.3	29.4	5.61
10	Tuttle-Giacobini-Kresak	1858 III	7	1989 b_1	1990.11	5.46	1.068	0.656	61.6	141.6	9.2	5.14
11	Tempel 1	1867 II	8	1989 I	1989.01	5.50	1.497	0.520	179.0	68.3	10.5	4.74
12	Wirtanen	1947 XIII	7	1991 s	1991.72	5.50	1.083	0.652	356.0	81.6	11.7	5.15
13	Clark	1973 V	4	1989 h	1989.91	5.51	1.556	0.501	208.9	59.1	9.5	4.68
14	Tempel-Swift[2])	1869 III	5	1908 II	1908.76	5.68	1.153	0.638	113.5	291.1	5.4	5.22
15	Howell	1981 X	2	1987 VI	1987.28	5.93	1.612	0.508	214.7	75.3	5.6	4.94
16	Hartley 1	1985 VII	2	1991 j	1991.37	6.02	1.818	0.451	174.1	40.4	24.9	4.80
17	Russell 1	1979 V	2	1985 IX	1985.51	6.10	1.612	0.517	0.4	230.1	22.7	5.06
18	Forbes	1929 II	7	1987 I	1987.00	6.26	1.475	0.566	262.7	22.9	4.7	5.32
19	Hartley 2	1985 V	2	1991 t	1991.70	6.26	0.953	0.719	174.9	226.1	9.3	5.84
20	De Vico-Swift[2])	1844 I	3	1965 VII	1965.31	6.31	1.624	0.524	325.4	24.4	3.6	5.21
21	Wild 2	1978 XI	3	1989 t	1990.96	6.37	1.578	0.541	41.6	135.6	3.2	5.30
22	Pons-Winnecke	1819 III	20	1989 VIII	1989.63	6.38	1.261	0.634	172.3	92.7	22.3	5.62
23	D'Arrest	1687	15	1989 II	1989.10	6.39	1.292	0.625	177.1	138.8	19.4	5.59
24	Du Toit-Neujmin-Delporte	1941 VII	4	1989 XIV	1989.80	6.39	1.715	0.502	115.3	188.3	2.9	5.17

Table 3. continued

No.	Name	Appearance			T	P	q	e	ω	Ω	i	Q
		first	number	last		[a]	[AU]		[deg]	[deg]	[deg]	[AU]
25	Schwassmann-Wachmann 2	1929 I	10	1987 XIX	1987.66	6.39	2.071	0.398	357.9	125.7	3.8	4.82
26	West-Kohoutek-Ikamura	1975 IV	3	1987 XV	1987.57	6.40	1.571	0.544	359.8	83.5	30.6	5.32
27	Kopff	1906 IV	13	1988 k	1990.05	6.46	1.585	0.543	162.8	120.3	4.7	5.35
28	Wolf-Harrington	1924 IV	8	1990 e	1991.26	6.54	1.616	0.538	186.9	254.2	18.4	5.38
29	Bus	1981 XI	2	1987 XXXIV	1987.97	6.54	2.193	0.373	24.6	181.5	2.6	4.80
30	Russell 4	1984 I	2	1989 g_1	1990.51	6.57	2.222	0.366	93.0	70.4	6.2	4.79
31	Churymov-Gerasimenko	1969 IV	4	1989 VI	1989.46	6.59	1.300	0.630	11.4	50.4	7.1	5.73
32	Giacobini-Zinner	1808 III	13	1991 m	1992.28	6.61	1.034	0.706	172.5	194.7	31.8	6.01
33	Biela[2])	1772	6	1852 III	1852.73	6.62	0.861	0.756	223.2	247.3	12.6	6.19
34	Kohoutek	1975 III	3	1987 XXVII	1987.83	6.65	1.775	0.498	175.7	269.0	5.9	5.30
35	Tsuchinshan 1	1965 I	4	1985 I	1985.01	6.67	1.508	0.575	22.8	96.2	10.5	5.58
36	Perrine-Mrkos[2])	1896 VII	5	1968 VIII	1968.84	6.72	1.272	0.643	166.1	240.2	17.8	5.85
37	Reinmuth 2	1947 VII	7	1987 XXVI	1987.82	6.72	1.936	0.456	45.5	296.0	7.0	5.19
38	Harrington	1953 VI	4	1987 XXVII	1987.83	6.84	1.596	0.557	233.0	119.0	8.7	6.61
39	Arend-Rigaux	1950 VII	6	1984 XXI	1984.92	6.84	1.446	0.599	328.9	121.6	17.8	5.76
40	Gunn[1])	1969 II	5	1989 XI	1989.73	6.84	2.472	0.314	196.9	67.9	10.4	4.74
41	Tsuchinshan 2	1965 II	4	1985 X	1985.55	6.85	1.794	0.502	203.2	287.6	6.7	5.42
42	Borrelly	1905 II	11	1987 XXXII	1987.96	6.86	1.357	0.624	353.3	74.7	30.3	5.86
43	Brooks 2	1889 V	13	1987 XXIV	1987.79	6.89	1.845	0.491	198.1	176.3	5.5	5.40
44	Wild 3	1980 VII	2	1987 XX	1987.67	6.90	2.292	0.367	179.6	72.0	15.5	4.95
45	Giglas	1978 XXII	2	1985 XV	1985.75	6.93	1.838	0.495	276.3	111.9	7.3	5.43
46	Johnson	1949 II	7	1990 h	1990.88	6.97	2.313	0.366	208.3	116.7	13.7	4.98
47	Finlay	1886 VII	11	1988 IX	1988.43	6.95	1.094	0.699	322.2	41.7	3.6	6.19
48	Taylor	1916 I	4	1990 n	1990.99	6.97	1.950	0.466	355.6	108.2	20.5	5.35
49	Longmore	1974 XIV	3	1988 XVIII	1988.78	7.00	2.409	0.341	195.7	15.0	24.4	4.91
50	Daniel	1909 IV	7	1985 XI	1985.59	7.07	1.651	0.552	10.8	68.5	20.1	5.72
51	Holmes	1892 III	7	1986 V	1986.20	7.08	2.168	0.412	23.3	327.3	19.2	5.20
52	Russell 2	1980 III	2	1987 XI	1987.50	7.10	2.151	0.418	245.4	44.4	12.5	5.24
53	Takamizawa	1984 VII	2	1991 h	1991.62	7.22	1.590	0.575	147.6	124.3	9.4	5.88
54	Shoemaker 1	1984 XVI	2	1991 p	1991.96	7.26	1.986	0.470	18.8	339.2	26.2	5.51
55	Reinmuth 1	1928 I	8	1988 VI	1988.36	7.29	1.869	0.503	13.0	199.1	8.1	5.65
56	Faye	1843 III	19	1991 n	1991.88	7.34	1.594	0.578	203.8	199.0	9.1	5.96
57	Shajn-Schaldach	1949 VI	4	1986 X	1986.40	7.46	2.331	0.390	216.5	166.2	6.1	5.31

Table 3. continued

No.	Name	Appearance			T	P	q	e	ω	Ω	i	Q
		first	number	last		[a]	[AU]		[deg]	[deg]	[deg]	[AU]
58	Ashbrook-Jackson	1948 IX	6	1986 II	1986.07	7.47	2.307	0.396	348.8	2.0	12.4	5.34
59	Russell 3	1982 IX	2	1989 d	1990.38	7.50	2.517	0.343	353.2	248.0	14.1	5.15
60	Harrington-Abell	1954 XIII	6	1990 m	1991.52	7.59	1.774	0.539	138.7	336.6	10.2	5.95
61	Metcalf-Brewington	1906 VI	2	1991 a	1991.02	7.77	1.592	0.595	208.1	187.1	13.0	6.25
62	Oterma[2])	1942 VII	3	1958 IV	1958.44	7.88	3.388	0.144	354.9	155.1	4.0	4.53
63	Kojima	1970 XII	3	1986 VII	1986.26	7.89	2.414	0.391	348.4	154.2	0.9	5.51
64	Gehrels 2	1973 XI	3	1989 n	1989.84	7.94	2.348	0.410	183.5	215.5	6.7	5.61
65	Arend	1951 X	5	1991 u	1991.59	8.02	1.857	0.536	46.9	355.6	19.9	6.15
66	Peters-Hartley	1846 VI	3	1990 d	1990.48	8.13	1.626	0.598	338.3	259.4	29.9	6.46
67	Gehrels 3	1977 VI	2	1985 IV	1985.42	8.14	3.442	0.149	231.3	242.4	1.1	4.65
68	Wolf	1884 III	13	1984 IX	1984.42	8.21	2.415	0.407	162.2	203.5	27.5	5.73
69	Schaumasse	1911 VII	8	1984 XXII	1984.93	8.26	1.213	0.703	57.4	80.4	11.8	6.96
70	Jackson-Neujmin	1936 IV	4	1987 VIII	1987.39	8.42	1.438	0.653	196.6	163.1	14.1	6.84
71	Whipple	1926 VIII	10	1986 XII	1986.48	8.49	3.077	0.261	202.0	181.8	9.9	5.25
72	Smirnova-Chernykh	1967 XV	3	1984 V	1984.14	8.51	3.557	0.146	90.9	77.0	6.6	4.78
73	Comas Sola	1927 III	8	1987 XVIII	1987.63	8.78	1.830	0.570	45.5	60.4	13.0	6.68
74	Kearns-Kwee	1963 VII	4	1989 u	1990.89	8.96	2.215	0.487	131.8	315.0	9.0	6.42
75	Denning-Fujikawa	1881 V	2	1978 XIX	1978.75	9.01	0.779	0.820	334.0	41.0	8.7	7.88
76	Lovas 1	1980 V	2	1989 XIII	1989.78	9.09	1.680	0.614	73.6	341.7	12.2	7.03
77	Swift-Gehrels	1889 VI	4	1991 c	1991.15	9.26	1.361	0.691	84.5	314.0	9.2	7.46
78	Neujmin 3 [2])	1929 III	3	1972 IV	1972.37	10.6	1.976	0.590	146.9	150.2	3.9	7.66
79	Väisälä 1	1939 IV	5	1982 V	1982.58	10.9	1.800	0.633	47.9	134.5	11.6	8.02
80	Klemola	1965 VI	2	1987 XIV	1987.57	10.9	1.773	0.640	154.5	175.8	11.0	8.09
81	Gale[2])	1927 VI	2	1938	1938.46	11.0	1.183	0.761	209.1	67.3	11.7	8.91
82	Boethin	1975 I	2	1986 I	1986.04	11.2	1.114	0.778	11.6	25.8	5.8	7.71
83	Slaughter-Burnham	1958 VI	3	1981 XVIII	1981.88	11.6	2.544	0.504	44.2	345.9	8.2	7.71
84	Van Biesbroeck	1954 IV	3	1989 h_1	1991.31	12.4	2.395	0.553	134.3	148.6	6.6	8.31
85	Sanguin	1977 XII	2	1989 z	1990.25	12.5	1.814	0.663	162.8	181.8	18.7	8.96
86	Wild 1	1960 I	2	1973 VIII	1973.50	13.3	1.981	0.647	167.9	358.2	19.9	9.24
87	Tuttle	1790 II	10	1980 XIII	1980.95	13.7	1.015	0.823	206.9	269.9	54.5	10.9
88	Chernykh	1978 IV	2	1991 o	1992.07	14.0	2.356	0.594	263.2	129.7	5.1	9.24
89	Schwassmann-Wachmann 1 [1])	1925 II	6	1989 XV	1989.82	14.9	5.772	0.045	49.9	312.1	9.4	6.31
90	Du Toit	1944 III	2	1974 IV	1974.25	15.0	1.294	0.787	257.2	22.1	18.7	10.9

Table 3. continued

No.	Name	Appearance			T	P	q	e	ω	Ω	i	Q
		first	number	last		[a]	[AU]		[deg]	[deg]	[deg]	[AU]
91	Kowal 1	1977 III	2	1991 i	1991.19	15.0	4.669	0.233	174.4	28.1	4.4	7.51
92	Gehrels 1	1973 I	2	1987 XVI	1987.61	15.1	2.989	0.510	28.5	12.9	9.6	9.23
93	Neujmin 1	1913 III	5	1984 XIX	1984.77	18.2	1.552	0.776	346.8	346.3	14.2	12.3
94	Crommelin	1818 I	5	1984 IV	1984.14	27.4	0.735	0.919	195.9	250.2	29.1	17.4
95	Tempel-Tuttle	1366	4	1965 IV	1965.33	32.9	0.982	0.904	172.6	234.4	162.7	19.6
96	Stephan-Oterma	1867 I	3	1980 X	1980.93	37.7	1.574	0.860	358.2	78.5	18.0	20.9
97	Westphal[2])	1852 IV	2	1913 VI	1913.90	61.9	1.254	0.920	57.1	347.3	40.9	30.0
98	Olbers	1815	3	1956 VI	1956.47	69.6	1.178	0.930	64.6	85.4	44.6	32.6
99	Brorsen-Metcalf	1847 V	3	1989 X	1989.70	70.5	0.479	0.972	129.6	310.9	19.3	33.7
100	Pons-Brooks	1812	3	1954 VII	1954.39	70.9	0.774	0.955	199.0	255.2	74.2	33.5
101	Halley	– 239	30	1986 III	1986.11	76.0	0.587	0.967	111.8	58.1	162.2	35.3
102	Herschel-Rigollet	1788 II	2	1939 VI	1939.60	155	0.748	0.974	29.3	355.3	64.2	56.9

[1]) Comets observed all around the orbit.

[2]) Comets missed at their last return. Some of them are probably lost for ever (Biela, Brorsen, Tempel-Swift, Neujmin 2). Perrine-Mrkos, Gale and Westphal may also have disappeared. Comet Oterma has been subjected to many perturbations by Jupiter, with the result that the period is about 19 years and perihelion distance more than 5 AU.

Table 4. Periodic comets of only one appearance before 1992 (references see text).

T = time of perihelion passage
P = period
q = perihelion distance
e = eccentricity
ω = argument of perihelion in the orbital plane
Ω = longitude of the ascending node
Q = aphelion distance
i = inclination of orbit to ecliptic
For details of definition see LB VI/2a, p. 202.

No.	Name	Appearance	T	P [a]	q [AU]	e	ω [deg]	Ω [deg]	i [deg]	Q [AU]
1	Helfenzieder	1766 II	1766.32	4.35	0.406	0.848	178.7	75.6	7.9	4.92
2	Blanpain	1819 IV	1819.89	5.10	0.892	0.699	350.2	79.2	9.1	5.03
3	Barnard 1	1884 II	1884.63	5.38	1.279	0.583	301.0	6.1	5.5	4.86
4	Brooks 1	1886 III	1886.43	5.44	1.325	0.571	176.8	54.5	12.7	4.86
5	Lexell	1770 I	1770.52	5.60	0.674	0.786	224.9	133.9	1.6	5.63
6	Mrkos	1991 k	1991.22	5.64	1.410	0.555	180.4	1.0	31.5	4.92
7	Piggot	1783	1783.89	5.89	1.459	0.552	354.6	58.0	45.1	5.06
8	Haneda-Campos	1978 XX	1978.77	5.97	1.101	0.665	240.4	131.6	6.0	5.48
9	Wild 4	1990 a	1990.51	6.15	1.989	0.408	170.5	21.5	3.7	4.72
10	Holt-Olmstead	1990 k	1990.73	6.16	2.043	0.392	2.6	14.6	14.9	4.68
11	Hartley 2	1985 V	1985.43	6.27	0.951	0.720	174.9	226.1	9.3	5.85
12	Singer-Brewster	1986 XI	1986.44	6.30	1.955	0.427	45.4	192.7	9.3	4.86
13	Tritton	1977 XIII	1977.82	6.35	1.438	0.580	147.7	300.0	7.0	5.42
14	Harrington-Wilson	1951 IX	1951.83	6.36	1.664	0.515	343.0	127.8	16.4	5.20
15	Spitaler	1890 VII	1890.82	6.37	1.818	0.471	13.4	45.9	12.8	5.06
16	Shoemaker-Levy 4	1991 f	1991.54	6.51	2.019	0.421	302.2	151.4	8.5	4.95
17	Kowal 2	1979 II	1979.04	6.51	1.521	0.564	189.4	247.2	15.8	5.45
18	Wiseman-Skiff	1986 XV	1986.89	6.52	1.505	0.569	171.8	271.0	18.2	5.47
19	Barnard 3	1982 V	1892.95	6.52	1.432	0.589	170.0	207.3	31.3	5.60
20	Mueller 2	1990 j	1990.86	6.56	2.083	0.406	171.1	218.1	7.1	4.95
21	Urata-Niijima	1986 XVI	1986.89	6.62	1.449	0.589	21.4	31.3	24.3	5.60
22	Giacobini	1896 V	1896.83	6.65	1.455	0.588	140.5	194.2	11.4	5.62
23	Schorr	1918 III	1918.75	6.67	1.884	0.469	279.2	118.3	5.6	5.21
24	Lovas 2	1986 XIII	1986.67	6.75	1.457	0.592	71.5	282.8	1.5	5.69

Table 4. continued

No.	Name	Appearance	T	P [a]	q [AU]	e	ω [deg]	Ω [deg]	i [deg]	Q [AU]
25	Hartley 3	1987 XII	1987.54	6.85	2.449	0.321	167.48	287.3	14.7	4.76
26	Swift	1895 II	1895.64	7.20	1.298	0.652	167.8	171.1	3.0	6.16
27	Ciffreo	1985 XVI	1985.83	7.21	1.702	0.544	357.9	53.1	13.1	5.76
28	Shoemaker-Levy 3	1991 e	1991.16	7.25	2.810	0.250	181.7	303.0	5.0	4.68
29	Kowal-Mrkos	1984 X	1984.43	7.32	1.951	0.483	338.1	248.5	3.0	5.59
30	Denning	1894 I	1894.11	7.42	1.147	0.698	46.3	85.1	5.5	6.46
31	Schuster	1978 I	1978.02	7.47	1.628	0.574	353.9	50.8	20.4	6.01
32	Skiff-Kosai	1976 XVI	1976.59	7.54	2.849	0.259	26.4	80.2	3.2	4.84
33	West-Hartley	1988 XVI	1988.76	7.59	2.129	0.449	102.6	46.1	15.4	5.60
34	Shoemaker 2	1984 XVII	1984.74	7.84	1.320	0.666	317.6	54.8	21.6	6.22
35	Shoemaker-Holt 2	1988 XI	1988.60	7.89	2.688	0.321	10.0	98.6	17.6	5.24
36	Helin-Roman-Crockett	1988 XIII	1988.70	8.13	3.470	0.141	10.2	91.4	4.2	4.62
37	Helin-Roman-Alu 2	1989 XVI	1989.83	8.31	1.942	0.527	202.7	202.5	7.5	6.26
38	Mueller 1	1987 XXXI	1987.92	8.45	2.747	0.338	30.3	4.0	8.8	5.55
39	Maury	1985 VI	1985.44	8.84	2.011	0.530	114.0	183.1	9.4	6.54
40	Mueller 3	1990 l	1990.58	8.65	3.000	0.288	226.0	137.3	9.4	5.34
41	Parker-Hartley	1987 XXXVI	1987.63	8.85	3.027	0.292	181.4	243.6	5.2	5.53
42	Shoemaker-Levy 2	1990 p	1990.73	9.28	1.844	0.582	140.1	235.2	4.6	6.99
43	Helin-Roman-Alu 1	1989 XXXVII	1989.86	9.50	3.709	0.173	216.4	72.8	9.8	5.26
44	Shoemaker-Holt 1	1987 VII	1988.39	9.59	3.049	0.324	210.6	213.8	4.4	5.98
45	Ge-Wang	1988 VIII	1988.40	11.39	2.516	0.503	176.0	179.8	11.7	7.61
46	IRAS	1983 XIV	1983.64	13.2	1.697	0.696	356.9	357.2	46.2	9.45
47	Helin	1987 XVII	1987.61	14.5	2.571	0.567	216.2	143.1	4.7	9.30
48	van Houten	1961 X	1961.32	15.6	3.957	0.367	14.4	22.9	6.7	8.54
49	Bowell-Skiff	1983 II	1983.20	15.7	1.945	0.689	169.0	345.6	3.8	10.6
50	Koval-Vavrova	1983 III	1983.25	15.9	2.609	0.588	19.5	201.8	4.3	10.1
51	Shoemaker 3	1985 XVIII	1985.96	16.9	1.794	0.728	14.8	96.6	6.4	11.4
52	Shoemaker-Levy 1	1990 o	1990.72	17.8	1.525	0.776	310.6	51.3	24.4	12.1
53	Hartley-IRAS	1984 III	1984.02	21.5	1.282	0.834	47.1	0.8	95.7	14.2
54	Levy	1991 q	1991.44	51.3	0.983	0.929	41.5	328.7	19.2	26.6
55	Pons-Gambart	1827 II	1827.43	57.5	0.807	0.946	19.2	319.3	136.5	29.0
56	Dubiago	1921 I	1921.34	62.3	1.115	0.929	97.4	66.5	22.3	30.3
57	de Vico	1846 I	1846.18	76.3	0.664	0.963	12.9	79.0	85.1	35.3

Table 4. continued

No.	Name	Appearance	T	P [a]	q [AU]	e	ω [deg]	Ω [deg]	i [deg]	Q [AU]
58	Bradfield 2	1988 XXIII	1988.93	81.5	0.420	0.978	194.7	27.7	83.1	37.3
59	Väisälä 2	1942 II	1942.13	85.4	1.287	0.934	335.2	171.6	38.0	37.5
60	Swift-Tuttle	(see Table 4a)	1862.64	120	0.963	0.960	152.8	138.7	113.6	47.7
61	Barnard 2	1889 III	1889.47	145	1.105	0.960	60.2	271.9	31.2	54.2
62	Mellish	1917 I	1917.27	145	0.190	0.993	121.3	88.0	32.7	55.1
63	Bradfield 1	1983 XIX	1983.99	151	1.357	0.952	219.2	356.2	51.8	55.5
64	Wilk	1937 II	1937.14	187	0.619	0.981	31.5	57.6	26.0	64.9

Table 4a. Updated list of rediscovered and discovered periodic comets in 1991-1992 (equinox J 2000.0)

Name		Number of appearances	T	P [a]	q [AU]	e	ω [deg]	Ω [deg]	i [deg]
Grigg-Skjellerup	1992	17	1992.56	5.10	0.995	0.664	359.3	213.3	21.1
Spacewatch	1990 XXIX	1	1990.97	5.59	1.541	0.511	87.1	153.4	10.0
Howell	1992c	3	1993.22	5.93	1.612	0.508	234.8	57.7	4.4
Kowal 2	1991 f_1	2	1991.84	6.39	1.500	0.564	189.5	247.8	15.8
Singer Brewster	1992 e	2	1992.82	6.43	2.026	0.414	46.6	192.6	9.2
Tsuchinshan 1	1991 c_1	5	1991.66	6.65	1.498	0.576	22.8	96.8	10.5
McNaught-Hughes	1991 y	1	1991.45	6.70	2.117	0.404	223.2	90.2	7.3
Shoemaker-Levy 7	1991 d_1	1	1991.82	6.72	1.629	0.542	91.7	313.0	10.3
Tsuchinshan 2	1991 e_1	5	1992.38	6.82	1.782	0.504	203.1	288.3	6.7
Giglas	1992 l	3	1992.67	6.96	1.847	0.493	276.4	111.9	7.3
Daniel	1992 o	8	1992.70	7.06	1.649	0.552	11.0	68.4	20.1
Ciffero	1992 s	2	1992.82	7.23	1.708	0.543	358.0	53.7	13.1
Schuster	1992 n	2	1992.68	7.26	1.846	0.493	355.7	49.9	20.1
Shoemaker-Levy 8	1992 f	1	1992.47	7.47	2.711	0.291	22.4	213.4	6.1
Ashbrook-Jackson	1992 j	7	1993.53	7.47	2.306	0.396	348.7	2.7	12.5
Shoemaker-Levy 6	1991 b_1	1	1991.78	7.57	1.132	0.706	333.1	37.9	16.9
Kojima	1992 z	4	1994.13	7.85	2.400	0.393	348.4	154.3	0.9
Gehrels 3	1992 v	3	1993.55	8.14	3.442	0.149	231.6	243.3	1.1
Schaumasse	1992 x	9	1993.17	8.22	1.202	0.704	57.5	81.1	11.8
Wolf	1992 m	14	1992.66	8.25	2.428	0.406	162.3	203.4	27.5
Brewington	1992 p	1	1992.42	8.65	1.560	0.630	45.2	342.9	18.1
Shoemaker-Levy 5	1991 z	1	1991.95	8.66	1.984	0.529	6.0	29.7	11.8
Mueller 4	1992 g	1	1992.13	9.06	2.656	0.389	43.3	145.6	30.0
Väisälä	1992 u	6	1993.46	10.9	1.800	0.633	47.4	135.1	11.6
Slaughter-Burnham	1992 w	4	1993.50	11.6	2.545	0.504	44.1	346.4	8.2
Tuttle	1992 r	11	1994.48	13.51	0.998	0.824	206.7	270.5	54.7
Swift-Tuttle1	1992 t	3	1992.95	135	0.958	0.964	153.0	139.4	113.4

Note: P/Swift-Tuttle 1992t = 1862 III = 1737 II.

Table 5. Least values of perihelion distances ("sun-grazers"). Symbols: see Table 3.

Comet		q [AU]	e	ω [deg]	Ω [deg]	i [deg]	P [a]
1981 XXI	Solwind 4	0.00450	1	77.7	356.9	143.8	
1989 q	SMM 9	0.00462	1	91.8	14.2	144.8	
1989 x	SMM 10	0.00476	1	87.5	8.9	144.8	
1979 XI	Solwind 1	0.00480	1	67.7	344.3	141.5	
1887 I	Great Southern Comet[1])	0.00483	1	83.5	3.9	144.3	
1963 V	Pereyra	0.00506	0.99995	86.2	7.2	144.5	903
1988 XVII	SMM 5	0.00513	1	88.1	9.6	144.8	
1988 X	SMM 3	0.00516	1	85.9	7.0	144.7	
1987 XXII	SMM 1	0.00538	1	85.6	0.5	144.3	
1880 I	Great Southern Comet[2])	0.00549	1	86.2	7.1	144.6	
1843 I	Great March Comet	0.00552	0.99991	82.6	2.8	144.3	513
1989 m	SMM 8	0.00557	1	84.7	5.5	144.6	
1988 XIX	SMM 7	0.00579	1	86.1	7.3	144.7	
1988 XXII	SMM 6	0.00590	1	91.1	13.3	144.8	
1988 XII	SMM 4	0.00591	1	82.3	2.5	144.4	
1981 XIII	Solwind 3	0.00612	1	68.4	345.3	141.7	
1680		0.00622	0.99998	350.6	275.9	60.6	
1987 XXV	SMM 2	0.00627	1	82.6	3.0	144.5	
1945 VII	Du Toit	0.00751	1	72.1	350.5	141.8	
1983 XX	Solwind 6	0.00753	1	78.6	358.0	144.0	
1882 II	Great September Comet	0.00775	0.99990	69.6	346.5	142.0	759
1965 VIII	Ikea-Seki	0.00778	0.99991	69.1	346.3	141.8	880
1981 I	Solwind 2	0.00792	1	65.4	341.4	140.7	
1970 VI	White-0rtiz-Bolelli	0.00887	1	61.3	336.3	139.0	
1984 XII	Solwind 5	0.01541	1	56.7	329.7	136.4	

[1]) Comet Thome.

[2]) Comet Gloud.

Solwind and SMM (Solar Maximum Mission) were satellites for solar wind and solar corona observations.

3.3.3.2 Photometry; polarimetry

Photometry

Brightness variation of comets:

Intrinsic brightness J_0 at $r = 1$ AU and $\Delta = 1$ AU is related to the observed brightness J by the equation $J = J_0 \Delta^{-2} r^{-n}$ (Δ = geocentric distance, r = heliocentric distance, both in [AU]) where n is derived for each comet from observation. However, the value of n is not constant for a given comet, sometimes changing abruptly. Mean values of n (ranging from 2 to 8) are given in Table 9 of LB VI/2a, p. 215.

For prediction of the apparent magnitude m the definition $m = H_y + 5 \log \Delta + y \log r$, with $y = 2.5\,n$ is often used. The mean value of y is assumed as 10 and H_y is assumed as a constant depending only on y and by some authors is defined as "absolute magnitude". In reality, however, the value of y is ranging from 5 to 28!

Visual estimation of total brightness of a comet is strongly depending on the observational techniques [91L1].

Intrinsic brightness variation (brightness outburst) is an unexpected flare-up in the brightness of the comet, with brightness increasing by 2 to 3 magnitudes. The brightness variation has usually an eruptive onset followed by a quasi-exponential decrease. Several cometary outbursts are accompanied by the expansion of a spherical halo [91H4]. Two examples are given in LB VI 2a, p. 215. The mean variation of absolute magnitude H of short-period comets is about $0^{m}.03$ to $0^{m}.2$ per revolution and it is assumed as an evidence of aging of periodic comets (discussions see in [91K3]).

The photometry and spectrophotometry of comets are extensively discussed for instance in [83A1, 87F, 89N, 91H1, 91J3, 91L1]. Photographic methods (equidensitometry, isophotometry) are entirely replaced by modern CCD technique.

IHW photometric system

The IHW photometric system was introduced in the photoelectric photometry of comets during the world-wide ground-based observational campaign (1982–1988) of Comet Halley organized by the International Halley Watch (IHW) [84A, 90O1]. The system consists of six interference filters for molecular bands and three for the continuum. Typical FWHM (full width at half maximum) of these filters is about 5 to 9 nm (see Table 6).

The IHW photometric standards are (a) bright stars, mostly spectral type B, which serve as flux standards, and (b) stars of spectral type G2V, used as solar analogs (Tables 7 and 8). The measurements of solar analogs are applied for the subtraction of flux of the solar reflected continuum underlying on the cometary molecular emissions. More details about the IHW system see in [84A, 90O1, 86P, 85V, 85W].

From pure molecule emission flux density $F_{\rm m}$ [$\rm J\,m^{-2}s^{-1}$] measured in the column of coma defined by the diaphragm of the photometer the total number of molecules $N_{\rm m}$ can be derived:

$$N_{\rm m} = 4\pi F_{\rm m} \Delta^2 r^2 / g$$

where the geocentric distance Δ is given in [m] and the heliocentric distance r in [AU]. The fluorescence efficiency g for typical molecular emissions at $r = 1$ AU is of the order of 10^{-19} to about 10^{-22} $\rm J\,s^{-1}$ molecule^{-1} [82A]. (Table 10, Fig. 4).

Polarimetric observations

The linear polarization in the near-visual and infrared spectral region is observed in the range of 10 to 20% and increases toward the tail, where it is oriented approximately perpendicular to the tail axis. The degree of polarization changes with scattering angle (Sun-comet-observer). In the continuum positive polarization is amounting to $20 \cdots 30\%$ at the scattering angle 90°, and inversion to negative polarization ($< -10\%$) was observed at 100° to 180° (back-scattering angles, see Fig. 5). In molecular emission the polarization ranges between 2 to 4%. The maximum of circular polarization in the continuum is about 2%. A critical evaluation of the data on 10 comets indicates that there are no universal polarization characteristics. The negative polarization, however, seems to be typical of the back-scattering domain. Discussion see in [87L1, 91M1, 88D, 89D].

See also subsect 3.3.3.7: The nature of cometary dust.

Table 6. IHW photometric system [84A, 90O1]. (FHWM = full width at half maximum)

Filter	Maximum transmission λ [nm]	FWHM [nm]	Spectral feature
OH	308.5	8	OH
UC	365.0	8	continuum
CN	387.1	5	CN ($\Delta v = 0$)
C3	406.0	7	C_3
CO	426.0	7	CO^+
BC	484.5	6.5	continuum
C2	514.0	9	C_2 ($v = 0$)
RC	684.0	9	continuum
H2O	700.0	8	H_2O

Table 7. IHW primary standards [90O1].

HD = Henry Draper catalogue
α = right ascension (equinox 1950.0)
δ = declination (equinox 1950.0)
V = visual magnitude
B-V = color index

HD	Other identification	α	δ	V	B-V	Spectral type
3379	53 Psc	$00^h34^m10.8^s$	$+14°57'24''$	$5^m.88$	$-0^m.15$	B2.5 IV
26912	μ Tau	04 12 49.0	+08 46 07	4.29	−0.06	B3 IV
52266	BD -5°1912	06 57 53.9	−05 45 21	7.23	−0.01	09 V
74280	η Hya	08 40 36.7	+03 34 46	4.30	−0.20	B4 V
89688	23 Sex	10 18 27.1	+02 32 31	6.66	−0.08	B2.5 IV
120086	BD -1°2858	13 44 44.2	−02 11 40	7.89	−0.18	B3 III
120315	η UMa	13 45 34.3	+49 33 44	1.86	−0.19	B3 IV
149363	BD -5°4318	16 31 47.9	−06 01 59	7.80	+0.01	B0.5 III
164852	96 Her	18 00 14.7	+20 49 56	5.28	−0.09	B3 IV
191263	BD +10°4189	20 06 15.1	+04 43 29	6.31	−0.12	B3 V
28099	Hyades: vB 64	$04^h23^m47.7^s$	$+16°38'07''$	8.12	+0.66	G6 V
29461	Hyades: vB 106	04 36 07.6	+14 00 29	7.96	+0.66	G5 V
30246	Hyades: vB 142	04 43 38.9	+15 22 59	8.33	+0.67	G5 V
44594	HR 2290	06 18 47.1	−48 42 50	6.60	+0.66	G3 V
105590	BD -11°3246	12 06 53.2	−11 34 36	6.56	+0.66	G2 V
186427	16 Cyg B	19 40 32.0	+50 24 03	6.20	+0.66	G2.5 V
191854	BD +43°3515 AB	20 08 33.7	+43 47 43	7.42	+0.66	G5 V

Table 8. Magnitudes of IHW standards in IHW photometric system [90O1]. HD catalogue numbers are used for identification of stars. For filter definition see Table 6.

(a) IHW primary flux standards.

	Spectral features:	OH	UC	CN	C_3	CO^+	BC	C_2	RC	H_2O^+
HD	V λ [nm]:	308.5	365.0	387.1	406.0	426.0	484.5	514.0	684.0	700.0
3379[1])	$5^{m}.88$	$5^{m}.88$	$5^{m}.88$	$5^{m}.88$	$5^{m}.88$	$5^{m}.88$	$5^{m}.88$	$5^{m}.88$	$5^{m}.88$	$5^{m}.88$
26912	4.27		4.55	4.42	4.42	4.41	4.33	4.32	4.22	4.22
52266	7.23		7.02	7.32	7.42	7.41	7.22	7.25	7.06	7.07
74280	4.29	4.38	4.20	4.25	4.27	4.29	4.26	4.30	4.32	4.33
89688	6.68	6.84	6.79	6.75	6.79	6.77	6.68	6.69	6.61	6.61
120086	7.89	7.72	7.70	7.83	7.89	7.86	7.84	7.87	7.87	7.89
120315	1.86	1.94	1.92	1.90	1.92	1.86	1.94	1.93	1.86	1.89
149363	7.80	7.80	7.71	7.94	8.02	8.00	7.80	7.81	7.62	7.59
164852	5.27	5.44	5.38	5.34	5.33	5.31	5.29	5.28	5.21	5.24
191263	6.33	6.42	6.38	6.36	6.35	6.35	6.34	6.34	6.31	6.33

[1]) The zero point of the system in all filters is arbitrarily defined by the V magnitude of HD 3379 (53 Psc).

(b) IHW primary solar analogs.

	Spectral features:	OH	UC	CN	C_3	CO^+	BC	C_2	RC	H_2O^+
HD	V λ [nm]:	308.5	365.0	387.1	406.0	426.0	484.5	514.0	684.0	700.0
28099	$8^{m}.12$		$9^{m}.58$	$9^{m}.84$	$9^{m}.27$	$9^{m}.14$	$8^{m}.39$	$8^{m}.33$	$7^{m}.53$	$7^{m}.49$
29461	7.96		9.44	9.70	9.12	8.98	8.24	8.20	7.40	7.33
30246	8.33	$10^{m}.65$	9.77	10.05	9.46	9.36	8.59	8.54		
44594	6.60		8.07	8.34	7.72	7.61	6.90	6.85		
105590	6.56		8.35	8.65	8.01	7.88	7.16	7.11	6.25	6.22
186427	6.20	8.70	7.68	7.96	7.35	7.23	6.50	6.46	5.62	5.64
191854	7.42	9.89	8.86	9.14	8.54	8.42	7.67	7.63	6.82	6.80
Solar colors[1])		2.12	1.19	1.46	0.87	0.75	0.00	–0.05	–0.86	–0.89

[1]) Average of stars HD 28099, 29461, 30246, 186427.

Table 9. Flux of zero-magnitude-star in the IHW photometric system. (Adapted from [89B]. Filter definition see Table 6).

Filter	Spectral feature	Flux[1]) [J s^{-1}]	Filter	Spectral feature	Flux [J s^{-1}nm^{-1}]
CN	CN ($\Delta v = 0$)	$5.3 \cdot 10^{-14}$	UC	continuum	$8.2 \cdot 10^{-15}$
C3	C_3	$1.4 \cdot 10^{-13}$	BC	continuum	$5.1 \cdot 10^{-15}$
C2	C_2	$6.8 \cdot 10^{-14}$	RC	continuum	$1.6 \cdot 10^{-15}$

[1]) The flux in the molecular emission feature is integrated over the band, thus, the dimension is [J s^{-1}].

Table 10. Fluorescence efficiency g for typical molecular emission features in coma at the heliocentric distance 1 AU.

Molecule	Spectral feature	g [J s^{-1} molecule^{-1}]
C_2	Swan; $\Delta v = 0$	$2.2 \cdot 10^{-20}$
C_3	$^1\Pi - ^1\Sigma_g^-$	$1.0 \cdot 10^{-19}$
CH	$A^2\Delta - X^2\Pi$, $\Delta v = 0$	$9.2 \cdot 10^{-21}$
CN	$B^2\Sigma^+ - X^2\Pi$, $\Delta v = 0$	$2.2 \cdot 10^{-20}$ [1])
OH	$A^2\Sigma^+ - X^2\Pi$, 0 – 0 band	$1.7 \cdot 10^{-22}$ [1])

[1]) Valid only for the heliocentric radial velocity of the comet $v_r = dr/dt = 0$. Values of g of CN and OH are very sensitive on v_r. Discussion see [82A].

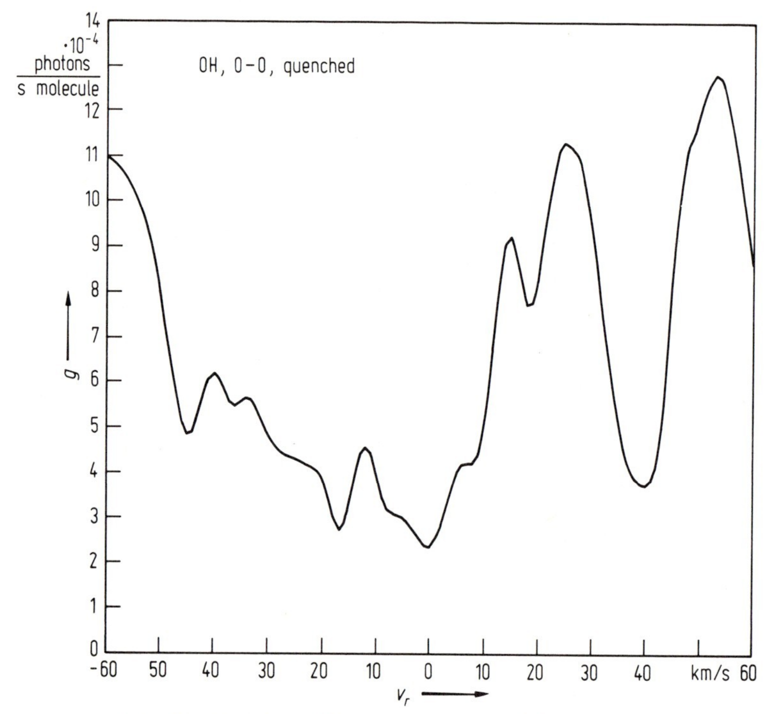

Fig. 4. Variation of the fluorescence efficiency g for OH with heliocentric radial velocity v_r of the comet. (Adapted from [90A]).

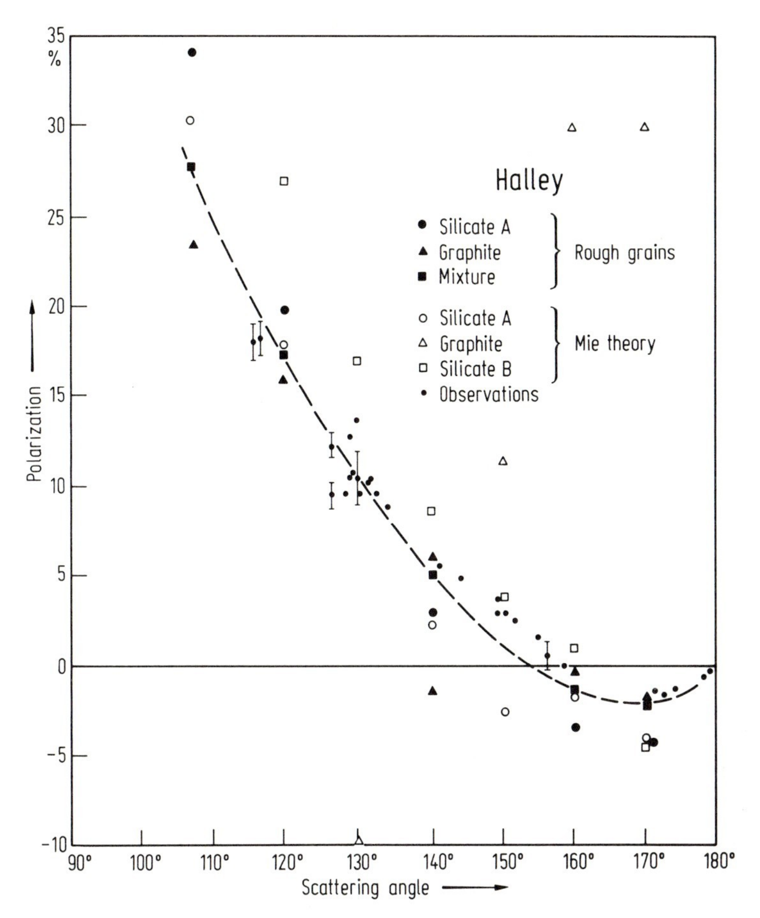

Fig. 5. Polarization observed for P/Halley at large scattering angle. For comparison model calculations are shown for rough and spherical particles (calculated by the Mie theory) of several materials. The material denoted silicate A is characterized by larger value of the dielectric part of the complex index of refraction in comparison to the refraction index of common silicates denoted as silicate B. The mixture is a combination of silicate A and graphite [87L].

3.3.3.3 Spectroscopic observations

Optical observations
The continuum is produced mainly by the sunlight scattered by the dust particles in the coma and tail. Only a very small fraction of the continuum is due to the solar radiation reflected by the nucleus. The energy distribution in the continuum is determined by size and optical characteristics (index of refraction) of small particles and the scattered light is often slightly redder than sunlight. Superimposed on the continuum background are emissions (molecular bands or atomic lines) due to neutral radicals or atoms (in coma) and ions (in tail). The resonance-fluorescence is a dominant radiation process. In small heliocentric distances often metallic lines are observed. Due to the heliocentric component of the comet's motion the Fraunhofer lines in the solar spectrum are Doppler-shifted relative to the comet: influence on the vibration-rotation line intensities in the molecular bands (Swings effect). Similar secondary effect (Greenstein effect) due to the internal motion (expansions) in the coma.

Infrared observations
Silicate-like characteristics of the cometary dust are confirmed by recent infrared observation. The temperature of small particles with diameters $< 2\mu m$ exceeded the black-body temperature.

The infrared observation, mainly of P/Halley from VEGA spacecraft, revealed the emission features superimposed on the background continuum in the spectral region between 2 to $5\mu m$. The strongest band is the H_2O $\nu 3$ band at $2.7\mu m$. Others are due to H_2CO, CO_2, CO, and emissions of complex molecules near $3.4\mu m$. In the far-infrared region the thermal radiation of dust particles dominates. In the near infrared (up to $3.5\mu m$) the continuum is the reflected sun light.

Discussions in [86T, 90A, 91E, 91H1, 91M1].

Radio observations
Radio observation, mainly of P/Halley in 1985-1986, confirmed the presence of 18-cm OH-transitions. Hydrogen cyanide HCN and formaldehyde H_2CO were detected. However, species which detections were claimed in the past (CH_3CN, NH_3) and the 22-GHz-line of H_2O were not confirmed. The study of the 18-cm-transitions of the OH radical is the most productive tool in the cometary astronomy (derivation of water production, velocity field in the coma, etc.). In millimeter wavelength ranges hydrogen sulfide H_2S and methanol CH_3OH were detected. Discussion in [91C2].

UV observations
Besides the typical emissions in the UV spectral region (H, OH), lines and bands of C, O, S, CO, CS, NH, H_2CO were detected, mostly from the International Ultraviolet Explorer (IUE) observations (see for instance [82F, 87F, 89R2] and discussion in [a]). At least in one case (comet 1983 VII) sulfur dimer S_2 has been observed [83A2]. The reflectivity of the dust indicates reddening of the scattered radiation at shorter wavelengths (visual and UV) [86J, 87F].

Update notes
Some spectroscopic results are given in Tables 11–13 and Fig. 6. Tables 12 and 13 are updated versions of Table 10 in LB VI/2a, p. 216. Molecular and atomic transitions observed in comets are given in LB VI/2a, Table 11, p. 217. However, in this table the tentative identification of He ($3^1P - 2^1S$) at 501.5 nm has to be deleted, while the tentative identification of the resonance lines Al (396.1 nm) and Si (390.4 nm) has been confirmed in the meantime. Lifetimes of upper states of cometary bands and production rates of major constituents in four comets are given in LB VI/2a, pp. 218 and 219. Water production rates and the gas-to-dust ratio in 13 comets are given in Table 14. Relative abundance of several atoms, isotopes and molecules in comets and other objects (for comparison) are given in Tables 15–21 and Figs. 7 and 8.

Review concerning the excitation and emission mechanisms in comets see in [90A].

Table 11. Spectral identification in comets (compiled by author).

	Spectral range
Coma (head) CN, ^{13}CN, C_2, $^{12}C^{13}C$, C_3, CH, NH, NH_2, OH, [OI], Na, Ca, Cu, Cr, Mn, Fe, Ni, K, Co, SH, CO^+, CH^+, $CO_2{}^+$, $N_2{}^+$, OH^+, Ca^+, H_2O^+ Reflected sunlight	Optical
H, C, O, S, OH, CO, CS, $CO_2{}^+$, CO^+, CN^+, C^+, S_2	UV
H_2O, H_2CO, CO_2, CO, (OCS) [1]) "C-H" feature near 3.4 μm of complex organic compounds, thermal emission of dust, silicate emission features near 9 μm	IR
OH, HCN, H_2CO, H_2CS, H_2S, OCS (CH_3CN, NH_3, H_2O) [2])	Radio
H^+, N^+, O^+, C^+, $S_2{}^+$, H_3O^+, $NH_4{}^+$, $NH_3{}^+$, $NH_2{}^+$, $CH_5{}^+$, $CH_4{}^+$, $CH_3{}^+$, $C_3H_3{}^+$, $C_3H_4{}^+$, C_3H^+, $CS_2{}^+$, CS^+, H_3S^+, $(H_2CO)_n$, H_2O, CO, N_2, C_2H_4 Many compounds probably organic, with mass up to 105 amu. "CHON" grains (dust particles with organic mantles or composed entirely of light elements, i.e. H,C,N,O compounds).	"In-situ" measurements by mass spectrometry during spacecraft flyby of P/Halley
Plasma tail CO^+, CH^+, $CO_2{}^+$, $N_2{}^+$, OH^+, H_2O^+, CN^+, OH^+, $NH_4{}^+$, SH^+	Optical and UV
Dust tail	Optical
Reflected sunlight	IR
Thermal emission of dust	

[1]) Tentative identification.
[2]) Identification claimed in the past, but not confirmed in P/Halley.

Table 12. Updated list of newly identified molecular infrared and radio emissions observed in comets (compiled by the author).

Species	Transitions or type of band	λ [μm]
CN	$A \Rightarrow X(2 \Rightarrow 1)$	0.917
	$A \Rightarrow X(1 \Rightarrow 0)$	0.941
	$A \Rightarrow X(0 \Rightarrow 0)$	1.10
	$A \Rightarrow X(1 \Rightarrow 0)$	1.46
C_2	$b \Rightarrow a(2 \Rightarrow 1)$	1.45
H_2O ?	$\nu_1 + \nu_2 + \nu_3$	2.44
H_2O	ν_3	2.67
H_2O ice	absorption feature	2.9
Unsaturated hydrocarbons	C-H stretch	3.28
Saturated hydrocarbons	C-H stretch	3.36
CO_2	ν_3	4.27
CO	1-0	4.64
OH ?	$X^2\Pi$ ($v = 3$)	0.973
OH	$^2\Pi_{3/2}$ ($J(5/2 - 3/2)$)	119
	$^2\Pi_{3/2}$ ($J(3/2)$)	18 cm (1612, 1665, 1667, 1721 MHz)
H_2CO	ν_5	3.50
	ν_1	3.62
	1_{10}-1_{11}	6.2 cm (4830 MHz)
	3_{12}-2_{11}	1.3 mm (226 GHz)
HCN	J (1-0)	3.37 mm (89 GHz)
	J (3-2)	1.13 mm (266 GHz)
H_2S	1_{10}-1_{01}	1.78 mm (169 GHz)
CH_3OH	A-ν_3	3.52
	J (3-2)	2.07 mm (145 GHz)
	J (2-1)	3.13 mm (96 GHz)

Table 13. Updated list of newly identified molecular emissions in ultraviolet spectral region (compiled by the author).

Species	Transition or type of band	λ [nm]
S_2	B-X	288.9 ··· 410.0
NH	$c\,^1\Pi - a\,^1\Delta(0.0)$	325.5 ··· 326.7
CO^+	$^2\Pi_{1/2u} - {}^2\Pi_{1/2g}(0.0)$	350.4 ··· 351.2
H_2CO	B-X	353.1 ··· 354.7

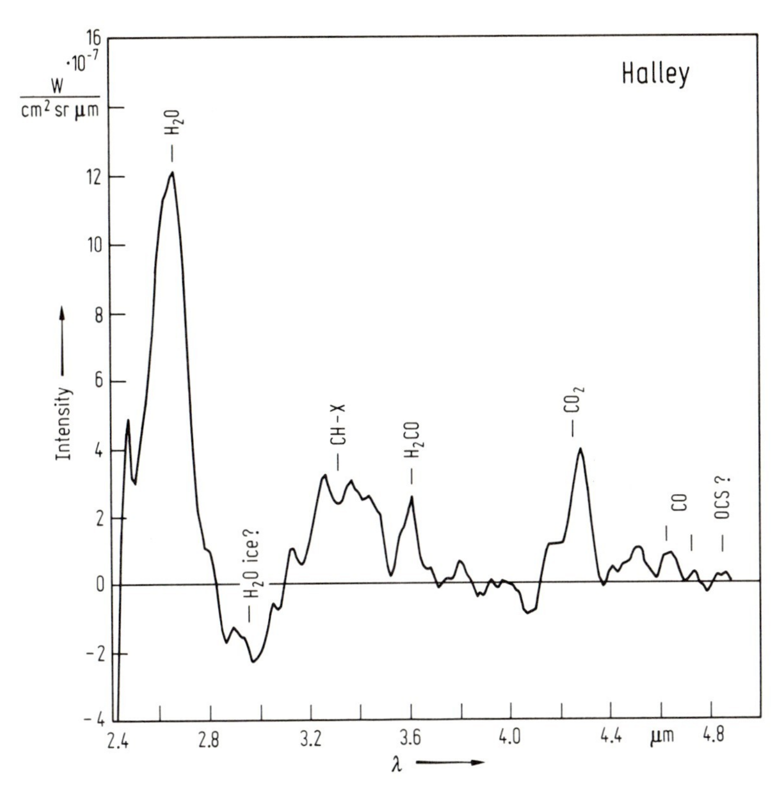

Fig. 6. Low-resolution infrared spectrum of comet Halley between 2.5 and 5µm obtained by the infrared spectrometer on the VEGA spacecraft near the time of the closest approach to the nucleus. Figure is from [88C], for discussion see [91E].

Table 14. Water production rate Q (in mass) and gas-to-dust ratio. (Compiled from [82F, 82N, 85H1, 85H2, 87F, 87H, 89R2, 91K4]). r = heliocentric distance.

Comet	r [AU]	Q (H_2O) [100 kg s^{-1}]	H_2O/dust
P/Encke	1.0	2	7
P/Grigg-Skjelerup 1982 IV	1.04	0.9	5
P/Giacobini-Zinner 1985 XIII	1.05	12	3
P/Crommelin 1984 IV	1.0	2	0.1
P/Halley 1986 III	1.0	150	3
Bennet 1970 II	1.0	150	10
Churyumov-Gerasimenko 1982 VIII	1.5	1	2
IRAS-Araki-Alcock 1983 VII	1.0	2	1
Kobayaschi-Berger-Milon 1975 IX	1.0	6	6
Kohoutek 1973 XII	1.0	90	20
Sukano-Saigusa-Fujikawa 1983 V	1.06	0.8	40
West 1976 VI	1.0	150	4
Wilson 1987 VII	1.0	18	6

Table 15. Ratios of stable isotopes.

Species	Solar system	Local ISM	Comets
H/D	$2 \cdot 10^{-5}$	$1.5 \cdot 10^{-5}$	$\geq 5 \cdot 10^{-4}$ [1,9])
		$1 \cdot 10^{-4} \cdots 1 \cdot 10^{-2}$ [2])	
$^{12}C/^{13}C$	89	43 ±4 [3])	89 ··· 115 [1,4])
		65 ±20 [5])	70 ··· 130 [6])
		12 ··· 110 [7])	< 100 [1,8,9])
$^{14}N/^{15}N$	270	≃ 400 [5])	> 200 ? [1,9])
$^{16}O/^{18}O$	490	≃ 400 [5])	< 450 [1,9])
$^{24}Mg/^{25}Mg$	7.8		variable [1,9])
$^{25}Mg/^{26}Mg$	0.9		< 2 [1,9])
$^{32}S/^{34}S$	22.6		22 [1,9])
$^{56}Fe/^{54}Fe$	15.8		15 [1,9])

Discussion and references in [89W, 91J1, 91K1, 91V].
ISM = interstellar matter.

[1]) Data for P/Halley only.
[2]) Range of observed values in dense ISM.
[3]) From visual spectra.
[4]) From ground-based observation of CN ($\Delta v = 0$) band.
[5]) From radio astronomical data.
[6]) From ground-based observation of C_2 spectra.
[7]) Range of observed values in dense ISM.
[8]) From in-situ mass spectrometry of dust.
[9]) From space experiment results.

Table 16. Carbon isotopic abundance ratio in comets.

Comet	$^{12}C/^{13}C$	Band	Ref.
Ikeya 1963 I	70 ±15	$C_2(\Delta v = +1)$	89W, 91V
Tago-Sato-Kosaka 1969 IX	100 ±20	$C_2(\Delta v = +1)$	89W, 91V
Kohoutek 1973 XII	$115\,^{+30}_{-20}$	$C_2(\Delta v = +1)$	89W, 91V
	$135\,^{+65}_{-45}$	$C_2(\Delta v = +1)$	89W, 91V
Kobayashi-Berger-Milon 1975 IX	$100\,^{-20}_{+30}$	$C_2(\Delta v = +1)$	89W, 91V
Halley 1986 III	65 ± 9	$CN(\Delta v = 0)$	89W
	89 ±17	$CN(\Delta v = 0)$	91J2
	100 ±15	$CN(\Delta v = 0)$	91K4

Table 17. Light elements in some bright comets and mean ratios of atom numbers in the volatile fractions [91D].

Element	Bright comets	P/Halley
H/O	1.8 ±0.4	1.9 ±0.4
C/O	0.2 ±0.1	0.20 ±0.05
N/O	0.10 ±0.05	0.10 ±0.05
S/O	0.003 ±0.0015	0.01 ±0.005

Table 18. Elemental ratios in atom numbers for H, C, N, O and S [91D].

Ratio	Sun	Comets	CI chondrites	Comet dust	Comet gas
H/O	1175	2.0	0.17	2.29	1.86
C/O	0.43	0.4	0.1	0.91	0.2
N/O	0.13	0.09	0.007	0.005	0.1
S/O	0.02	0.03	0.07	0.08	0.003

Table 19. Abundances of molecules (relative to H_2O) in the coma of P/Halley [91Y1].

Species	X/H_2O	Species	X/H_2O
H_2O	1	NH_3	$0.01 \cdots 0.02$
CO	$0.15 \cdots 0.2$ [1]	N_2	< 0.02
	$\geq 0.07 \cdots 0.05$ [2]	HCN	0.001
CO_2	$0.02 \cdots 0.04$	H_2C	(0.04)
CH_4	$0.005 \cdots 0.02$	POM [3]	(0.1)

[1]) At distances larger than 15 000 km from the nucleus.

[2]) At distances less than 10 000 km from the nucleus.

[3]) POM = polyoxymethylene (polymerized $(H_2CO)_n$).

Table 20. Carbon and oxygen abundances (relative to H_2O) of some carbon-bearing parent molecules in P/Halley compiled from infrared observations and spacecraft VEGA experiments [91E].

Species	C/H_2O	O/H_2O
H_2O		1
CO_2	0.03 ±0.01	0.06 ±0.02
H_2CO	0.04 ±0.02	0.04 ±0.02
CO	0.05 ±0.02	0.05 ±0.02
OCS	0.007 ±0.003	0.007 ±0.003
HCN	0.001 ±0.0005	
CH_4	0.02 ±0.01	
Total	0.148 ±0.064	

Note: Upper limits of abundances of carbon-bearing molecules in P/Halley obtained from the GIOTTO spacecraft mass spectrometer relative to H_2O are: CO 0.20, CO_2 0.035, CH_4 0.07, and from the ion mass spectrometer: CH_4 0.02 [86B, 86K].

Table 21. Average atomic abundance of elements in the comet Halley. (Compiled from [89A2, 91J2]). Atomic numbers normalized to Mg = 100

Element	Halley		Solar system	CI-chondrites
	dust	dust and ice		
H	2025	4062	$2.6 \cdot 10^6$	492
C	814	1010	940	70.5
N	42	95	291	5.6
O	890	2040	2216	712
Na	10	10	5.34	5.34
Mg	100	100	100	100
Al	6.8	6.8	7.91	7.91
Si	185	185	93.1	9.31
S	72	72	46.9	47.9
K	0.2	0.2	0.35	0.35
Ca	6.3	6.3	5.69	5.69
Ti	0.4	0.4	0.223	0.223
Cr	0.9	0.9	1.26	1.26
Mn	0.5	0.5	0.89	0.89
Fe	52	52	83.8	83.8
Co	0.3	0.3	0.21	0.21
Ni	4.1	4.1	4.59	4.59

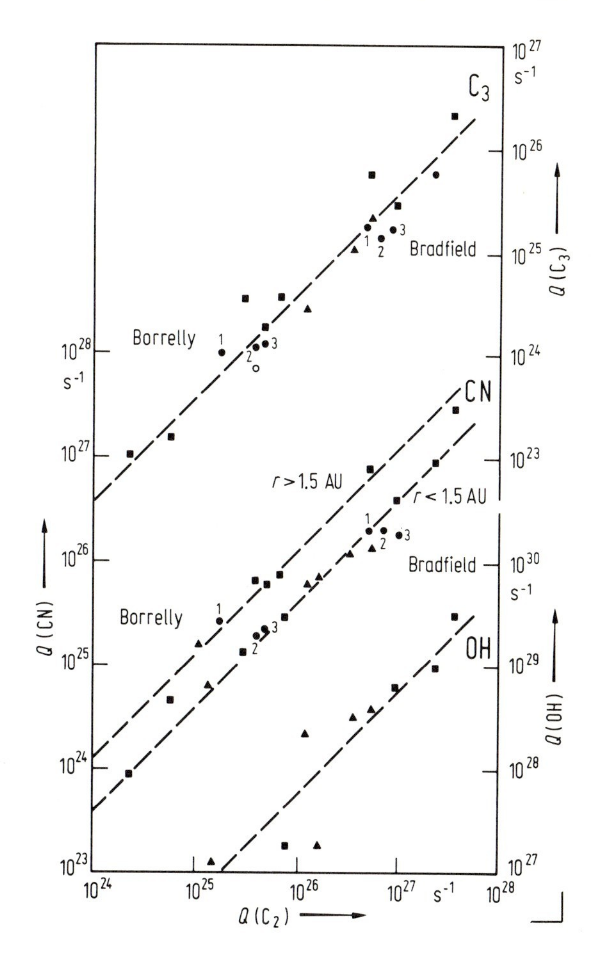

Fig. 7. Correlation abundance of OH, CN, C_2 and C_3 among comets. The production rates Q of the comets P/Borrelly and Bradfied 1987 XXIX for different sets of scale lengths of parent molecules marked by dots with digits 1, 2, 3, are compared with results for other comets. Squares are pre- perihelion observations, triangles are post-perihelion data. Correlation for CN depends slightly on the heliocentric distance r. [89B].

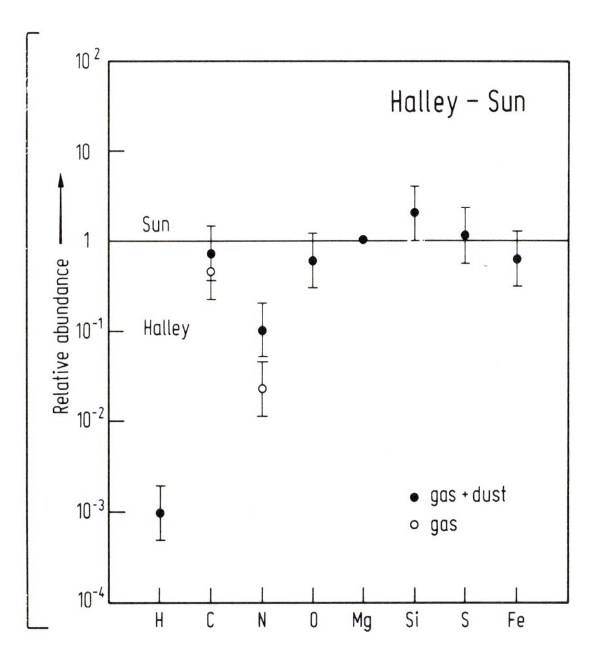

Fig. 8. Abundances of elements in P/Halley relative to the sun. Note the depletion of hydrogen and nitrogen [90W].

3.3.3.4 Nucleus

Structure and composition

The classical "icy conglomerate" model, proposed in 1950 by Whipple is still – although in modified version – the best approach to the structure and composition of real cometary nuclei which are mixtures of frozen-gas ices and meteoric material partly in the form of dust grains. The gas-to-dust ratio is poorly known. Some investigators placed this ratio about 1. In volatile constituents water ice is dominant, but other molecular compounds may be significant. The composition of refractory material resembles the composition of carbonaceous chondrites type 1. In the dust Fe seems to be deficient and Si overabundant. The dust particles contain a considerable amount of organic compounds (CHON particles). Otherwise, the elemental abundance appears to be equivalent to the solar system abundance. Tensile strength of the cometary material (estimated from the splitting of cometary nuclei) is very weak (less than 10^5 Nm^{-2}). Under solar radiation the sublimation process on the cometary nucleus is activated. However, the outgasing fraction of the nucleus surface, i.e. the active area, is for most cases below 10%, and even 1% only. It is assumed that the phase transitions from crystalline to amorphous ice, as well as the clatharate hydrates, may play an important role in the nucleus [91K2]. The erosion of active regions is about 4 to 6 m per revolution for nuclei with bulk density 1000 kg m^{-3} and about 25 m per revolution if the density is assumed to be about 200 kg m^{-3}. A typical cometary nucleus seems to be a loosely packed structure with high porosity. Discussion in [86G2, 89J1, 91M2, 91S4, 91W2, 91R]. General review about the cometary nuclei see in [90K].

Shape and dimensions

The size of cometary nuclei is estimated mostly from the brightness of the comet at large heliocentric distances where the dust and gas production is negligible and only the sun radiation reflected by the bare comet nucleus is observed [82S, 87S2, 82W].

Another approach to the shape and size determination can be the radar observation. The radar-echo signal was received at the Arecibo Observatory from comets Encke, Halley, Grigg-Skjellerup, and IRAS-Araki-Alcock [85O].

Direct images of the cometary nucleus were obtained by the GIOTTO and VEGA space missions in 1986 to P/Halley (Figs. 9 and 10). The nucleus showed to be quite irregular, although it can be very crudely approximated by a prolate spheroid with dimensions of axes 15.3 : 7.2 : 7.2 km. The size determination is not more accurate than ±0.5 km. The volume of the nucleus is about 365 km^3 [86W1, 87K, 90K, 91S4]. The prolate shape of cometary nuclei is assumed also for some other comets from photoelectric studies. These dimensions of cometary nuclei are larger than the dimensions derived from previous studies that were based on intrinsic magnitudes at large heliocentric distances and on relatively high values of an assumed albedo. The range of radii 4 to 6 km is very likely typical of observed comets (Tables 22 and 23). However, the size distribution of cometary nuclei is unknown and much smaller cometary nuclei as well as much larger comets may exist; e.g. the Great Comet of 1729 that could be seen with the naked eye at 4-AU-distance from the Earth and Sun.

Albedo

The cometary nuclei are "black" with very low geometric albedo. This was confirmed by the space missions and by comparisons of visual and infrared photometry. The albedo of the P/Halley nucleus is about 0.035 to 0.045 with somewhat larger variation of the surface reflectivity. The geometric albedo of most cometary nuclei appears to range from 0.02 to 0.1 with a typical value of about 0.04 (see Table 23).

Fig. 10. Three-dimensional model of comet Halley's nucleus reconstructed from near-nucleus images taken by VEGA 1 (63 images), VEGA 2 (3 images) spacecraft and single view provided by Halley Multicolor Camera (HMC) experiment from GIOTTO spacecraft. In (c) the long axis points to the reader, in (a) it is rotated to the right by 40°, and in (b) it is rotated to the left by 40°. Some significant features are (1) protrusion, (2) elevation, (3) "chain of hills", and (4) "cave"; concave-convex depression and elevation. The dimensions of the axes are 15.3, 7.2, 7.2 km with the mean error ±0.5 km. (From [91S3]). →

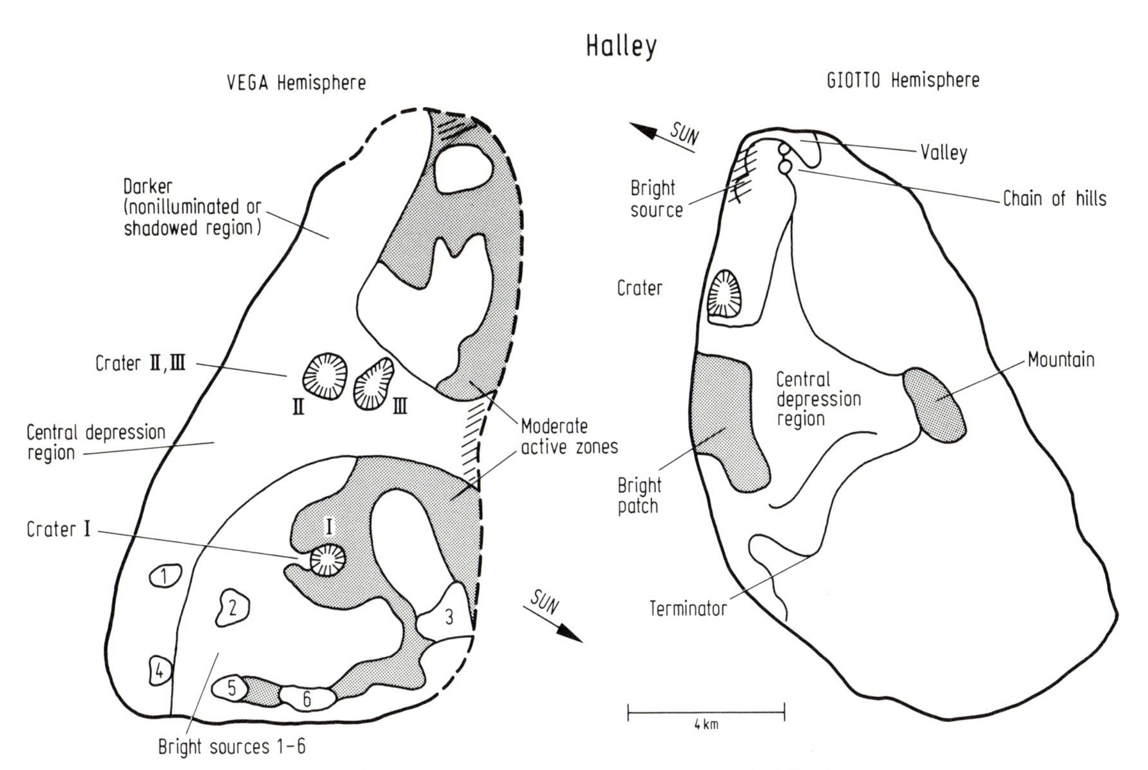

Fig. 9. Schematic picture of the nucleus of Halley's comet reconstructed from VEGA and GIOTTO imaging experiments [91M2].

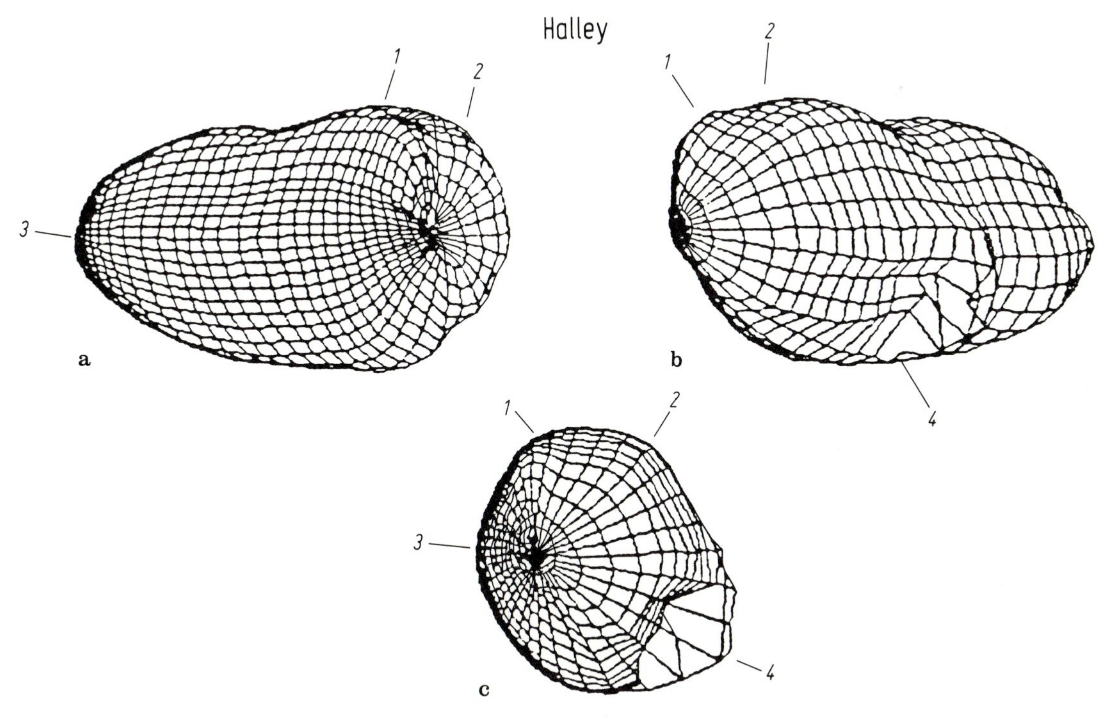

Fig. 10. (For caption see previous page.)

Table 22. Estimated radii R of cometary nuclei for some short-periodic comets (with assumed albedo 0.055 and assumed spherical shape) [87S2]. V_0 = visual magnitude of the nucleus reduced to unit heliocentric and geocentric distances ($r = \Delta = 1$ AU).

Comet	V_0	R [km]
Halley	$13^{m}_{.}5$	5.6
Oterma	14.5	3.5
van Houten	11.4	14.8
Gunn	13.3	6.3
Kearns-Kwee	13.8	4.9
Ashbrook-Jackson	14.0	4.4
Whipple	13.8	4.8
Brooks 2	15.9	1.8
van Biesbroeck	13.0	7.2
Slaughter-Burnham	14.8	3.0
Schwassmann-Wachmann 2	14.1	4.3
Comas Sola	13.3	6.3
Faye	14.0	4.4
Tempel 2	15.4	2.3
Encke	15.0	2.9

Directly observed nucleus of the comet Halley is here accepted as a standard object with projected surface area 100 km^2 and with upper limit of the albedo $A = 0.055$.

Table 23. Visual geometric albedo, water production (at heliocentric distance r) and active fraction of the surface of some cometary nuclei (compiled by the author).

Comet	R [1]) [km]	Albedo	H_2O production [molecules s^{-1}]	r [AU]	Fraction of active surface
P/Halley	6	0.04	$6 \cdot 10^{29}$	0.8	0.3
P/Arend-Rigaux	5.2	0.03	$2 \cdot 10^{26}$	1.6	0.008
P/Neujmin 1	10.4	0.02	$1 \cdot 10^{26}$	1.7	0.001
P/Tempel 2	5.6	0.02	$2 \cdot 10^{27}$	1.7	0.01
P/Encke	<2	0.03 ?	$6 \cdot 10^{28}$	0.76	0.17

[1]) R is an "average" radius

Temperature

Surface temperature depends of the albedo and vaporization rate. The steady state for a rotating nucleus can be written

$$0.25F_0(1 - A_v)r^{-2} = \sigma T^4(1 - A_i) + Z(T)L(T),$$

where F_0 is the solar flux density at heliocentric distance $r = 1$ AU; A_v and A_i are mean albedos in the visual and infrared spectral regions, respectively, T is the surface temperature, Z is the vaporization rate [molecules $m^{-2}s^{-1}$], L is the latent heat per molecule, and σ denotes the Stefan-Boltzmann constant. For $Z = 0$, the surface temperature $T \cong T_0 r^{-1/2}$, where $T_0 \cong 280$ K if A_v as well as A_i are < 0.1. The surface temperature of a low-albedo nucleus covered by ice ranges from 195 to 215 K at $r = 1$ AU [82D]. The central temperature for a typical periodic comet is estimated in the range from 68 K (P/Halley) to 183 K (P/Encke). The central temperature of a cometary nuclei may be partly determined also by exothermic processes if some radioactive long-lived isotopes are present [91R].

Gas production rates

Gas production rates from a cometary nucleus can be derived from the photometric and spectrophotometric observations or by the solution of the energy balance and Clausius-Clapeyron equation. The vaporization rate of H_2O at heliocentric distance $r = 1$ AU from the ice surface of a steadily rotating nucleus with visual and infrared albedos A_v, $A_i < 0.1$ is about 10^{22} molecules $m^{-2}s^{-1}$ (see Fig. 11).

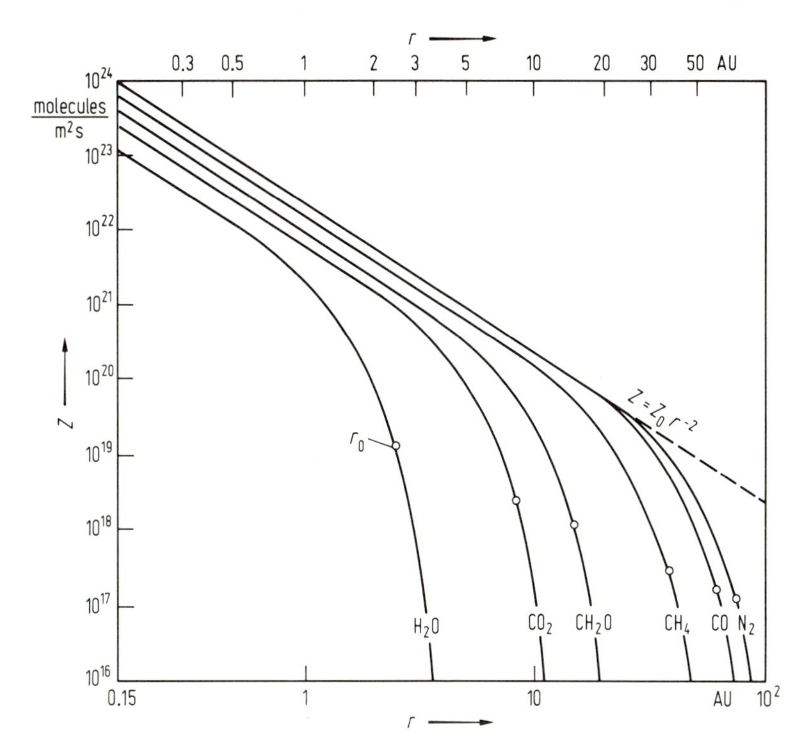

Fig. 11. Vaporization rate Z as a function of heliocentric distance r for various frozen gases; rotating nucleus, with visual albedo A_v being twice the infrared albedo A_i. Beyond the distance r_0 the vaporization rate becomes negligible [82D].

Density and mass

The bulk density of the cometary nuclei is low. Values derived from nongravitational effects [89R1] are between 280 ±100 to 650 ±180 kg m^{-3}. However, some authors find 800 to 2000 kg m^{-3} for P/Halley. The real value is more likely about 500 kg m^{-3}. Consequently, the mass of P/Halley nucleus is about $2 \cdot 10^{14}$ kg, and masses of observed comets range from 10^{13} to 10^{17} kg. Mass loss is 0.1 to 1% per revolution. Internal mass distribution is unknown. Discussion and references concerning the formation of cometary nuclei see in [a, 90H2].

Rotation of cometary nuclei

Although there are about 60 reported determinations of rotational periodicities of cometary nuclei, only few appear to be reliable, and even they do not necessarily describe the true rotational state (see Table 24). The determination of the spin state, i.e. rotation period and orientation of the spin axis, is based on the search for periodicities in time-series of some continuously varying observed property of comets, mainly variation of brightness [81S]. The difficulties in attempts to determine the spin parameters of cometary nuclei can be demonstrated on Halley's comet. Although most of the published results indicate that the nucleus rotates with the period of about 53 h around the shortest axis (short-axis mode), there are strong evidences also for a period of 178 h. According to [89S] the nucleus of P/Halley rotates as a slightly asymmetric top, the orientation of angular momentum vector being $b = -54°, l = 39°$ in the ecliptic system, with mean error about 15°. In this case it is assumed that the nucleus rotates around the short axis, its long axis "noding" with period of about 178 h. However, extensive search to determine all rotational states of the nucleus of P/Halley [91S1] shows that in the most likely modes for the rotational state the long axis executes a precessional motion around the space-fixed total angular momentum vector with a period near 3.7 days, while performing a rotational period around itself with period of about 7.3 days. Review in [91B2].

Table 24. Rotation of cometary nuclei derived for some periodic comets.

Comet	Period [h]	Projected axial ratio	Ref.
d'Arrest	5.17 ± 0.01		78F
Neujmin 1	25.34		86W2
	12.67	≅1.45:1	88A
	12.68		88J
Encke	22.43	2:1	87J2
	15.08 [1])		90L
Arend-Rigaux	13.47 [2])	1.6:1	88M
Tempel 2	8.95 ± 0.01		89J2
	8.9 [3])	1.9:1	89A1
Halley [4])	53	2.1:1.1:1	91B2
	52.8 ± 1.2		89S

[1]) Or 7.54 h, or 22.62 h, respectively.
[2]) Visual geometric albedo 0.028, effective radius of the nucleus 5.15 ± 0.2 km.
[3]) Visual geometric albedo 0.022, effective radius of the nucleus 5.9 km and approximate size 16×8.5×8.5 km. Comprehensive model for the nucleus of P/Tempel 2 see [91S2].
[4]) According to [91B1] the total spin period is 68.16 h and the nucleus precesses with a period of 88.56 h.

Table 25. Rotation-pole determination for P/Halley.

Apparition	Time span	North pole (equinox 1950.0)		Obliquity of equator to orbit	Ref.
		α [deg]	δ [deg]		
1835/36 [1])		225 ± 25	−70 ± 10	40 ± 10	82W
1910 [1])	May-June	357	−49	30	86S
1986 [2])	Nov. 85	34 ··· 43	−15 ··· −40	30 ··· 56	86G2
[3])	March 86	55 ± 15	−36 ± 15		89S
[3])	March 86	50 ± 10	−40 ± 5	30^{+5}_{-10}	86W2
[4])	Apr.-June 86	20	−66	9	86S

Method of determination:
[1]) from recurring jets;
[2]) from stationary jets;
[3]) from nucleus images;
[4]) from sunward spike.

3.3.3.5 Coma

Phenomenology

Coma (and nucleus) make up the head of a comet. General review see in [90A]. A comet nucleus behaves like a fountain that ejects water upward, but the gravitational field is here substituted by the solar radiation pressure. The matter ejected from the nucleus is pushed in the anti-sunward direction and the observed shape of the coma exhibits some asymmetry and in many cases also jets, rays and fans are displayed.

Inner, molecular coma is dominated by parent molecules released by sublimation from the nucleus. They are photodissociated by the solar UV radiation. The typical lifetime of most of the parent molecules is 10^4 s at the heliocentric distance $r = 1$ AU. If the expansion velocity is approximately 1 $\mathrm{km\,s^{-1}}$, then the dimension (scale-length) of the inner coma is about 10^4 km at $r = 1$ AU.

Chemical reactions in the coma are also considered [91H3].

Table 26. Scale lengths of molecules in the cometary atmosphere at heliocentric distance $r = 1$ AU (compiled by the author).

Species	$v\tau$ [km][1])	Species	$v\tau$ [km][1])
CN	$3.0 \cdot 10^4$	HCN	$8 \cdot 10^4$
C_2	$5.7 \cdot 10^4$	C_2H_2	$4 \cdot 10^5$
NH_2	$5.0 \cdot 10^4$	NH_3	$7 \cdot 10^3$
CH	$6.0 \cdot 10^3$	CH_4	$2 \cdot 10^6$
H_2O	$8.0 \cdot 10^4$	OH	$3 \cdot 10^5$

[1]) $v\tau$ = scale length; τ = photodissociation lifetime.
Expansion velocity v is slightly variable with heliocentric distance r and production rate. For $r \cong 1$ AU is $v \cong 1$ $\mathrm{km\,s^{-1}}$. From 18-cm OH radio line observation of P/Halley the following mean values of v were found: 0.4 $\mathrm{km\,s^{-1}}$ (at $r = 2$ AU); 0.8 $\mathrm{km\,s^{-1}}$ (1.5 AU); 1.2 $\mathrm{km\,s^{-1}}$ (1 AU); 1.8 $\mathrm{km\,s^{-1}}$ (0.6 AU) [91C2].

Visible coma is dominated by the products of photodissociation, i.e. radicals (like CN, C_2, OH), most of them have fluorescent bands in the visible spectral range. The size of the visible coma varies proportionally with the heliocentric distance: from 10^4 km at $r = 0.3$ AU up to 10^6 km at $r > 3$ AU. The typical dimension of the visible coma of medium bright comets is about 10^5 km ($r = 1$ AU).

UV coma is dominated by atomic H easily detectable by the Lyα emission line. The dimension of this hydrogen coma is up to several 10^7 km.

Dust coma pervades the inner and visible coma. Expansion velocity of the dust depends on the dust grain mass. The terminal velocities at $r = 1$ AU are about 100 $\mathrm{m\,s^{-1}}$ for dust particles with mass 10^{-8} kg, and 1 $\mathrm{km\,s^{-1}}$ for all particles with mass $< 10^{-13}$ kg. The dynamics of the dust coma is determined by hydrodynamical interaction between gas and dust and by the solar radiation pressure [91G, 91H2].

Spiral-shaped jets of CN and C_2 in the coma of P/Halley and some other comets were discovered by the new observational CCD technique. These jets are seen to persist several weeks and extended to the projected distance up to $5 \cdot 10^4$ km from the nucleus. The strength of the jets is well correlated with the fluctuation in the overall production rate of radicals [86A, 90A].

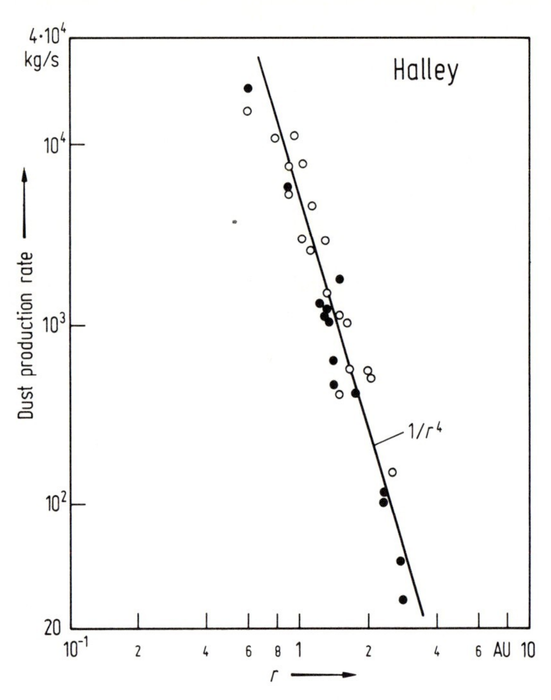

Fig. 12. Variation of the dust production rate with heliocentric distance r of P/Halley. The straight line represents a r^{-4} dependence. Dots and crosses refer to the infrared observation for pre-perihelion and post-perihelion distances, respectively [91K4].

Table 27. Dust production rate Q_d estimated for some comets and reduced to the heliocentric distance $r = 1$ AU [1]).

Comet	Q_d [$kg\,s^{-1}$]	Ref.
P/Halley [2])	5400	91K4
	5500	85H1
Bennet 1970 II	1500	91K4
	1180	82N
Kohoutek 1973 XII	460	91K4
	500	82N
P/Encke	28	91K4
	11	82N
Sugano-Saigusa-Fujikawa 1983 V	2	91K4
	<1.2	87H

[1]) Assumed density of dust particles is 1000 $kg\,m^{-3}$

[2]) The maximum dust production rate of comet Halley measured in-situ at the time of the Giotto encounter was $2.55 \cdot 10^4$ $kg\,s^{-1}$ [86K].

Anisotropic ejection of fan-shaped forms and jets observed in visual light in many comets can be explained by varying the geometrical conditions in the orientation of rotating nuclei in respect to the Sun (discussion in [87S1, 90A]).

Note
For further far-infrared and radio observations see also subsection 3.3.3.3: Spectroscopic observations.

3.3.3.6 Tails

For general description of the plasma (ion, type I) tail and the dust (type II) tail see LB VI/2a, pp. 223, 224. New reviews for dust tail [a,i].

Plasma tail in-situ measurements
Spacecraft encounters with comets P/Giacobini-Zinner, P/Halley and Grigg-Skjellerup revealed a great variety of collective plasma phenomena accompanying the interaction of solar wind (see subsect. 3.3.5.2) with comets, especially numerous plasma instabilities. It turns out that a relatively small number of them generates Alfvén wave turbulence near comets. The measurements have provided detailed data on the bow-shock and information on the bow-shock structure, magnetic field (40 to 100 nT), and change of magnetic field along the spacecraft trajectory through coma and tail. Discussions and reviews see in [a, b, c, d, m, n, o, 90I, 91C1, 91F2]. A schematic picture of the plasma tail and solar wind-comet interaction morphology is shown in Fig. 13.

Table 28. Plasma population in the vicinity of comets [91C1].

Population	Type of distribution	Energy
Protons		
Solar wind	Maxwellian	2 eV, total energy 1 keV
Cometary plasma		
Hot	Ring/shell	20 keV
Warm	Shell	100 keV
Heavy cometary ions		
Hot	Ring/shell	20 keV
Warm	Shell	1 keV
Cold	Maxwellian	1 eV
Electrons		
Solar wind	Maxwellian	$10 \cdots 30$ eV
Photoelectrons	Non-maxwellian	20 eV
Cometary cold	Maxwellian	1 eV
Accelerated	Beams	≥ 1 keV
Cometary neutrals		
Cold	$v = 1\ \mathrm{km\,s^{-1}}$	≤ 0.01 eV
Warm (dissociative recombination and charge transfer)	$v = 1 \cdots 30\ \mathrm{km\,s^{-1}}$	
Very warm (charge transfer with warm ions)	$v = 30 \cdots 100\ \mathrm{km\,s^{-1}}$	
Hot (charge transfer with hot ions)	$v \geq 400\ \mathrm{km\,s^{-1}}$	

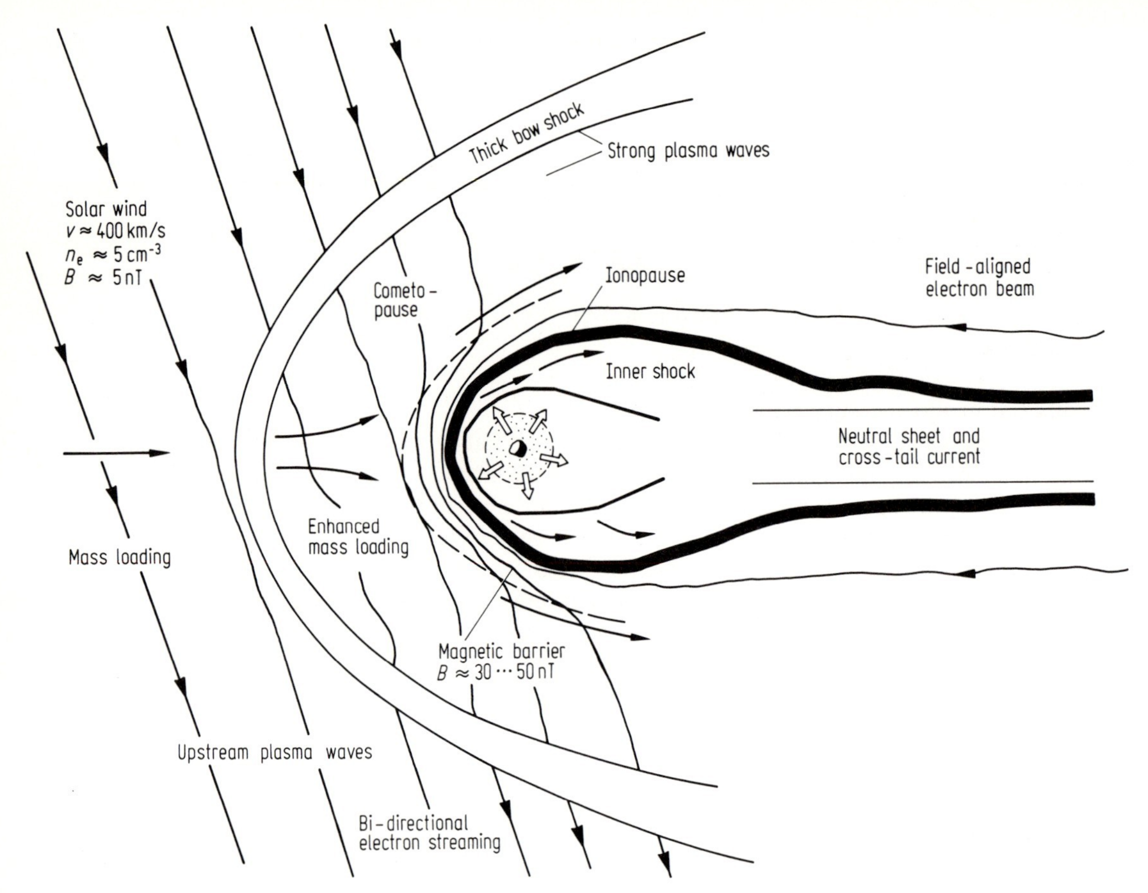

Fig. 13. Schematic representation of the global morphology of the solar wind interaction with the cometary atmosphere. (From [91F2]).

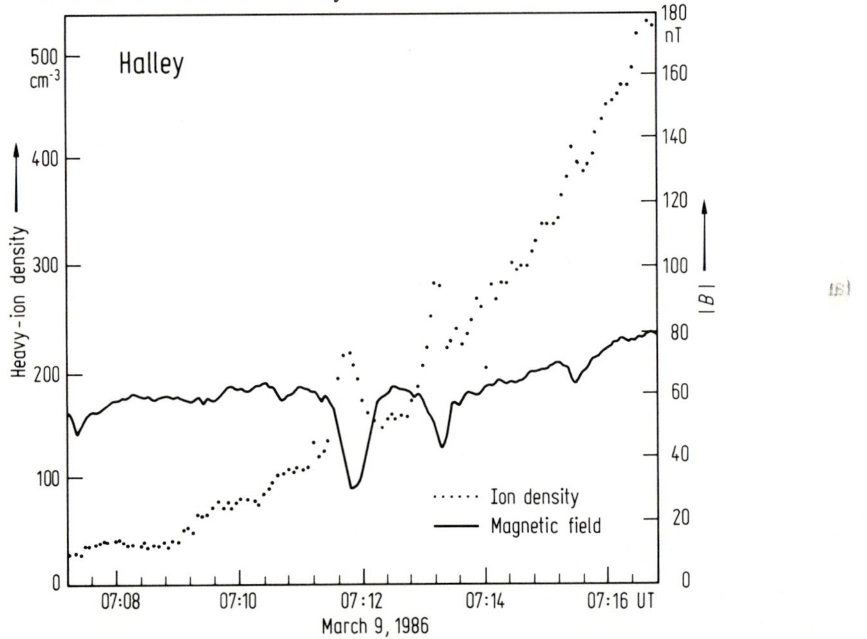

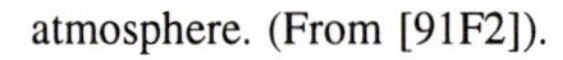

Fig. 14. Ion density and magnetic field data measured near magnetic barrier region of comet Halley by instruments on spacecraft VEGA 2. Ion condensation is caused by the minor instability [91C1].

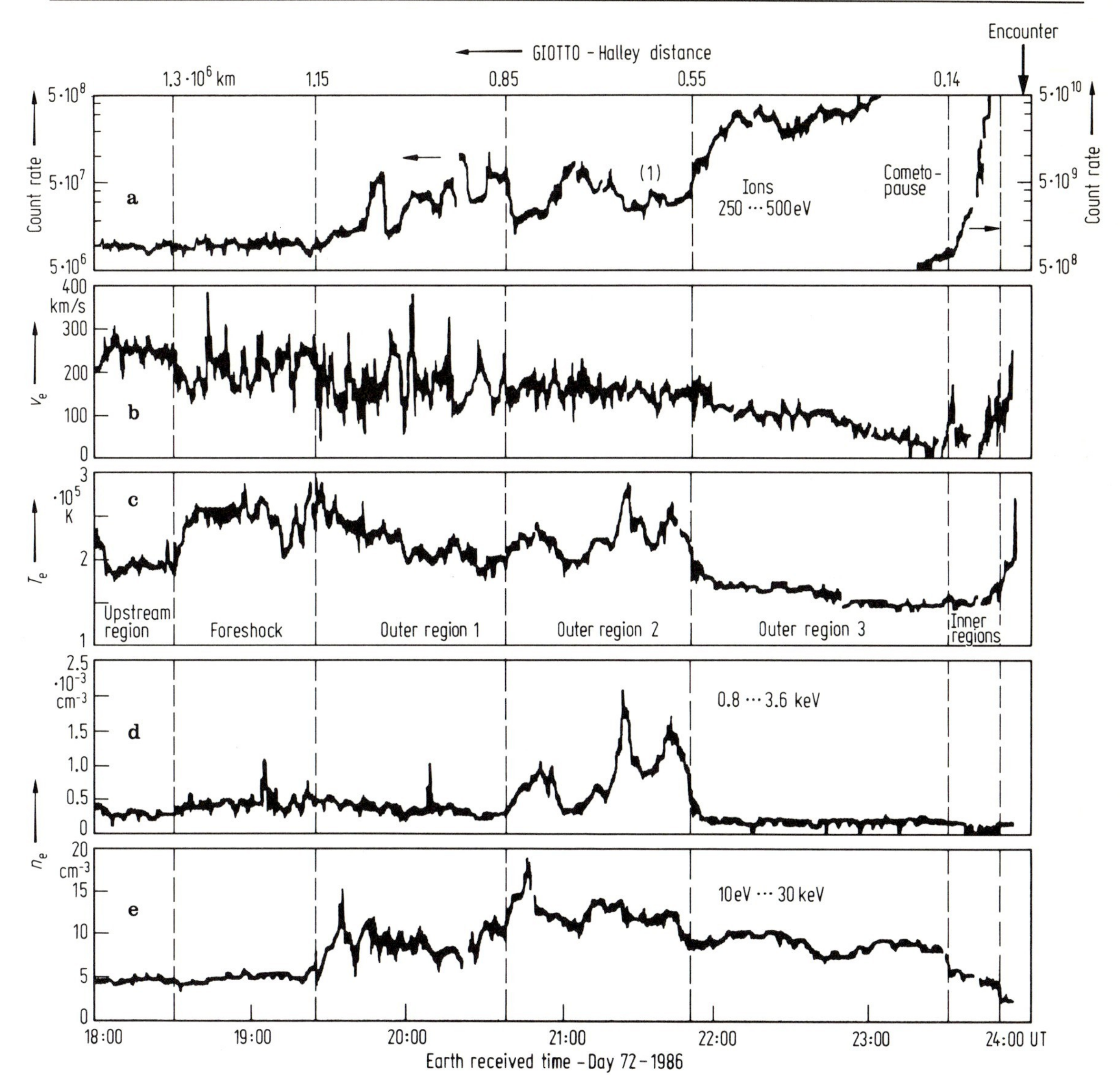

Fig. 15. Electron parameters measured by GIOTTO on its inbound trajectory at comet Halley. From top to bottom: (a) total electron count rate, (b) electron velocity, (c) electron temperature, (d) density of electrons in energy range 0.8 ··· 3.6 keV, (e) density of electrons in the range 10 eV to 30 keV. Near the cometopause ions with energy 250 to 500 eV were dominant. According to [91F2].

Dust tail

The dust tail contains very fine small particles from 10^{-18} to 10^{-20} kg which in the plasma environment of a comet are necessarily electrically charged. This charging has both physical and dynamical effects on the cometary dust, being particularly important for the smallest grains observed in the dust size spectrum [87B]. The disruption of larger grains into smallest particles by electric charges is one of the possible mechanisms which may lead to additional production of submicron particles in coma and tail. Discussions and reviews in [90G, 91H2].

More in-situ data of comet Halley are given in Figs. 14–16. For the energy flux in the continuum of the coma, tail and antitail of comet Kohoutek 1973 XII, see Fig. 15 in LB VI/2a, p. 223.

3.3.3.7 The nature of cometary dust

Cometary dust is studied by various methods which allow the analysis of the spectral distribution and polarization of the scattered solar radiation, thermal emission, coma and tail morphology, radar echoes, meteors associated with comets and data from in-situ measurements (reviews [87J1, 89J1, 90G, 91L2, 91M1]). For the nature of the cometary dust from ground-based observations see also LB VI/2a, p. 225. The ground-based observations of recent comets, namely of P/Halley, show that the reflectivity of the cometary dust is characterized by reddening in the ultraviolet (and partly in visual) wavelengths.

The polarization of light scattered by the cometary dust is determined by the same physical parameters as the colour, i.e. by the refractive index and size distribution of grains. The value of the maximum polarization (and the corresponding scattering angle is mostly controlled by the imaginary part of the refractive index (i.e. absorption properties of the grain) [87L1]. Also the grain roughness has a very strong influence on the polarization curve (i.e. dependence of the polarization on the phase angle). For discussion see [91M1].

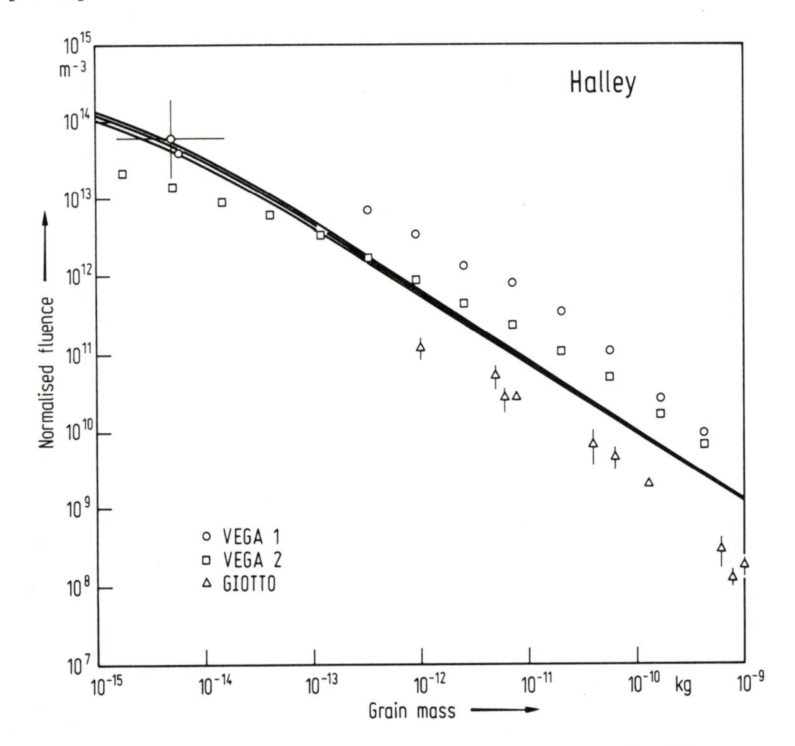

Fig. 16. Fluency of dust measured in Comet Halley's coma normalized by the varying distances of closest approach to the nucleus. Data from VEGA 1 (circles), VEGA 2 (squares) and GIOTTO (diamonds). Solid lines indicate fountain model distribution. (From [91M1]).

By the in-situ measurement of dust at comet Halley in 1986 information has been gained on particles at the extreme ends of the size distribution, where ground-based observation are ineffective. Very small grains (< 0.1μm size) are more abundant than previously estimated. The detectable size distribution extending down to the molecular dimensions. A large fraction (about 30%) of small to medium-sized grains (in the mass range $3 \cdot 10^{-19} \cdots 3 \cdot 10^{-13}$ kg) are dominated by light elements and carbon-based chemical composition ("CHON" particles, with large amount of organic compounds dominated by H, C, O and N). The bulk density of these grains is about 1000 kg m^{-3} (1 g cm^{-3}). These grains may also be sources for some volatiles or molecules observed in the coma. The composition of about 35% of grains is most likely consistent with chondritic meteorites [87L2]. The geometric albedo of the cometary grains is relatively low (0.02 to 0.1).

The nature of cometary dust from in-situ measurements:

Aspect	Basic information
Chemical composition	Great variability, from high content of metals to silicate composition dominated by light elements, "CHON" particles with organic molecular content. Isotopic ratio of $^{12}C/^{13}C$ variable. Correlation between Mg/C and Si/C ratios. About 35% of particles can be interpreted as minerals similar to that encountered in carbonaceous chodrites. The "CHON" particles represent about 30% of the dust population.
Size, structure and masses	The size from 0.5 to 100μm is probably dominant. The measured full mass range is 10^{-5} to 10^{-19} kg. Mass distribution index < 1 with a constant density function. The structure of "CHON" particles is fluffy and density considerably less than 1000 kg m^{-3}.

3.3.3.8 Laboratory studies and space experiments relevant to comets

Laboratory studies are mostly concerned with simulations of the environment on and in the cometary nuclei: physical properties of frozen volatiles, irradiation effects on comets and cometary debris and photochemical reactions. The large-scale experiments are made in Germany (project KOSI = KOmetenSImulation). These experiments include study of heat transport in ice-dust porous mixtures, material modification and chemical fractionation, detailed studies of gas release from mixtures of volatile ices. These experiments are reviewed in [a, i, k].

First space missions to comets have been realized in 1985–1986. Following a series of highly complicated orbital maneuvers about Earth and the Moon, the International Sun-Earth Explorer 3 (ISEE-3), launched in 1978, was renamed International Cometary Explorer and went through the tail of P/Giacobini-Zinner on September 11, 1985 at the distance from nucleus of 7800 km.

Four space agencies – the European Space Agency (ESA), the Intercosmos of USSR Academy of Sciences, the Japanese Institute of Space and Aeronautical Science (ISAS) and the National Aeronautics and Space Administration (NASA) sent six spacecrafts to Halley's comet. ESA launched GIOTTO, Intercosmos launched VEGA 1 and VEGA 2, ISAS launched SAGIGAKE and SUISEI, and NASA used its ICE spacecraft. The results are presented in [a, b, d, i, k, m, n, o].

Table 29. Cometary missions since 1985.

Name	Target	Encounter date	Flyby distance [km]
ICE	P/Giacobini-Zinner	Nov. 10, 1985	$7.8 \cdot 10^3$
	P/Halley	March 25, 1986	$2.8 \cdot 10^7$
VEGA 1	Venus and P/Halley	March 6, 1986	$8.9 \cdot 10^3$
SUISEI	P/Halley	March 8, 1986	$1.5 \cdot 10^5$
VEGA 2	Venus and P/Halley	March 9, 1986	$8 \cdot 10^3$
SAGIGAKE	P/Halley	March 11, 1986	$7 \cdot 10^6$
GIOTTO	P/Halley	March 14, 1986	596
GIOTTO	P/Grigg-Skjellerup	July 10, 1992	200

The reactivation of GIOTTO spacecraft in May 4, 1992 (Project Giotto Extended Mission) has provided new data concerning the interaction of the solar wind with comet P/Grigg-Skjellerup during the encounter with this comet on June 10, 1992. Preliminary results are discussed in [92S].

For the discussion concerning implications of comet research including the history of the solar system and the origin of life see [90H1, 90H2, h].

Acknowledgement

The contributions of M.F. A'Hearn, C. Arpigny, H. Boehnhardt, H. Fechtig, B.G. Marsden and L. Kresák are gratefully acknowledged.

References for 3.3.3

Listed are only references not included in Landold-Börnstein NS Vol. VI/2a 1981.

Proceedings and monographs

(In special references referred as Proceedings).

a Comets in the Post-Halley Era, (Newburn, R.L., Neugebauer, M., Rahe, J., eds.). Dordrecht, The Netherlands: Kluver Academic Publishers (1991).
b Symposium on the Diversity and Similarity of Comets, ESA SP-278, (Rolfe, E.J., Battrick, B., eds.) Brussels 1987.
c 20th ESLAB Symposium on the Exploration of Halley's Comet, ESA SP-250, (Battrick, B., Rolfe, E.J., Reinhard, R., eds.) Paris 1986.
d Halley's Comet, Astron. Astrophys. **187** (1987), Nos. 1-2 (special volume).
e Asteroids, Comets, Meteors II (Lagerkvist, C.-I., Lindblad, B.A., Rickman, H., Lundstedt, H., eds.) Uppsala 1986.
f Asteroids, Comets, Meteors III (Lagerkvist, C.-I., Rickman, H., Lundstedt, H., eds.) Uppsala 1990.
g Comets (Wilkening, L.L., ed.). Tucson: University of Arizona Press (1982).
h Comets and Origin of Life (Ponnamperuma, C., ed.). Dordrecht, The Netherlands: D. Reidel Publ. Co. (1981).
i Physics and Chemistry of Comets (Huebner, W.F., ed). Heidelberg: Springer-Verlag (1990).
j Physical Processes in Comets, Stars and Active Galaxies (Hillebrandt, W., Mayer-Hofmeister, E., Thomas, H.C., eds.). Heidelberg: Springer-Verlag (1987).
k The Comet Nucleus Sample Return Mission, ESA SP-249, Canterbury, UK, 1986.
l Cometary Exploration (Gombosi, T.I., ed.). Budapest: Hungarian Acad. Sci. (1982).
m Workshop on Observations of Recent Comets (Huebner, W.F., Rahe, J., Wehinger, P.A., Konno, I., eds.) Southwest Research Institute, San Antonio, Texas, USA 1990.
n Nature **321** (1986) May, special issue (Comet Halley).
o Science **232** (1986) April 18, special issue (Comet Giacobini-Zinner).

Catalogues

C86B Belyaaev, N.A., Kresák, L., Pittich, E.M., Pushkarev, A.N.: Catalogue of Short-Period Comets. Bratislava: Slovak Academy of Sciences (1986).

C86M Marsden, B.G.: Catalogue of Cometary Orbits (5th edition). Cambridge, Mass.: Smithsonian Astrophys. Observatory (1986).

C89M Marsden, B.G.: Catalogue of Cometary Orbits (6th edition). Cambridge, Mass.: Smithsonian Astrophys. Observatory (1989). (This catalogue appears in updated edition every few years)

Special references

78F Fay, T.D., Wisniewski, W.Z.: Icarus **34** (1978) 1.

80A A'Hearn, M.F., Millis, R.L.: Astron. J. **85** (1980) 1528.

81S Sekanina, Z.: Annual Rev. Earth Planet. Sci. **9** (1981) 113.

82A A'Hearn, M.F.: see Proceeding [g] p. 433.

82D Delsemme, A.H.: see Proceeding [g] p. 85.

82F Festou, M.C., Feldman, P.D., Weaver, H.A., Keller, H.U.: 3rd European IUE Conf. ESA SP-176 (1982) p. 445.

82N Ney, E.P.: see Proceeding [g] p. 323.

82S Svoreň, J.: see Proceeding [l] p. 31.

82W Whipple, F.L.: see Proceeding [l] p. 95.

83A1 A'Hearn, M.F., in: Solar Photometry Handbook (Gennet, R., ed.). Richmond, VA: Willman-Bell (1983) p. 31.

83A2 A'Hearn, M.F., Feldman, P.D., Scleicher, D.G.: Astrophys. J. (Letters) **274** (1983) L99.

83E Everhart, E., Marsden, B.G.: Astron. J. **88** (1983) 135.

84A A'Hearn, M.F., Vanysek, V., Campins H.: International Halley Watch Letter No. 4 (1984) 21.

84L Lüst, Rhea: Astron. Astrophys. **141** (1984) 94.

85H1 Hanner, M.S., Tedesko, E., Veeder, G.J., Lester, D.F., Witteborn, F.G., Bregman, J.D., Gradie, J., Lebofsky, L.: Icarus **64** (1985) 11.

85H2 Hanner, M.S., Knacke, R., Sekanina, Z., Tokunaga, A.T.: Astron. Astrophys. **152** (1985) 177.

85O Ostro, S.J.: Publ. Astron. Soc. Pac. **97** (1985) 877.

85V Vanysek, V., Wolf, M.: Bull. Astron. Inst. Czech. **36** (1985) 267.

85W Wisniewski, W.Z., Zellner, B.: Icarus **63** (1985) 333.

86A A'Hearn, M.F., Birch, P.V., Klinglesmith III, D.A., Hoban, S., Bowers, C., Martin, R.: see Proceeding [c] p. 483.

86B Balsiger, H., Altwegg, K., Buhler, F., Geiss, J., Ghielmetti, A.G., Goldstein, B.E., Goldstein, R., Huntress, W.T., Ip, W.-H., Lazarus, A.J., Meier, A., Neugebauer, M., Rettenmund, U., Rosenbauer, H., Schwenn, R., Sharp, R.D., Shelley, E.G., Ungstrup, E., Young, D.T.: Nature **321** (1986) 330.

86D Delsemme, A.H., in: The Galaxy and the Solar System (Smoluchowski, R., Bahcall, J.N., Matthews, M.S., eds.). Tucson: Univ. of Arizona Press (1986) p. 173.

86G1 Greenberg, J.M.: see Proceeding [e] p. 221.

86G2 Grün, E., Graser, U., Kohoutek, L., Thiele, U., Massone, L., Schwehm, G.: Nature **321** (1986) 144.

86J Jewitt, D.C., Meech, K.J.: see Proceeding [c] Vol. II, p. 47.

86K Krankowski, D., Lammerzahl, P., Herrwerth, I., Woweries, J., Eberhardt, P., Dolder, U., Herrmann, U., Schulte, W., Berthelier, J.J., Illiano, J.M., Hodges, R.R., Hoffman, J.H.: Nature **321** (1986) 326.

86P Pfau, W., Stecklum, B.: Astron. Nachr. **307** (1986) 64.

86S Sekanina, Z., Larson, S.M., Emerson, G., Helin, E.F., Schmidt, R.E.: Proceeding [c] Vol. II, p. 177.

86T Togunaga, A.T., Golisch, W.F., Griep, D.M., Kaminski, C.D., Hanner, M.S.: Astron. J. **92** (1986) 1183, and Astron. J. **96** (1988) 1971.

86W1 Wilhelm, K., Cosmovici, C.B., Delamere, W.A., Huebner, W.F., Keller, H.U., Reitsema, H., Schmidt, H.U., Whipple, F.L.: see Proceeding [c] Vol. II p. 367.

86W2 Wisniewski, W.Z.: IAU Circular No. 4603 (1986).

87B Boehnhardt, H., Fechtig, H.: Astron. Astrophys. **187** (1987) 824.

87E Everhard, E., Marsden, B.G.: Astron. J. **93** (1987) 753.

87F Feldman, P.D., Festou, M.C., A'Hearn, M.F., Arpigny, C., Butterworth, P.S., Cosmovici, C.B., Danks, A.C., Gilmozzi, R., Jackson, W.M., McFadden, L.A., Patriarchi, P., Schleicher, D.G., Tozzi, G.P., Wallis, M.K., Waver, H.A., Woods, T.N.: Astron. Astrophys. **187** (1987) 325.

87G Grün, E., Massone, L., Schwehm, G.: see Proceeding [b] p. 305.

87H Hanner, M.S., Newburn, R.L., Spinrad, H., Veeder, G.J.: Astron. J. **94** (1987) 1081.

87J1 Jessberger, E.K., Kissel, J., Fechtig, H., Krueger, F.R.: see Proceeding [j] p. 36.

87J2 Jewitt, D., Meech, K.: Astron. J. **93** (1987) 1542.

87K Keller, H.U.: see Proceeding [b] p. 447.

87L1 Lamy, P.L., Grün, E., Perrin, J.M.: Astron. Astrophys. **187** (1987) 767.

87L2 Langevin, Y., Kissel, J., Bertaux, J-L., Chassefiere, E.: Astron. Astrophys. **187** (1987) 761.

87N Nakano, S.: Minor Planet Circular No. 12577 (1987).

87S1 Sekanina, Z.: see Proceeding [b] p. 315 and p. 323.

87S2 Svoren, J.: see Proceeding [b] p. 707.

88A A'Hearn, M.F.: Annu. Rev. Earth Planet. Sci. **16** (1988) 273.

88C Combes, M., Moroz, V.I., Crovisier, J., Encrenaz, T., Bibring, J.-P., Grigoriev, A.V., Sanko, N.F., Coron, N., Crifo, J.F., Gispert, R., Bockelée-Morvan, D., Nikolsky, Y.V., Krasnopolsky, V.A., Owen, T., Emerich, C., Lamarre, J.M., Rocard, F.: Icarus **76** (1988) 404.

88D Dollfus, A., Bastien, P., Le Borgne, J.-F., Levasseur-Regourd, A.C., Mukai, T.: Astron. Astrophys. **206** (1988) 348.

88J Jewitt, D., Meech, K.: Astrophys. J. **328** (1988) 974.

88M Millis, R.L., A'Hearn, M.F., Campins, H.: Astrophys. J. **324** (1988) 1194.

89A1 A'Hearn, M.F., Campins, H., Schleicher, D.G., Millis, R.L.: Astrophys. J. **347** (1989) 1155.

89A2 Anders, E., Grevesse, E.: Geoechim. Cosmochim. Acta **53** (1989) 197.

89B Boehnhard, H., Drechsel, H., Vanysek, V., Waha, L.: Astron. Astrophys. **220** (1989) 286.

89D Dollfus, A.: Astron. Astrophys. **213** (1989) 469.

89J1 Jessberger, E.K., Rahe, J., Kissel, J., in: Origin and Evolution of Planetary and Satellite Atmospheres (Atreya, S.K., Pollak, J.B., Matthews, M.S., eds.). Tucson, AZ: Univ. of Arizona Press (1989).

89J2 Jewitt, D., Luu, J.: Astron. J. **97** (1989) 1766.

89M Marsden, B.G.: Harvard Smithsonian Astrophys. Obs. Prepr. Ser. 2905 (1989).

89N Newburn, R.L., Spinrad, H.: Astron. J. **97** (1989) 552.

89R1 Rickman, H.: Adv. Space Res. **9** (1989) 59.

89R2 Roettger, E.E., Feldman, P.D., Festou, M.C., A'Hearn, M.F., McFadden, L.A., Gilmozzi, R.: Icarus **80** (1989) 303.

89S Sagdeev, R.Z., Szegö, K., Smith, B.A., Larson, S., Merenyi, E., Kondor, A., Toth, I.: Astron. J. **97** (1989) 546.

89W Wyckoff, S., Lindholm, E., Wehinger, P.A., Peterson, B.A., Zucconi, J.M., Festou, M.C.: Astrophys. J. **339** (1989) 488.

90A A'Hearn, M.F., Festou, M.C.: see Proceeding [i] p. 69.

90G Grun, E., Jessberger, E.: see Proceeding [i] p. 113.

90H1 Huebner, W.F.: see Proceeding [i] p. 1.

90H2 Huebner, W.F., McKay, Ch.P.: see Proceeding [i] p. 305.

90I Ip, W-H., Axford, I.: see Proceeding [i] p. 177.

90K Keller, H.U.: see Proceeding [i] p. 13.
90L Luu, J., Jewitt, D.: Icarus **86** (1990) 69.
90O1 Osborn, W.H., A'Hearn, M.F., Carsenty, U., Millis, R.L., Schleicher, D.G., Birch, P.V., Moreno, H., Gutierrez-Moreno, A.: Icarus **88** (1990) 228.
90O2 Oort, J.H.: see Proceeding [i] p. 235.
90R Rickman, H., Huebner, W.F.: see Proceeding [i] p. 245.
90W Wyckoff, S.: see Proceeding [m] p. 28.
91B1 Belton, M.J.S., Julian, H.W., Mueller, B.E.A.: Int. Conf. on Asteroids, Comets, Meteors (ACM IV) Flagstaff, AZ, 1991 (abstracts).
91B2 Belton, M.J.S.: see Proceeding [a] p. 691.
91C1 Cravens, T.E.: see Proceeding [a] p. 1211.
91C2 Crovisier, J., Schloerb, F.P.: see Proceeding [a] p. 149.
91D Delsemme, A.H.: see Proceeding [a] p. 377.
91E Encrenaz, T., Knacke, R.: see Proceeding [a] p. 107.
91F1 Fernandes, J.A., Ip, W.-H.: see Proceeding [a] p. 487.
91F2 Flammer, K.R.: see Proceeding [a] p. 1125.
91G Gombosi, T.I.: see Proceeding [a] p. 991.
91H1 Hanner, M.S., Tokunaga, A.T.: see Proceeding [a] p. 67.
91H2 Horanyi, M., Mendis, D.A.: see Proceeding [a] p. 1093.
91H3 Huebner, W.F., Boice, D.C.: Schmidt, H.U., Wegmann, R.: see Proceeding [a] p. 907.
91H4 Hughes, D.W.: see Proceeding [a] p. 825.
91J1 Jaworski, W.A., Tatum, J.B.: Astrophys. J. **377** (1991) 317.
91J2 Jessberger, E.K., Kissel, J.: see Proceeding [a] p. 1075.
91J3 Jewitt, D.: see Proceeding [a] p. 19.
91K1 Kleine, M., Wyckoff, S., Wehinger, P.A., Petrson, B.A.: Bull. Am. Astron. Soc. (abstracts) **23** (1991) 1166.
91K2 Klinger, J.: see Proceeding [a] p. 227.
91K3 Kresák, L.: see Proceeding [a] p. 607.
91K4 Krishna Swamy, K.S.: Astron. Astrophys. **241** (1991) 260.
91L1 Larson, S.M., Edberg, S.J., Levy, D.H.: see Proceeding [a] p. 209.
91L2 Lien, D.J.: see Proceeding [a] p. 1005.
91M1 McDonnell, J.A.M., Lamy, P.L., Pankiewicz, G.S.: see Proceeding [a] p. 1043.
91M2 Möhlmann, D., Kuhrt, E.: see Proceeding [a] p. 761.
91R Rickman, H.: see Proceeding [a] p. 733.
91S1 Samarasinha, N.H., A'Hearn, M.F.: Icarus **93** (1991) 194.
91S2 Sekanina, Z.: Astron. J. **102** (1991) 350.
91S3 Sekanina, Z.: see Proceeding [a] p. 769.
91S4 Szegö, K.: see Proceeding [a] p. 713.
91V Vanysek, V.: see Proceeding [a] p. 879.
91W1 Weissman, P.R.: see Proceeding [a] p. 463.
91W2 Whipple, F.L.: see Proceeding [a] p. 1259.
91Y1 Yeomans, D.K.: see Proceeding [a] p. 3.
91Y2 Yamamoto, T.: see Proceeding [a] p. 361.
92S Schwehm, G.H.: ESA Bulletin No. 72 (1992) 61.
93F Festou, M.C., Rickman, H., West, R.M.: Astronomy Astrophys. Review **4** (1993) 363.

3.3.4 See LB VI/2a

3.3.5 Interplanetary particles and magnetic field

3.3.5.1 Gases of non-solar origin in the solar system

3.3.5.1.1 Introduction

The state of the undisturbed very local interstellar medium (VLISM) with its density, composition, temperature and relative velocity can be determined from measurements of interstellar absorption lines in the light of nearby stars [LB/NS VI/1, subsect. 8.4.1.2, 87C, 90F]. The solar system is moving relative to the VLISM. While the plasma component of the interstellar medium is excluded from the solar system by the interplanetary magnetic field, the neutral interstellar gas flows through the system [70B].

Close to the sun the neutral particles of the VLISM perform Keplerian trajectories under the influence of solar gravitation and radiation pressure. When approaching the sun the neutral gas is subject to ionization by solar ultraviolet radiation, charge exchange with the solar wind and electron collisions. The combination of these processes leads to a characteristic spatial distribution of each neutral species in the heliosphere. Schematic views of the distribution in the ecliptic plane are given for H and He in Figs. 1a and 1b, respectively. The loss processes lead to a density depletion for all species in the vicinity of the sun. For H the radiation pressure generally is comparable to or even exceeds the gravitation thus leading to a wake of the neutral gas on the downwind side of the sun. For He and heavier species the radiation pressure is negligible and thus the gas is focused in a cone on the downwind side of the sun. For reviews see [72A, 74F, 77H, 78T, 90A, 90L, 90M].

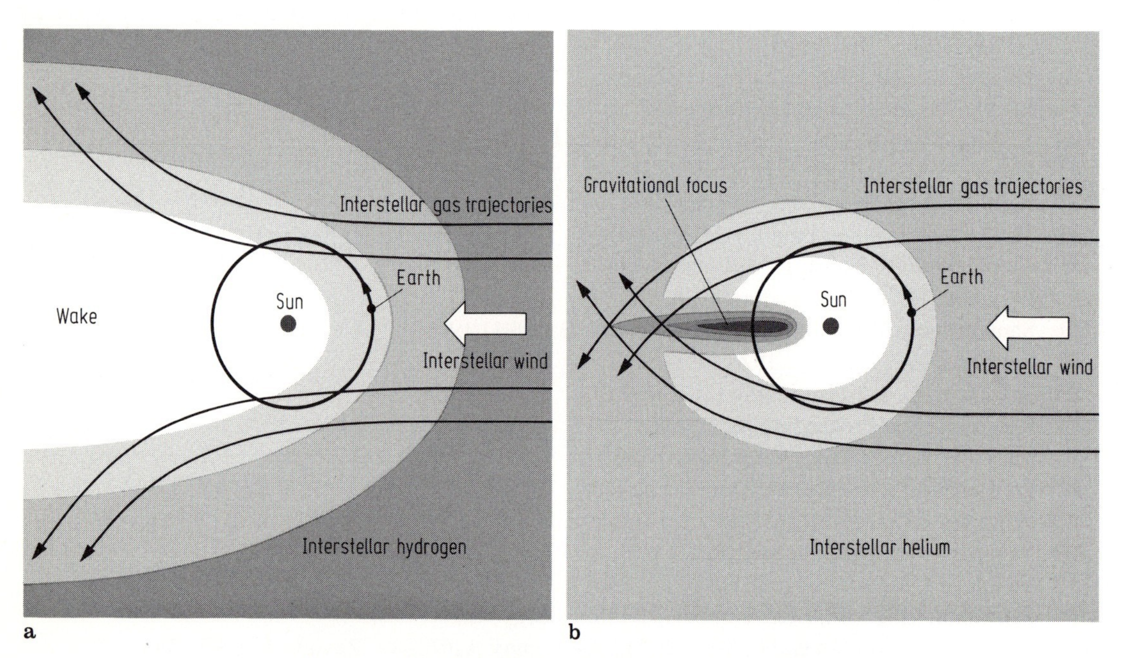

Fig. 1. Cut through the distribution of interstellar neutral gas in the inner solar system in the ecliptic plane. (a) Distribution of neutral hydrogen: the radiation pressure of the sun dominates. A wake is created downwind of the sun. (b) Distribution of helium: the solar gravitation dominates. The interstellar wind is focused on the downwind side of the sun.

3.3.5.1.2 Determination of the interstellar gas distribution in the heliosphere

The distribution of the interstellar gas flow through the heliosphere is determined by model calculations [77H, 78T, 79W, 90L] and by two independent observational techniques listed below. As a result of the comparison of the calculations with the measurements the local interstellar density n_0 and temperature T_0, as well as the speed v_{rel}, heliocentric ecliptical longitude λ and latitude β of the interstellar wind flow in the solar system are derived (for the coordinate system see Fig. 2).

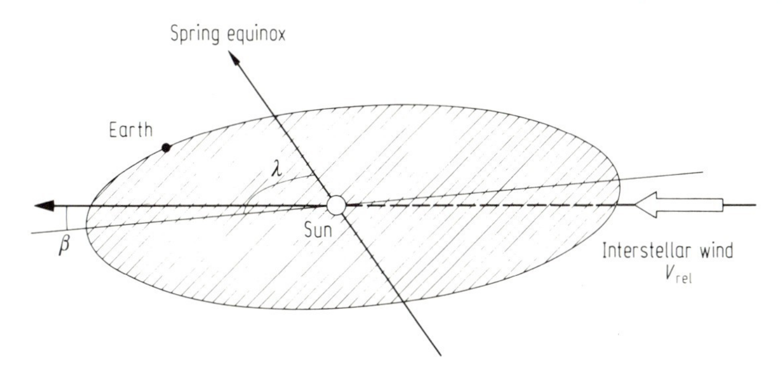

Fig. 2. The interstellar wind in relation to the Earth's orbit. v_{rel} is the relative velocity, λ the longitudinal, and β the latitudinal direction of the interstellar gas flow with respect to the ecliptic.

1) Observation of the directional distribution of the backscattered light of solar emission lines. This sky map of the scattered light is compared with the results of a model calculation of the neutral gas density distribution and the corresponding line-of-sight integral of the light emission in order to derive the VLISM parameters. These techniques have been applied to H using the H I 1216 Å line and to He using the He I 584 Å line. In addition, v_{rel} and T_0 have been derived directly from the line profile of the scattered light for H.

2) Observation of the spatial flux distribution of pickup ions in the solar wind from spacecraft. Upon ionization the interstellar particles are subject to the combined forces of the interplanetary magnetic and $\mathbf{V}_{sw} \times \mathbf{B}$ electric fields and swept away as pickup ions with the solar wind. Model calculations of the pickup ion distribution in the solar wind are given by [76V, 85M, 87I, 88M]. The energy flux density of the pickup ions is proportional to the production of these ions (density $\times$ ionization rate) upstream of the observer. From an earthbound spacecraft the neutral gas distribution in the ecliptic inside the Earth's orbit can be determined during the course of a year. This technique has been applied to He.

Table 1 gives the most recent results for the VLISM parameters. A compilation of the earlier results may be found in [86C, 87A]. The derived values for the parameters are model dependent. In addition, the backscatter results are dependent on the uncertainties in the total intensity and the shape [88C] of the solar spectral lines, while the pickup ion results depend on the uncertainties in the ionization rate [88M, 90R].

The neutral gas distribution for each species varies with the ionization rate and therefore with solar activity [87F] and the solar wind flux [85L]. Except for the focusing cone on the downwind side of the sun the interstellar gas distribution is well described by a cold gas approximation [76V, 77H, 78T]

$$n(r,\theta) = \frac{n_0}{\sin\theta}\left\{\frac{\partial b_1}{\partial r}\exp[-r_i\theta/b_1] + \frac{\partial b_2}{\partial r}\exp[-r_i(2\pi-\theta)/b_2]\right\}$$

with

$$\begin{Bmatrix} b_1 \\ b_2 \end{Bmatrix} = \left[\left(\frac{r}{2}\sin\theta\right)^2 + (1-\mu)\frac{GM_s}{v_{rel}^2}r(1-\cos\theta)\right]^{1/2} \pm \frac{r}{2}\sin\theta$$

where r is the distance from the sun and θ the angular separation from the velocity vector of the sun with respect to the VLISM. To include the cone structure in the model and to allow a determination

Table 1. Parameters of the interstellar wind in the heliosphere as measured for H and He. For definition of the symbols see text.

	Method	n_0 [cm^{-3}]	T_0 [K]	v_{rel} [km/s]	λ	β	Ref.
Hydrogen	Backscatter	0.03 ··· 0.075	7000 ··· 9000	19 ··· 21	70° ··· 72°	4.5° ··· 10.5°	85B, 86C, 87A, 88W, 90A
Helium	Backscatter	0.004 ··· 0.012	5000 ··· 9000	19 ··· 24	71.5° ··· 77.5°	3° ··· 9°	86C, 87A, 88C, 90A
	Pickup ions	0.005 ··· 0.007	7000 ··· 12500				88M, 90M, 90R

of the temperature a more elaborate treatment is necessary [77H, 78T, 79W]. GM_s/r^2 represents the gravitational acceleration of the sun and μ the relative contribution of the radiation pressure. $r_i = r_0^2 \nu_{ion}/v_{rel}$ is a characteristic penetration distance of the interstellar gas with ν_{ion} being the ionization rate at the reference distance r_0. Ionization rates at 1 AU and the typical penetration distances r_i are compiled for various species in Table 2 for low to moderate solar activity.

Table 2. Ionization rates ν_{ion} at 1 AU and penetration distances r_i from the sun for various species of the interstellar gas for v_{rel} = 20 km/sec.

Species	Reaction	ν_{ion} [10^{-7} sec^{-1}]		r_i [AU]	Ref.
		partial	total		
Hydrogen			7.8	5.9	
	$H + h\nu \rightarrow H^+ + e$	0.88			77H
	$H + H^+ \rightarrow H^+ + H$	6.6			77H
	$H + He^{2+} \rightarrow H^+ + He^+$	0.033			77H
	$H + e \rightarrow H^+ + 2e$	0.27		adapted from	89R
Helium			0.995	0.76	
	$He + h\nu \rightarrow He^+ + e$	0.8			73B, 90R
	$He + H^+ \rightarrow He^+ + H$	0.045			90R
	$He + e \rightarrow He^+ + 2e$	0.15			89R
Nitrogen			8.8	6.6	
	$N + h\nu \rightarrow N^+ + e$	2.8			73B, 68M
	$N + H^+ \rightarrow N^+ + H$	6			72A
Oxygen			6.0	4.5	
	$O + h\nu \rightarrow O^+ + e$	3.3			73B
	$O + H^+ \rightarrow O^+ + H$	2.7			72A
Neon			2.93	2.2	
	$Ne + h\nu \rightarrow Ne^+ + e$	2.6			73B, 68M
	$Ne + H^+ \rightarrow Ne^+ + H$	0.33			72A
Argon			6.1	4.6	
	$Ar + h\nu \rightarrow Ar^+ + e$	2.8			73B, 68M
	$Ar + H^+ \rightarrow Ar^+ + H$	3.3			72A

The photoionization rates are based on photon flux densities of about $2 \cdot 10^{10}$ photons cm^{-2}s^{-1} for 910 ··· 375 Å and about $3.6 \cdot 10^{10}$ photons cm^{-2}s^{-1} for 665 ··· 150 Å [73B], which are appropriate for low to moderate solar activity. Most recent measurements of the UV flux which affects the ionization of He have confirmed that it varies linearly with the F10.7-cm solar radio flux. The photon flux density for 575 ··· 50 Å varies typically from $3.6 \cdot 10^{10}$ to $5.5 \cdot 10^{10}$ photons cm^{-2}s^{-1} for F10.7-cm radio flux densities from 0.01 to 0.02 W cm^{-2}Hz^{-1} [89F, 90O]. To adjust for varying solar activity, the UV flux densities and ionization rates may be scaled accordingly. The charge exchange rates are based on solar wind flux densities of about $3.3 \cdot 10^8$ particles cm^{-2}s^{-1} for H^+ and about $1.5 \cdot 10^7$ particles cm^{-2}s^{-1} for He^{2+}. Electron ionization rates correspond to a solar-wind electron distribution which consists of a cold core component with a number density and temperature of about 6.7 cm^{-3} and $1.5 \cdot 10^5$ K, respectively, and a hot halo component with a density and temperature of about 0.3 cm^{-3} and $7 \cdot 10^5$ K, respectively. All values are given for 1 AU.

The extrapolation of the measured interstellar gas parameters to the undisturbed conditions outside the solar system depend on assumptions about the influence of the interface between the heliosphere and the interstellar medium [83R, 84F, 87B]. Depending on the fractional ionization of the interstellar medium, part of the neutral gas may be coupled to the interstellar plasma flow and prevented from entering the heliosphere.

The charge exchange between interstellar neutrals and solar wind ions creates a population of neutral particles with the solar wind velocity which may be used as an additional diagnostic of the interaction between the interstellar gas and the solar wind [90G1, 90G2]. Further indirect information on the composition of the interstellar neutral gas in the heliosphere can be obtained from the composition of the anomalous component of cosmic rays – most probably pickup ions which were accelerated at the solar wind termination shock (see subsect. 3.3.5.3.2).

References for 3.3.5.1

General Reference

a Physics of the Outer Heliosphere (Grzedzielski, S., Page, D.E., eds.). COSPAR Colloquia Series **1** (1990).

Special References

68M McGuire, E.J.: Phys. Rev. **175** (1968) 20.
70B Blum, P.W., Fahr, H.J.: Astron. Astrophys. **4** (1970) 280.
72A Axford, W.I., in: Solar Wind (Sonett, C.P., Coleman, P.J., jr., Wilcox, J.M., eds.). NASA-SP-308 (1972) 609.
73B Banks, P.M., Kockarts, G.: Aeronomy, Part A, New York: Academic Press (1973).
74F Fahr, H.J.: Space Sci. Rev. **15** (1974) 483.
76V Vasyliunas, V.M., Siscoe, G.L.: J. Geophys. Res. **81** (1976) 1247.
77H Holzer, T.E.: Rev. Geophys. Space Phys. **15** (1977) 467.
78T Thomas, G.E.: Annu. Rev. Earth Planet. Sci. **6** (1978) 173.
79W Wu, F.M., Judge, D.L.: Astrophys. J. **231** (1979) 594.
83R Ripken, H.W., Fahr H.J.: Astron. Astrophys. **122** (1983) 181.
84F Fahr, H.J., Ripken, H.W.: Astron. Astrophys. **139** (1984) 551.
85B Bertaux, J.L., Lallement, R., Kurt, V.G., Mironova, E.N.: Astron. Astrophys. **150** (1985) 1.
85L Lallement, R., Bertaux, J.L., Kurt, V.G.: J. Geophys. Res. **90** (1985) 1413.
85M Möbius, E., Hovestadt, D., Klecker, B., Scholer, M., Gloeckler, G., Ipavich, F.M.: Nature **318** (1985) 426.
86C Chassefiere, E., Bertaux, J.L., Lallement, R., Kurt, V.G.: Astron. Astrophys. **160** (1986) 229.
87A Ajello, J.M., Stewart, A.I., Thomas, G.E., Graps, A.: Astrophys. J. **317** (1987) 964.
87B Bleszynski, S.: Astron. Astrophys. **180** (1987) 201.
87C Cox, D.P., Reynolds, R.J.: Annu. Rev. Astron. Astrophys. **25** (1987) 303.
87F Fahr, H.J., Rucinski, D., Nass, H.U.: Ann. Geophys. **5A** (1987) 255.
87I Isenberg, P.A.: J. Geophys. Res. **92** (1987) 1067.
88C Chassefiere, E., Dalaudier, F., Bertaux, J.L.: Astron. Astrophys. **201** (1988) 113.
88M Möbius, E., Klecker, B., Hovestadt, D., Scholer, M.: Astrophys. Space Sci. **144** (1988) 487.
88W Wu, F.M., Gangopadhyay, P., Ogawa, H.S., Judge, D.L.: Astrophys. J. **331** (1988) 1004.
89F Feng, W., Ogawa, H.S., Judge, D.L.: J. Geophys. Res. **94** (1989) 9125.
89R Rucinski, D., Fahr, H.J.: Astron. Astrophys. **224** (1989) 290.
90A Axford, W.I.: see General Reference [a] p. 7.
90F Frisch, P.C.: see General Reference [a] p. 19.
90G1 Gruntman, M.A., Grzedzielski, S., Rucinski, V.B.: see General Reference [a] p. 355.
90G2 Grzedzielski, S., Rucinski, D.: see General Reference [a] p. 367.
90L Lallement, R.: see General Reference [a] p. 49.
90M Möbius, E.: see General Reference [a] p. 345.
90O Ogawa, H.S., Canfield, L.R., McMullin, D., Judge, D.L.: J. Geophys. Res. **95** (1990) 4291.
90R Ratkiewicz, R., Rucinski, D., Ip, W.H.: Astron. Astrophys. **230** (1990) 227.

3.3.5.2 Interplanetary plasma and magnetic field (solar wind)

A wealth of valuable measurements has been performed using various spaceprobes cruising through large parts of the heliosphere: the two *Helios* probes collected data in the distance range between 0.29 and 1.0 AU from the sun for more than 12 years [90S2, 91S]. The *Pioneer 10/11* and the *Voyager 1/2* spaceprobes keep travelling into the outer heliosphere. Pioneer 10 is presently collecting and transmitting data from a solar distance of more than 52 AU [90G]! The *Ulysses* probe has finally left the ecliptic plane and started its journey over the sun's poles [92W].

The theoretical concepts of coronal expansion and solar wind have not changed substantially in the past 10 years since the crucial questions of coronal heating and solar wind acceleration could not yet be answered. It appears now that several different and maybe competing processes are involved, since the various types of solar wind are so intrinsically different:

1. The *fast-type solar wind* usually found in *high-speed streams* emerges from *coronal holes* which are representative of the inactive sun. At times around minimum solar activity large coronal holes cover both polar caps of the sun, with occasional extensions down to the equatorial regions. The plasma properties remain fairly steady throughout a high-speed stream and do not vary much between different fast streams.
2. The *slow solar wind* of the *interstream type* is also typical of times around activity minimum. Its sources are constrained to the warped activity belt of about 40° in latitude which encircles the sun close the equator, thus separating the polar coronal holes from each other. This is the regime of bright *coronal streamers*. The strikingly low helium content ($< 2\%$) indicates a larger release height in the gravitationally stratified solar atmosphere.
3. The *slow solar wind* of the *maximum type* is found to emerge at high solar activity from substantially larger areas often located far from coronal streamers. It is highly variable and usually contains a significant fraction of helium (some 4%), thus indicating its release at lower altitudes in the corona. Otherwise, the properties are similar to the interstream-type slow wind.
4. The *coronal mass ejecta* in conjunction with transient events in the corona may exhibit some of the following signatures: drastic increase in the helium content (up to 40%), occurrence of high fluxes of Fe^{16+} or He^{+} ions, depressions of ion and electron temperatures, bi-directional streaming of electrons and suprathermal particles, high magnetic field strength with low variance, very low plasma-beta, large-scale rotation of the magnetic field vector indicating the occurrence of *magnetic clouds*.

The basic differences and similarities between the first three types show up in more detail in the data collection of Table 1 (from [90S1, 90M] in [90S2]). The data were collected by the Helios solar probes between December 1974 and December 1976, i.e. around solar activity minimum in early 1976. After coarse binning of the data average numbers for slow ($< 400\,\mathrm{km\,s^{-1}}$) and fast ($> 600\,\mathrm{km\,s^{-1}}$) solar wind, plus the grand averages were calculated. Very few transients occurred in that period. All parameters involving the proton density n_p were normalized to 1 AU assuming a r^{-2} dependence. Only those temperature data obtained between $r = 0.9$ and 1.0 AU were included. For reference, the values assembled in [77F] on the basis of IMP (Interplanetary Monitoring Platform) 6/7/8 data at 1 AU taken in the 3.5 years before July 1974 are listed in parallel. (Note that all quantities involving proton densities obtained by the IMP satellites were multiplied by a factor of 0.70 which was determined from direct cross calibrations). For the magnetic field strength and the absolute values of its vector components the dependence on distance r[AU] to the sun is given in the form $M\,r^{-s}$.

The two data sets show some minor differences. These may be due to the slightly different limits applied for the IMP data ($< 350\,\mathrm{km\,s^{-1}}$ for slow and $> 650\,\mathrm{km\,s^{-1}}$ for fast wind, respectively). Also, the selected time periods do not coincide. After all, the differences indicate the uncertainties typically inherent to these data.

The variations of the average plasma parameters with distance from the sun were studied in much detail in recent years. In the *inner heliosphere* where the solar wind is streaming off freely the actual profiles depend strongly on the type of solar wind in question [90S1, 91M, 90P]. With increasing distance from the sun solar wind streams of different speeds (from different coronal sources) interact with each other. Finally, all solar wind plasma will have been processed by *stream-stream interactions*. Thus, in the *outer heliosphere* the original signatures are lost. Naturally, the radial profiles are strongly affected as well [84B, 92G].

With the advent of new instrumentation, in particular of satellite-born *coronagraphs* and *EUV-telescopes* and *spectrometers*, the diagnostics of transient phenomena in the solar atmosphere dramatically improved. Most prominent was the discovery of *huge coronal mass ejections* (CMEs) which in turn may drive *interplanetary shock waves* through the ambient solar wind. Several recent reviews document the enormous progress in this field: [85R, 86S, 87K, 88H, 88K, 92J].

The understanding of the role that *solar flares* play in such transient processes has also improved. There is evidence that flares are just consequences of a more fundamental process rather than causes themselves, since at times a CME starts rising some tens of minutes before an eventually associated flare erupts [91H]. Flares appear to be side effects of larger-scale magnetic changes which in parallel or even earlier may trigger CMEs.

There are many cases of CMEs with no associated flare. Also, the number of flares (including the smaller ones) exceeds that of notable CMEs. Any CME faster than $400\,\mathrm{km\,s^{-1}}$ close to the sun gives rise to a shock wave observable at any point within the angular extent of the CME [85S]. This association works whether a flare was involved or not.

The properties of 998 CMEs observed between March 1979 and December 1981 were examined in a statistical study by [85H]. Table 2 summarizes the results.

Table 1. Average values of basic solar wind parameters. The Helios data were taken from various references as indicated. The 1-AU-data (from the IMP satellites) stem from the collection in [77F].

v_p — proton bulk speed
n_p — proton density
$F_\mathrm{p} = n_\mathrm{p} v_\mathrm{p}$ — proton flux density
$M_\mathrm{p} = n_\mathrm{p} m_\mathrm{p} v_\mathrm{p}^2$ — proton momentum flux density
$A = n_\mathrm{p}/n_\alpha$ — proton to helium ion density ratio
$E_\mathrm{k} = 1/2 n_\mathrm{p} m_\mathrm{p} v_\mathrm{p}^3$ — kinetic energy flux density
$E_\mathrm{g} = n_\mathrm{p} v_\mathrm{p} G m_\mathrm{p} M_\mathrm{s}/R_\mathrm{s}$ — potential energy flux density
$E_\mathrm{thp} = 5/2 n_\mathrm{p} v_\mathrm{p} k T_\mathrm{p}$ — proton enthalpy flux density
$E_\mathrm{the} = 5/2 n_\mathrm{e} v_\mathrm{e} k T_\mathrm{e}$ — electron enthalpy flux density
Q_e — electron heat flux density
Q_p — proton heat flux density
E_a — Alfvén wave energy flux density
$E_\mathrm{t} = E_\mathrm{k} + E_\mathrm{g}$ — total energy flux density
T_p — proton temperature
$T_\mathrm{p}(\mathrm{perp})/T_\mathrm{p}(\mathrm{par})$ — proton temperature anisotropy
T_e — electron temperature
$T_\mathrm{e}(\mathrm{perp})/T_\mathrm{e}(\mathrm{par})$ — electron temperature anisotropy
T_α — helium ion temperature
$T_\alpha(\mathrm{perp})/T_\alpha(\mathrm{par})$ — helium ion temperature anisotropy
T_α/T_p — helium ion to proton temperature ratio
L_p — angular momentum flux (particles)
L_m — angular momentum flux (magnetic field)
$|B|$ — magnetic field strength
$|B_r|$ — magnetic field, radial component
$|B_\theta|$ — magnetic field, out-of-ecliptic component
$|B_\phi|$ — magnetic field, azimuthal component

Parameter	Slow wind (< 400 km s^{-1})		Fast wind (> 600 km s^{-1})		All data		
	Helios	1-AU	Helios	1-AU	Helios	1-AU	Ref.
v_p [km s^{-1}]	(348)	(327)	(667)	(702)	481	468	83S
n_p [cm^{-3}]	10.7	8.3	3.0	2.73	6.80	6.1	83S
F_p [10^8 cm^{-2}s^{-1}]	3.66	2.7	1.99	1.90	2.86	2.66	83S
M_p [10^8 dyn cm^{-2}]	2.12		2.26		2.15	2.03	83S
A	0.025	0.038	0.036	0.048	0.032	0.047	83S
E_k [erg cm^{-2}s^{-1}]	0.37	0.25	0.76	0.79	0.52	0.49	83S
E_g [erg cm^{-2}s^{-1}]	1.17	0.87	0.95	0.60	0.91	0.85	83S
E_{thp} [10^{-3} erg cm^{-2}s^{-1}]	11	3.0	23	16	16	11	83S
E_{the} [10^{-3} erg cm^{-2}s^{-1}]		11		7		12.6	83S
Q_e [10^{-3} erg cm^{-2}s^{-1}]	6.0	2.7	8.0	2.2		4.3	90P
Q_p [10^{-3} erg cm^{-2}s^{-1}]	0.08	0.029	0.3	0.16		0.13	84M
E_α [10^{-3} erg cm^{-2}s^{-1}]		0.7	14	6.7		5.7	83S
E_t [erg cm^{-2}s^{-1}]	1.55	1.18	1.43	1.51	1.45	1.46	83S
T_p [10^3 K]	55	34	280	230		120	82M1
$T_p(\text{perp})/T_p(\text{par})$	1.7		1.2			1.5	82M1
T_e [10^3 K]	190	130	130	100		140	90P
$T_e(\text{perp})/T_e(\text{par})$	1.2		1.6			1.18	90P
T_α [10^3 K]	170	110	730	1420		580	82M2
$T_\alpha(\text{perp})/T_\alpha(\text{par})$	1.4		1.3			1.3	82M2
T_α/T_p	2.9	3.2	3	6.2		4.9	82M2
L_p [10^{30} dyn cm sr^{-1}]	2.14		−0.56		0.2 ··· 0.3		83P
L_m [10^{30} dyn cm sr^{-1}]	0.14		0.15		0.15		83P
	M	s	M	s	M	s	
$\lvert B \rvert$ [nT]	3.45	1.64	3.28	1.86	3.29	1.84	90M
$\lvert B_r \rvert$ [nT]	2.47	2.02	2.77	1.97	2.77	1.97	90M
$\lvert B_\theta \rvert$ [nT]	1.95	1.32	1.44	1.34	1.59	1.32	90M
$\lvert B_\phi \rvert$ [nT]	2.96	1.07	2.19	1.13	2.43	1.14	90M

Table 2. Average properties of 998 CMEs observed between March 1979 and December 1981 [85H].

	Average	Range
Speed of the leading edge	$470\,\mathrm{km\,s^{-1}}$	$50 \cdots 1680\,\mathrm{km\,s^{-1}}$
Total ejected mass	$4.1 \cdot 10^{15}$ g	$2 \cdot 10^{14} \cdots 4 \cdot 10^{16}$ g
Total kinetic energy	$3.5 \cdot 10^{30}$ erg	$1 \cdot 10^{29} \cdots 6 \cdot 10^{31}$ erg
Average angular span	45°	$2° \cdots 360°$
Average occurrence rate	1.8 per day	

References for 3.3.5.2

77F Feldman, W.C., in: The Solar Output and its Variations (White, O.R., ed.). Boulder, USA: Colorado Associated University Press (1977) 351.

82M1 Marsch, E., Mühlhäuser, K.-H., Schwenn, R., Rosenbauer, H., Pilipp, W., Neubauer, F.M.: J. Geophys. Res. **87** (1982) 52.

82M2 Marsch, E., Mühlhäuser, K.-H., Rosenbauer, H., Schwenn, R., Neubauer, F.M.: J. Geophys. Res. **87** (1982) 35.

83P Pizzo, V., Schwenn, R., Marsch, E., Rosenbauer, H., Mühlhäuser, K.-H., Neubauer, F.M.: Astrophys. J. **271** (1983) 335.

83S Schwenn, R., in: Solar Wind Five (Neugebauer, M., ed.), NASA Conf. Publ. 2280, Pasadena, California, USA (1983) 489.

84B Burlaga, L.F.: Space Sci. Rev. **39** (1984) 255.

84M Marsch, E., Richter, A.K.: J. Geophys. Res. **89** (1984) 6599.

85H Howard, R.A., Sheeley jr., N.R., Koomen, M.J., Michels, D.J.: J. Geophys. Res. **90** (1985) 8173.

85R Richter, A.K., Hsieh, K.C., Luttrell, A.H., Marsch, E., Schwenn, R., in: Collisionless Shocks in the Heliosphere: Reviews of Current Research (Tsurutani, B.T., Stone, R.G., eds.) **35**, American Geophysical Union, Washington DC, USA (1985) 33.

85S Sheeley jr., N.R., Howard, R.A., Koomen, M.J., Michels, D.J., Schwenn, R., Mühlhäuser, K.-H., Rosenbauer, H.: J. Geophys. Res. **90** (1985) 163.

86S Schwenn, R.: Space Sci. Rev. **44** (1986) 139.

87K Kahler, S.: Rev. Geophys. **25** (1987) 663.

88H Hundhausen, A.J., in: Proc. Sixth International Solar Wind Conference (Pizzo, V.J., Holzer, T.E., Sime, D.G., eds.) **1**, National Center for Atmospheric Research, Boulder, USA (1988) 181.

88K Kahler, S.W., in: Proc. Sixth International Solar Wind Conference (Pizzo, V.J., Holzer, T.E., Sime, D.G., eds.) **1**, National Center for Atmospheric Research, Boulder, USA (1988) 215.

90G Grzedzielski, S., Page, D.E. (eds.): Physics of the Outer Heliosphere, COSPAR Colloquia Series **1**. Oxford: Pergamon Press (1990).

90M Mariani, F., Neubauer, F.M., in: [90S2] p. 183.

90P Pilipp, W.G., Miggenrieder, H., Mühlhäuser, K.-H., Rosenbauer, H., Schwenn, R.: J. Geophys. Res. **95** (1990) 6305.

90S1 Schwenn, R., in: [90S2] p. 99.

90S2 Schwenn, R., Marsch, E. (eds.): Physics of the Inner Heliosphere, I. Large-Scale Phenomena. Berlin: Springer-Verlag (1990).

91H Harrison, R.A.: Adv. Space Res. **11** (1991) 25.

91M Marsch, E., in: [91S] p. 45.

91S Schwenn, R., Marsch, E. (eds.): Physics of the Inner Heliosphere, II. Particles, Waves and Turbulence. Berlin: Springer-Verlag (1991).

92G Gazis, P.R., Barnes, A., Mihalov, J.D., Lazarus, A.J., in: Solar Wind Seven, COSPAR Colloquia Series **3** (Marsch, E., Schwenn, R., eds.). Oxford: Pergamon Press (1992) 179.
92J Jackson, B.V., in: Solar Wind Seven, COSPAR Colloquia Series **3** (Marsch, E., Schwenn, R., eds.). Oxford: Pergamon Press (1992) 623.
92W Wenzel, K.P., Marsden, R.G., Page, D.E., Smith, E.J.: Astron. Astrophys. Suppl. Ser. **92** (1992) 207.

3.3.5.3 Energetic particles in interplanetary space

3.3.5.3.1 Modulation of galactic cosmic rays

See LB VI/2a p. 251

Additional direct measurements of the radial gradient: At relativistic energies (protons) $\approx 2\%$ per AU between 1 and 40 AU [90L].

3.3.5.3.2 Anomalous component of low energy cosmic rays

See LB VI/2a p. 252

Hump of the cosmic ray energy spectrum of predominantly N, O, Ne in the energy range $3 \cdots 20$ MeV/nucleon with unusual composition (C/O < 0.1); flat energy spectrum of helium between $10 \cdots 80$ MeV/nucleon at quiet times. In the outer solar system possibly also H and C observed [90C1, 90C2].

New measurements:
Heliocentric radial gradient of anomalous oxygen between 1 and 22 AU:
14% per AU at $7 \cdots 25$ MeV/ nucleon [90C3].

Heliocentric latitudinal gradient of anomalous ^{4}He and ^{16}O:
$(+2.1 \pm 0.3)\%$ per degree at $11 \cdots 20$ MeV/nucleon for $1975 \cdots 1977$ [89M],
-4.3% per degree for He at $20 \cdots 25$ MeV/nucleon for 1987 [90C3],
-5.6% per degree for O at $7 \cdots 25$ MeV/nucleon for 1987 [90C3].

Ionization state, indirectly inferred from (1) modulation effects in the heliosphere [77M, 80K], (2) propagation in the earth's magnetic field [89O, 90B]: ionic charge of oxygen, neon ≤ 2.

Trapped anomalous component ions in the magnetosphere [90A].

As a possible source is also discussed:
Interstellar neutral particles ionized in and picked up by solar wind and further accellerated in the outer solar system at the termination shock [81P, 86J].

Numerical solution of the convection-diffusion equation including particle drift effects and accelleration at the termination shock of the heliosphere [85P, 86J].

Composition of the local interstellar medium derived from anomalous cosmic ray composition [90C2].

3.3.5.3.3 see LB VI/2a

3.3.5.3.4 Coronal propagation and injection

See LB VI/2a p. 255

Variations of the number of the solar flare particle events and intensity-time profiles with respect to heliolongitude are also explained by shock accelleration with different shock geometries [88C].

New calculation of coronal diffusion coefficients [89W].

Solar injection time constants: Flare particles are injected near the preferred connection region not instantaneously but over some finite time interval (typical values are 1 hr for 10 MeV protons [89W]. Compilation of time constants [89W].

3.3.5.3.5 Solar flare particle composition and charge state

Large intensity solar particle events: Average abundance of elements from He to Ni in large solar particle events is organized with first ionization potential (FIP), low FIP elements are overabundant (factor 3 to 5) with respect to local galactic or photospheric abundances [81M]. At energies below 10 MeV/nucleon large variations from flare to flare [71A, 86M2] and within each individual flare [74V, 76O].

The variation of the composition within each event, before the time of maximum intensity, is possibly due to a rigidity-dependent interplanetary mean free path [74V, 78S].

^{3}He-rich flares: Abnormally large abundances of ^{3}He in weak flares ($^3\mathrm{He}/^4\mathrm{He} \approx 0.01 \cdots \geq 1$) [75H1, 75H2, 75S].

He-rich flares show heavy-ion enrichment, the enrichment increasing with nuclear charge number [86M1], the event-to-event variations in ^{3}He and heavy ions are uncorrelated [86M1].

^{3}He-rich flares are correlated with solar electrons [85R].

Explanation of the overabundance of ^{3}He is given in terms of resonant [78F] and non-resonant [83V] heating of the ^{3}He by ion cyclotron waves.

Ionization states: Direct measurement of ionization state abundance of solar flare particles (C, N, O, Ne, Mg, Si, Fe) below 2 MeV/nucleon [77S, 81H, 84L]. Significant differences of Fe and Si mean ionic charge Q in large flares ($Q \approx 14$ for Fe) and small (^{3}He-rich) flares ($Q \approx 20$ for Fe) [84K, 87L]. Difference interpreted as due to higher temperature of $\approx 10^7$ K in the source region [87L].

Isotopic abundances of solar flare nuclei [79D, 79M]: $^{20}\mathrm{Ne}/^{22}\mathrm{Ne} \approx 7.7$.

3.3.5.3.6 Corotating energetic particle events

See LB VI/2a p. 256

Further measurements of the heliocentric gradient [81C].

Energy spectra of energetic particles in corotating events [79G].

References for 3.3.5.3

71A Armstrong, T.P., Krimigis, S.M.: J. Geophys. Res. **76** (1971) 4230.
74V Van Allen, J.A., Venkatarangan, P., Venkatesan, D.: J. Geophys. Res. **79** (1974) 1.
75H1 Hovestadt, D., Klecker, B., Gloeckler, G., Fan, C.Y.: Proc. 14th Int. Cosmic Ray Conf. **5** (1975) 1613.
75H2 Hurford, G.J., Stone, E.C., Vogt, R.E.: Proc. 14th Int. Cosmic Ray Conf. **5** (1975) 1624.
75S Serlemitsos, A.T., Balasubrahmanyan, V.K.: Astrophys. J. **198** (1975) 195.
76O O'Gallagher, J.J., Hovestadt, D., Klecker, B., Gloeckler, G., Fan, C.Y.: Astrophys. J. **209** (1976) L97.
77M McKibben, R.B.: Astrophys. J. **217** (1977) L113.
77S Sciambi, R.K., Gloeckler, G., Fan, C.Y., Hovestadt, D.: Astrophys. J. **214** (1977) 316.
78F Fisk, L.A.: Astrophys. J. **224** (1978) 1048.
78S Scholer, M., Hovestadt, D., Klecker, B., Gloeckler, G. Fan, C.Y.: J. Geophys. Res. **83** (1978) 3349.
79D Dietrich, W.F., Simpson, J.A.: Astrophys. J. **231** (1979) L91.
79G Gloeckler, G., Hovestadt, D., Fisk, L.A.: Astrophys. J. **230** (1979) L191.
79M Mewaldt, R.A., Spalding, J.D., Stone, E.C., Vogt, R.E.: Astrophys. J. **231** (1979) L97.
80K Klecker, B., Hovestadt, D., Gloeckler, G., Fan, C.Y.: Geophys. Res. Letters **7** (1980) 1033.
81C Christon, S.P.: J. Geophys. Res. **86** (1981) 8852.
81H Hovestadt, D., Gloeckler, G., Höfner, H., Klecker, B., Fan, C.Y., Fisk, L.A., Ipavich, F.M., O'Gallagher, J.J., Scholer, M.: Adv. Space Res. **1** (1981) 61.
81M Meyer, J.P.: Proc. 17th Int. Cosmic Ray Conf. **3** (1981) 145.
81P Pesses, M.E., Jokipii, J.R., Eichler, D.: Astrophys. J. **246** (1981) L85.
83V Varvoglis, H., Papadopoulos, K.: Astrophys. J. **270** (1983) L95.
84K Klecker, B., Hovestadt, D., Gloeckler, G., Ipavich, F.M., Scholer, M., Fisk, L.A.: Astrophys. J. **281** (1984) 458.
84L Luhn, A., Klecker, B., Hovestadt, D., Gloeckler, G., Ipavich, F.M., Scholer, M., Fan, C.Y., Fisk, L.A.: Adv. Space Res. **4** (1984) 161.
85P Potgieter, M.S., Fisk, L.A., Lee, M.A.: Proc. 19th Int. Cosmic Ray Conf. **4** (1985) 180.
85R Reames, D.V., van Rosenvinge, T.T., Lin, R.P.: Astrophys. J. **292** (1985) 716.
86J Jokipii, J.R.: J. Geophys. Res. **91** (1986) 2929.
86M1 Mason, G.M., Reames, D.V., Klecker, B., Hovestadt, D., von Rosenvinge, T.T.: Astrophys. J. **303** (1986) 849.
86M2 McGuire, R.E., von Rosenvinge, T.T., McDonald, F.B.: Astrophys. J. **301** (1986) 938.
87L Luhn, A. Klecker, B., Hovestadt, D., Möbius, E.: Astrophys. J. **317** (1987) 951.
88C Cane, H.V., Reames, D.V., von Rosenvinge, T.T.: J. Geophys. Res. **93** (1988) 9555.
89M McKibben, R.B.: J. Geophys. Res. **94** (1989) 17021.
89O Oschlies, K., Beaujean, R., Enge, W.: Astrophys. J. **345** (1989) 776.
89W Wibberenz, G., Kecskemety, K., Kunow, H., Somogyi, A., Iwers, B., Logachev, YU.I., Stolpovskii, V.G.: Solar Phys. **124** (1989) 353.
90A Adams, J.H., Tylka. A.J.: Proc. 21th Int. Cosmic Ray Conf. **6** (1990) 172.
90B Biswas, S., Durgaprasad, N., Mitra, B., Singh, R.K.: Astrophys. J. **359** (1990) L5.
90C1 Christian, E.R., Cummings, A.C., Stone, E.C.: Proc. 21th Int. Cosmic Ray Conf. **6** (1990) 186.
90C2 Cummings, A.C., Stone, E.C.: Proc. 21th Int. Cosmic Ray Conf. **6** (1990) 202.
90C3 Cummings, A.C., Mewaldt, R.A., Stone, E.C., Webber, W.R.: Proc. 21th Int. Cosmic Ray Conf. **6** (1990) 206.
90L Lopate, C., McKibben, R.B., Pyle, K.R., Simpson, J.A.: Proc. 21th Int. Cosmic Ray Conf. **6** (1990) 128.

3.4 Abundances of the elements in the solar system

3.4.1 Introduction

In the past it has been assumed that the Sun, the planets and all other objects of the solar system formed from a gaseous nebula with well-defined chemical and isotopic composition. Recent findings of comparatively large and widespread variations in oxygen isotopic compositions have cast doubt upon this assumption (see [88T] and references). Additional evidence of incomplete mixing in the primordial solar nebula is provided by isotopic anomalies of a variety of elements in refractory inclusions of carbonaceous chondrites ([80B] and references therein) and by the detection of huge isotope anomalies of some heavy elements in meteoritic silicon-carbide (SiC) grains (e.g., [88Z]). A good example for such isotope anomalies is given in Fig. 1 (data from [92R]), where the unusal isotopic composition of Nd in a SiC-grain from the Murchison carbonaceous chondrite is shown. The Nd-isotopic compositions of all other solar-system materials analysed (i.e., terrestrial, lunar and meteoritic samples) are indistinguishable within the scale of Fig. 1 and all points would fall on the line designated "average solar system". The basic agreement of the measured composition and the calculated s-process composition (see subsect. 3.4.9 on nucleosynthetic components) demonstrates the presence of material of distinct nucleosynthetic origins at the time of accretion of meteorite parent bodies, about $4.55 \cdot 10^9$ years ago. However, such isotope anomalies are confined to a very small fraction (much less than 1%) of the bulk of a meteorite, i.e., this material is truly exotic. As it is, in addition, likely that the more widespread oxygen isotope anomalies were produced by fractionation processes within the solar nebula, it is still a reasonable working hypothesis that the bulk of the matter of the solar system formed from a chemically and isotopically uniform reservoir, the primordial solar nebula. The composition of this nebula, the average solar system composition, is well known and carries the signatures of a variety of nucleosynthetic processes in stellar environments. The elemental composition of the solar system is similar to that of many other stars in particular with respect to the relative abundances of the non-gaseous elements. In this sense one can speak of cosmic abundances ([88W] and cited literature).

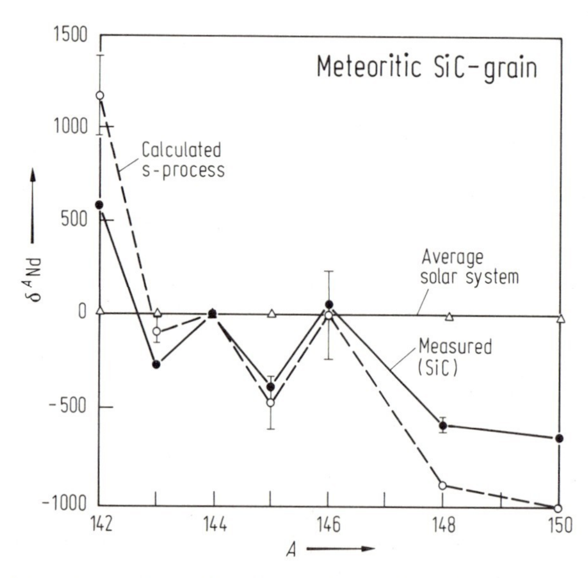

Fig. 1. Nd-isotope anomaly in a meteoritic SiC-grain (Murchison). The deviation of the Nd-isotopic composition from normal is given in per-mil (δ). All ratios are normalized to ^{144}Nd. Full symbols are measured ratios. Data from [92R]. Error bars are in most cases smaller than symbol sizes. Calculated s-process productions are indicated (see text). All previous analyses of Nd-isotopes in terrestrial, lunar or meteorite samples fall on the line marked "average solar system" which is used for normalization.

More than 99.8% of the present solar system matter is contained in the Sun. A representative composition of the solar system therefore implies a knowledge of the composition of the Sun. The most reliable solar elemental abundances are obtained from studies of absorption-line spectra of the solar photosphere. The quality of these data has continually improved and the most recent compilation by Anders and Grevesse [89A] lists 37 elements, determined in the photosphere of the Sun, with errors below 25% (0.1 dex in logarithmic scale). Of the solid materials available for analysis, i.e., primarily meteorites, there is one group of primitive meteorites, CI-chondrites (see below) that comes compositionally very close to the Sun, except for H, Li, Be, C, N, O, and rare gases. This remarkable coincidence justifies the use of the generally more precise meteorite data for estimating solar system abundances, in particular in those cases in which solar data are non-existent or are difficult to obtain. A better understanding of the importance of the CI-meteorites requires a few remarks on meteorites, in general (see [88P]).

3.4.2 – 3.4.4 see LB VI/2a

3.4.5 Cosmochemical classification of meteorites

Differences in volatile-element contents among meteorites reflect the importance of physical processes in the early solar nebula. In cosmochemistry it is usual practice to arrange elements according to volatility. A convenient measure for volatility is the condensation temperature. Condensation temperatures are calculated assuming a cooling gas of solar composition at a given pressure under equilibrium conditions. The temperature at which 50% of an element is contained in solid phases and 50% is still remaining in the gas is called the 50% condensation temperature of the element. Condensation temperatures are reasonably well known for major elements, but are, in some cases, rather uncertain for trace elements because of a lack of knowlege of relevant thermodynamic data. A compilation of condensation temperatures can be found in [85W]. According to their condensation temperatures four groups of elements can be distinguished:

1) The refractory component makes up about 5% of the condensible matter. Major elements are: Al, Ca and Ti. Trace elements are classified into lithophiles, associated with silicates and oxides: Be, Sc, V, Sr, Y, Zr, Nb, Ba, REE (= Rare Earth Elements), Hf, Ta, Th and U, and siderophiles, contained in the metal phase: Mo, Ru, Rh, W, Re, Os, Ir and Pt.
2) Mg-silicates, condensing as the minerals forsterite (Mg_2SiO_4), and enstatite ($MgSiO_3$) and metal (FeNi-alloy) represent the major fraction of condensible matter. The condensation temperatures of forsterite and FeNi-metal are very similar. Elements condensing with these components are: Si, Mg, Cr and Fe, Ni, Co and Pd.
3) Moderately volatile elements with condensation temperatures below Mg-silicates and FeNi and above, but including, sulfur as FeS. Lithophile elements: Li, Mn, P, Rb, K, Na, F, Zn. Siderophile elements: Au, As, Ag, Ga, Sb, Ge, Sn and Te. Chalcophile elements, associated with sulfides: Cu, Se and S.
4) Highly volatile elements with condensation temperatures below FeS. Lithophile elements: Cl, Br, I, Cs, Tl. Chalcophile elements: Pb, In, Bi, Cd and Hg. Rare gases (He, Ne, Ar, Kr and Xe) should also be included here.

3.4.6 Classification of solar system materials

Meteorites recovered from the surface of the Earth have only spent a few million years as individual bodies in space as determined from the abundance of certain cosmic-ray-produced isotopes, mostly rare gases. Before that meteorites were buried in the interior of larger planetesimals. Two different types of such planetesimals may be distinguished, those which were once molten and thus differentiated into an iron core and a silicate mantle and those which were not heated sufficiently to reach melting temperatures. Meteorites from unmelted planetesimals are called *chondritic* meteorites because most of them contain mm-sized spherules, so-called chondrules, which were molten in space, before accretion onto the meteorite parent body. Chondritic meteorites have preserved the bulk composition of their parent planetesimal and their composition should therefore reflect the average solar system composition. There are, however, some chemical variations among chondritic meteorites indicating processes that occurred in the solar nebula such as incomplete condensation, evaporation, preferred accumulation and separation of metal by magnetic forces, differential movement of fine vs. coarse grained material, etc.

A variety of meteorite types is derived from differentiated planetesimals, *iron*-meteorites from the core (or segregated metal-pods), *stony-irons* from the core-mantle boundary and *basaltic* meteorites from the crust of the parent planetesimal. These meteorites are, obviously, less useful in deriving average solar system abundances.

3.4.7 The composition of chondritic meteorites

3.4.7.1 General remarks

The definition of undifferentiated meteorites implies that compositional differences among them reflect processes in the solar nebula, during or before condensation, and/or before accretion. The large variety of chondritic meteorites is an important source of information for processes and conditions in the early solar system. Chondritic meteorites comprise three major groups:

(1) *carbonaceous* chondrites, generally rather oxidized with little or no metallic iron;
(2) *ordinary* chondrites with iron as metal and oxide in silicates;
(3) *enstatite* chondrites with very little oxidized iron.

The large number of meteorites recovered from Antarctica and dry desert areas has recently led to the recognition of additional groups of chondritic meteorites with properties intermediate between (1) and (3).

The most important chemical differences among the groups of chondritic meteorites are:

(a) variations in refractory element contents (e.g., Ca), although relative abundances among refractory elements appear to be constant;
(b) some variations in Mg/Si ratios;
(c) variable total Fe-contents;
(d) variations in volatile element contents.

In Fig. 2 variations of several diagnostic elements, representing the above defined cosmochemical groups, are shown through the traditional groups of chondritic meteorites, from oxidized to reduced, as indicated by the O/Si-ratios. For comparison, the solar photospheric abundances are shown. It is found that CI-chondrites, a subgroup of the group of carbonaceous chondrites, fit the solar abundances best (in CI, C stands for carbonaceous and I for Ivuna, a member of this group; however, many authors still use the old designation as C1-meteorites). As shown in Fig. 2 the CI-chondrites have solar refractory element (e.g., Ca) to Si ratios, solar Mg/Si and Fe/Si ratios and are not depleted in moderately volatile elements (Na, Zn and S). The lower Mn/Si ratio in the solar photosphere is one of the few cases where the match between solar and meteoritic abundances is not within 20% (see Fig. 3). The CI-chondrites will thus be used as the prime source for deriving average solar system abundances.

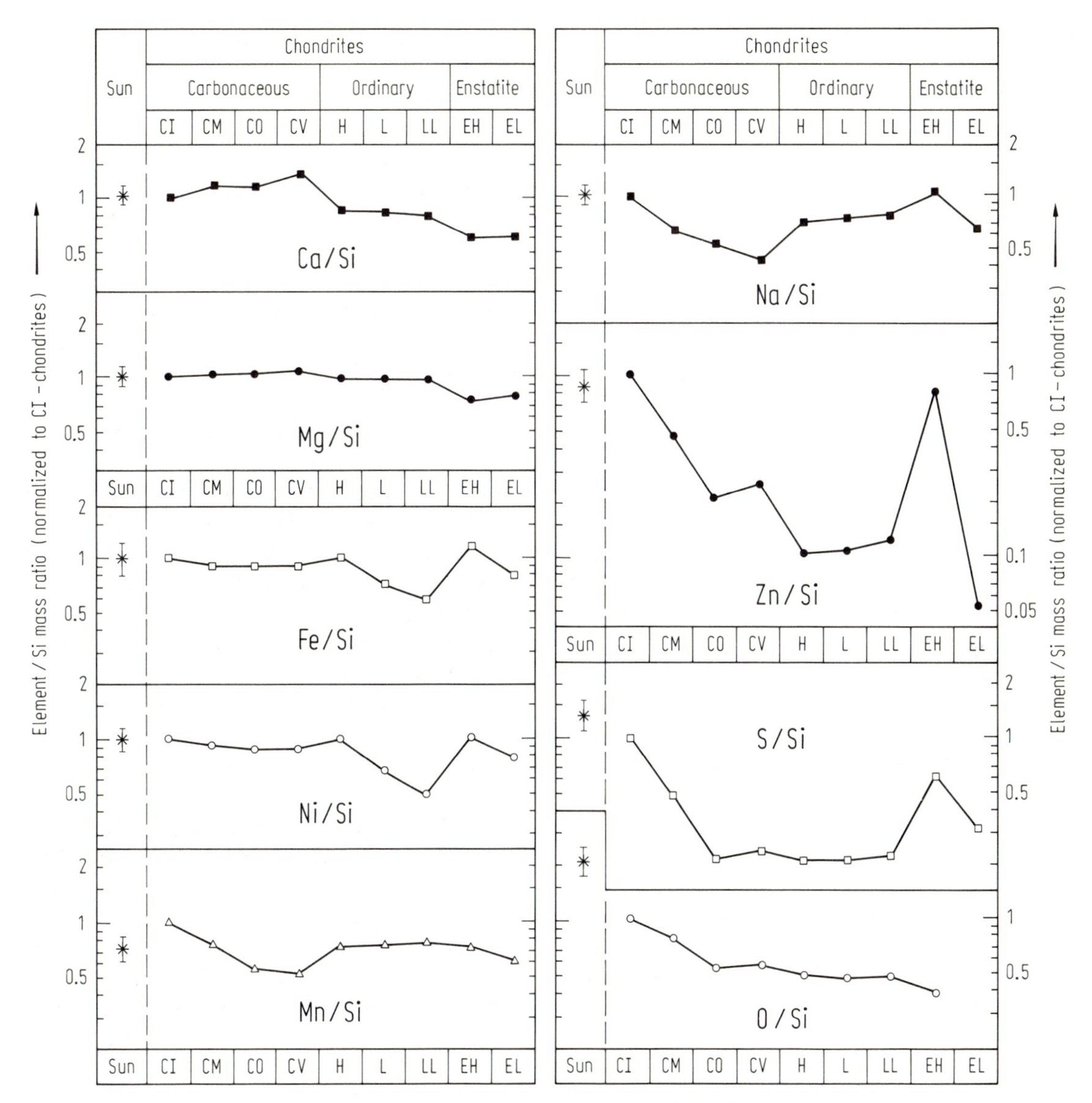

Fig. 2. Element/Si mass ratios of characteristic elements in various groups of chondritic (undifferentiated) meteorites (for meteorite classification see text and LB VI/2a, subsect. 3.3.2.2.4, p.191). Meteorite groups are arranged according to decreasing oxygen contents as reflected in O/Si-ratios. The best match between solar abundances and meteoritic abundances is with CI-meteorites. Refractory elements (Ca), metals and volatiles fit, except Mn. Oxygen (O) is not fully condensed, even in CI-meteorites.

3.4.7.2 Composition of CI-chondrites

Among the several thousand recovered meteorites there are only five CI-meteorites (Orgueil, Ivuna, Alais, Tonk, Revelstoke). Most analyses have been performed on the Orgueil meteorite, simply because Orgueil is the largest CI-meteorite and material is easily available for analysis. However, problems with sample size, sample preparation, and mobility of elements reflecting chemical inhomogeneities within the Orgueil meteorite make a comparison of data obtained by different authors difficult and contribute significantly to abundance uncertainties.

A new CI-abundance table has been prepared for this contribution. In setting up this table the following procedure was adopted. The average of a set of four different Orgueil and two Alais analyses performed at the University of California, Los Angeles (UCLA), by Kallemeyn and Wasson [81K] and the average of a similar set of six Orgueil, one Ivuna and one Alais analyses from the Cosmochemistry Department at the Max-Planck-Institut für Chemie in Mainz [93S] are listed in Table 1 (both data sets are by instrumental neutron activation analysis, INAA). The excellent agreement between the two, completely independent, sets of data suggests high accuracy of the analyses. Elemental abundances in Table 1 are therefore considered to be as good and in some cases superior to earlier analyses of CI-meteorites. For the solar system abundance table 20 elements were taken from the compilation in Table 1. For all these elements the agreement between the two data sets is better than 10%, except for Na and Br. Variations in Na and Br are unlikely to be of analytical nature and may be caused by variable concentrations in different samples of CI-chondrites, i.e., inhomogeneous distribution of both elements in CI-chondrites. We have also taken the concentrations of Mg, Ca and Al from Table 1, although wet chemical analyses are normally considered more accurate for major elements. The Mg, Ca, and Al concentrations obtained by INAA are, however, very similar to those found by other analysts [63M, 90J]. The Fe content listed in Table 1 (18.23%) is slightly lower than literature data, e.g., the compilation by Anders and Grevesse [89A] gives 18.51% for Orgueil and 19.04% for mean CI-chondrites. However, since INAA-procedures involve no processing (sieving, drying, etc.) of the fragile, volatile-rich CI-meteorites they should provide accurate results in absolute terms. The Si-content listed in Table 2 was obtained from a Mg/Si ratio of 0.90 typical of carbonaceous chondrites of type 1 (CI) and 2 (CM). As refractory elements occur in the same relative abundances in chondritic meteorites, ratios among them are often more accurately known than CI-abundances. The abundances of REE (rare earth elements) were taken from [78E]. The few REE elements listed in Table 1 fit well with those given by Evenson et al. [78E]. These authors, however, have determined abundances of all REE and, in addition, the accuracy of their analyses is certainly higher than by INAA. The principal agreement of the INAA-REE determination with the [78E] data indicates the proper absolute level of REE, i.e., their correct ratios to CI-major elements (Fe, Ca, etc.). Elements that were not determined by INAA were, in some cases, taken from the compilation of Anders and Grevesse [89A]. Other sources are indicated in Table 1.

The only element which was never analysed in CI-chondrites is Rh. Its abundance is extrapolated from data on other chondritic meteorites [82A]. Mercury abundances in CI-chondrites are extremely variable. The high contents of Hg in all samples of Orgueil analysed in this laboratory [93S] suggests contamination. A single analysis of Ivuna (0.31 ppm) was therefore used in the present compilation. A value of 0.39 ppm was reported by [67R] in Ivuna for Hg released above 450°C, while corresponding Orgueil Hg-data of these authors scattered by more than a factor 70. The value of 0.31 ppm is in agreement with the Hg content inferred from nuclear abundance systematics (see Table 4b). Recent Orgueil analyses for Th, U and Pb were used [93R, 93C]. However, the large variability of U in CI-chondrites and also in other meteorites still leaves considerable uncertainty on the precise solar U-content and in particular on the solar Th/U-ratio.

The last column of Table 2 contains the Anders and Grevesse [89A] data for comparison. The largest discrepancies between this and the present compilation (Table 2) are found for S (-19%), Hg (17%), Se (12%), P (-11%), Au (8%), Sr (-7.8%). The sulfur content is lower in the present compilation because only Orgueil S-data where considered while [89A] also considered other CI-meteorites with higher S-contents. New analyses of CI-chondrites confirm the lower S-content [93D]. The higher Se content is based on INAA data of Mainz and Los Angeles. Both values agree, but are higher than other CI-contents [89A]. Some of these data were obtained by radiochemical neutron activation analysis where larger errors are involved. An even better agreement between the UCLA and Mainz data is found for Au, whereas lower Au contents are also the result of radiochemical neutron activation procedures. Differences in P and Sr reflect differences in the selection of literature data.

Noble gases in meteorites are mixtures of a variety of components with different isotopic compositions, and noble gases are, in general, depleted by several orders of magnitude compared to inferred solar abundances. They are, therefore, not included in Table 2.

Table 1. Chemical composition of CI-chondrites. Mass abundance a in [ppm], unless otherwise stated. All data by instrumental neutron activation analysis (INAA).
(1) Average: 6 samples Orgueil + 1 sample Alais and Ivuna; Max-Planck-Institut Mainz MPI-MZ, Cosmochemistry [93S].
(2) Average: 4 samples Orgueil + 2 samples Alais; University of California Los Angeles, UCLA [81K].
(3) Difference between MPI-MZ and UCLA relative to average mass abundance.
(4) Average of of MPI-MZ and UCLA mass abundances.

Element		a_{MPI-MZ}		a_{UCLA}		$\frac{a_{MPI-MZ}-a_{UC}}{a_{av.}}$	$a_{av.}$
		(1) [ppm]	s.d. [%]	(2) [ppm]	s.d. [%]	(3) [%]	(4) [ppm]
11	Na	5238	7.6	4725	4.3	10.3	4982
12	Mg	9.58%	2.4	9.63%	5.2	–0.5	9.61%
13	Al	8600	4.9	8700	7.5	–1.2	8650
19	K	534	7.3	553	8.1	–3.5	544
20	Ca	9600	7.0	9428	5.8	1.8	9510
21	Sc	5.94	1.5	5.87	5.6	1.1	5.90
23	V	53.2	8.5	55.5	6.9	–5.2	54.3
24	Cr	2646	1.5	2647	2.8	0.0	2646
25	Mn	1903	5.9	1963	2.9	–3.1	1933
26	Fe	18.29%	1.4	18.18%	2.6	0.6	18.23%
27	Co	502	1.8	510	2.6	–1.6	506
28	Ni	1.118%	3.8	1.035%	5.8	7.7	1.077%
30	Zn	335	6.6	311	5.2	7.5	323
31	Ga	9.67	7.7	9.76	1.6	–1.0	9.71
33	As	1.75	14	1.87	6.2	–6.9	1.81
34	Se	21.2	2.2	21.3	3.8	–0.3	21.3
35	Br	3.29	27	3.70	12.5	–11.7	3.50
57	La	0.267	21	0.226	4.4	16.9	0.247 *
62	Sm	0.149	5.6	0.141	4.3	5.5	0.145 *
63	Eu	0.056	4.9	0.058	5.7	–3.3	0.057 *
66	Dy	0.253	14				0.253 *
70	Yb	0.162	12	0.163	4.6	–0.9	0.162 *
71	Lu	0.028	36	0.024	5.2	15.9	0.026 *
76	Os	0.496	8.6	0.475	4.4	4.2	0.486
77	Ir	0.468	1.8	0.450	3.4	3.9	0.459
79	Au	0.150	7	0.154	7.5	–3.1	0.152

* Not used for average in Table 2;
s.d. – standard deviation.

Table 2. Elemental abundances in CI-chondrites. Mean mass abundances a in [ppm] unless stated otherwise, and atomic abundances N relative to 10^6 Si atoms.
(1) Average of CI-meteorites, mainly Orgueil;
(2) Anders and Grevesse 1989 [89A].

Element		a [ppm] (1)	Estimated accuracy [%]	N [N(Si)=10^6]	Source	N [N(Si)=10^6] (2)
1	H	2.02%	10	527	[63M]	2.79E+10 *
3	Li	1.49	10	56.4	[89A]	57.1
4	Be	0.0249	10	0.73	[89A]	0.73
5	B	0.87	10	21.2	[89A]	21.2
6	C	3.22%	10	7.05E+05	[63M]	1.01E+07 *
7	N	0.318	10	5.97E+04	[89A]	3.13E+06 *
8	O	46.5%	10	7.64E+06	[63M]	2.38E+07 *
9	F	58.2	15	805	[89A]	843
11	Na	4982	5	5.70E+04	(a)	5.74E+04
12	Mg	9.61%	3	1.04E+06	(a)	1.074E+06
13	Al	8650	3	8.43E+04	(a)	8.49E+04
14	Si	10.68%	3	1.00E+06	(b)	1.00E+06
15	P	1105	10	9.38E+03	(c)	1.04E+04
16	S	5.25%	10	4.31E+05	[63M]	5.15E+05
17	Cl	698	15	5176	[89A]	5240
19	K	544	5	3658	(a)	3770
20	Ca	9510	3	6.24E+04	(a)	6.11E+04
21	Sc	5.90	3	34.5	(a)	34.2
22	Ti	441	5	2421	[79S]	2400
23	V	54.3	5	280	(a)	293
24	Cr	2646	3	1.34E+04	(a)	1.35E+04
25	Mn	1933	3	9251	(a)	9550
26	Fe	18.23%	3	8.58E+05	(a)	9.00E+05
27	Co	506	3	2257	(a)	2250
28	Ni	1.077%	3	4.82E+04	(a)	4.93E+04
29	Cu	131	10	542	[82R]	522
30	Zn	323	10	1299	(a)	1260
31	Ga	9.71	5	36.6	(a)	37.8
32	Ge	32.6	10	118	[89A]	119
33	As	1.81	5	6.35	(a)	6.56
34	Se	21.3	5	70.8	(a)	62.1
35	Br	3.50	10	11.5	(a)	11.8
37	Rb	2.32	5	7.14	[84B]	7.09
38	Sr	7.26	5	21.8	[84B]	23.5
39	Y	1.57	5	4.64	[86J]	4.64
40	Zr	3.87	5	11.2	[86J]	11.4
41	Nb	0.246	5	0.696	[86J]	0.698
42	Mo	0.928	5	2.54	[82R]	2.55
44	Ru	0.714	10	1.86	[89A]	1.86
45	Rh	0.134	20	0.342	[89A]	0.344
46	Pd	0.556	10	1.37	[89A]	1.39
47	Ag	0.197	10	0.480	[89A]	0.486

Table 2 (continued)

Element		a [ppm] (1)	Estimated accuracy [%]	N [N(Si)=10^6]	Source	N [N(Si)=10^6] (2)
48	Cd	0.680	10	1.59	[89A]	1.61
49	In	0.0778	10	0.178	[89A]	0.184
50	Sn	1.68	10	3.72	[89A]	3.82
51	Sb	0.133	10	0.287	[89A]	0.309
52	Te	2.27	10	4.68	[89A]	4.81
53	I	0.433	20	0.90	[89A]	0.90
55	Cs	0.188	5	0.372	[84B]	0.372
56	Ba	2.41	5	4.61	[84B]	4.49
57	La	0.245	5	0.464	[78E]	0.4460
58	Ce	0.638	5	1.20	[78E]	1.136
59	Pr	0.0964	10	0.180	[78E]	0.1669
60	Nd	0.474	5	0.864	[78E]	0.8279
62	Sm	0.154	5	0.269	[78E]	0.2582
63	Eu	0.0580	5	0.100	[78E]	0.0973
64	Gd	0.204	5	0.341	[78E]	0.3300
65	Tb	0.0375	10	0.0620	[78E]	0.0603
66	Dy	0.254	5	0.411	[78E]	0.3942
67	Ho	0.0567	10	0.0904	[78E]	0.0889
68	Er	0.166	5	0.261	[78E]	0.2508
69	Tm	0.0256	10	0.0398	[78E]	0.0378
70	Yb	0.165	5	0.251	[78E]	0.2479
71	Lu	0.0254	5	0.0382	[78E]	0.0367
72	Hf	0.107	5	0.158	[86J]	0.154
73	Ta	0.014	10	0.020	[86J]	0.0207
74	W	0.095	7	0.136	[82R]	0.133
75	Re	0.0383	7	0.0541	[82R]	0.0517
76	Os	0.486	5	0.672	(a)	0.675
77	Ir	0.459	3	0.628	(a)	0.661
78	Pt	0.994	10	1.34	[82R]	1.34
79	Au	0.152	5	0.203	(a)	0.187
80	Hg	0.31	20	0.41	(d)	0.34
82	Tl	0.143	10	0.184	[89A]	0.184
82	Pb	2.53	10	3.21	[93R]	3.15
83	Bi	0.111	15	0.140	[89A]	0.144
90	Th	0.0298	5	0.0338	[93R]	0.0335
92	U	0.0078	10	0.0086	(e)	0.0090

* Solar system (not CI-chondrite)

(a) Data from Table 1;
(b) Assuming Mg/Si=0.90;
(c) Average [63M] and [90J];
(d) Single Alais analysis [93S], since Orgueil samples are contaminated with Hg;
(e) Average [93R] and [93C].

For H, C, O and S only Orgueil analyses were considered; in cases where the Anders and Grevesse abundance data [89A] were used there may be small differences between Si-normalized atomic abundances because of slightly different Si and because Orgueil data were preferred over average CI.

3.4.8 Meteorites and the composition of the solar photosphere

In Table 3 the composition of the solar photosphere as obtained by absorption spectroscopy is given. Abundances are normalized to 10^{12} H atoms, the usual practice in astronomy. The Si-normalized abundances are converted to this scale by adding 1.55 (see [89A]) to the log of the Si-normalized abundances and are listed in the third column of Table 3. The photospheric data and corresponding error bars are from the Anders and Grevesse [89A] compilation except for C, N, and Fe for which newer values are included [90G, 91G, 91H]. The rare gases listed in the photospheric abundance column are derived from solar wind data (He, Ne, Ar) [89A] or are from s-process calculations (Kr, Xe). The lower photospheric Fe-abundance, recently reported [91H], is the most important change in this table. The disagreement between solar and meteoritic Fe-abundance has finally vanished. There are 38 high-quality determinations of photospheric abundances, i.e., with standard deviations below 25% (0.1 dex units in logarithmic scale). Seven elements differ by more than 25% from meteorite-derived abundances. The solar abundances of C, N and O are higher because these elements are incompletely condensed in CI-meteorites. The lower solar abundances of Li and Be reflect fusion processes in the interior of the Sun. Only Mn and Pb show unexplained deviations significantly beyond those expected from statistical errors. In Fig. 3 solar/meteoritic abundance ratios are plotted. For 29 elements the agreement between solar and meteoritic abundances is better than ±10%. Most elements outside the 10% limit have such large uncertainties in their photospheric abundances that agreement with meteorite data is within error bars. If the difference is larger, error bars are indicated in Fig. 3. The remarkable agreement between

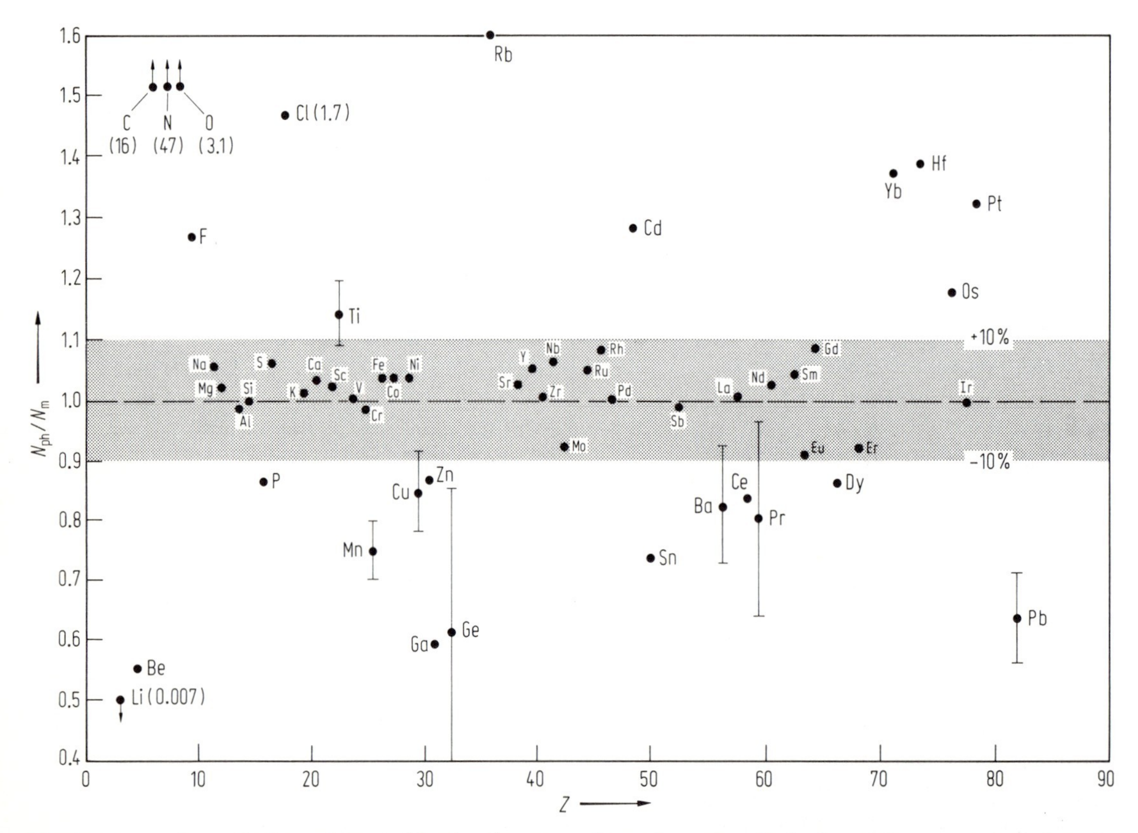

Fig. 3. Comparison of solar and meteoritic abundances (see Table 3). The elements C, N and O are incompletely condensed in meteorites. Li and Be are consumed by fusion processes in the interior of the sun. Solar and meteoritic abundances agree in most cases within 10%. Error bars are only indicated in cases where they do not overlap with $N_{ph}/N_m = 1$. The clustering of points above $N_{ph}/N_m = 1$ suggests a slightly higher conversion factor for meteoritic to solar abundances of 1.559 instead of 1.55 used here (see [89A]).

solar photospheric abundances and meteorite-derived abundances applies, however, only to CI-chondrites (Fig. 2). Any other type of chondrite would show a significantly higher disagreement with photospheric abundances, in particular with respect to volatile elements.

Table 3. Comparison of meteorite-derived solar-system atomic abundances with photospheric atomic abundances, normalized to 10^{12} H atoms. Standard deviations of $\log N$ are indicated. Meteorite data: Table 2; Solar data: Anders and Grevesse 1989 [89A], except C [90G], N [91G] and Fe [91H]. For solar wind derived He, Ne, Ar see [89A], Kr and Xe from s-process calculations (Table 4b). Values in parentheses are uncertain.

Element		$\log N(N(\mathrm{H}) \equiv 10^{12})$		$(N_{ph} - N_m)/N_m$
		Meteorites N_m	Photosphere N_{ph}	[%]
1	H		12.00	
2	He		10.99 ± 0.035	
3	Li	3.30 ± 0.04	1.16 ± 0.1	-100
4	Be	1.41 ± 0.04	1.15 ± 0.1	-45
5	B	2.88 ± 0.04	(2.6 ± 0.3)	
6	C	7.40 ± 0.04	8.60 ± 0.05	1500
7	N	6.33 ± 0.04	8.00 ± 0.05	4600
8	O	8.43 ± 0.04	8.93 ± 0.035	210
9	F	4.46 ± 0.06	4.56 ± 0.3	30
10	Ne		8.09 ± 0.1	
11	Na	6.31 ± 0.02	6.33 ± 0.03	6
12	Mg	7.57 ± 0.01	7.58 ± 0.05	3
13	Al	6.48 ± 0.01	6.47 ± 0.07	-1
14	Si	7.55	7.55 ± 0.05	
15	P	5.52 ± 0.04	5.45 ± (0.04)	-15
16	S	7.18 ± 0.04	7.21 ± 0.06	6
17	Cl	5.26 ± 0.06	5.5 ± 0.3	70
18	Ar		6.56 ± 0.10	
19	K	5.11 ± 0.02	5.12 ± 0.13	2
20	Ca	6.35 ± 0.01	6.36 ± 0.02	4
21	Sc	3.09 ± 0.01	3.10 ± (0.09)	3
22	Ti	4.93 ± 0.02	4.99 ± 0.02	15
23	V	4.00 ± 0.02	4.00 ± 0.02	1
24	Cr	5.68 ± 0.01	5.67 ± 0.03	-2
25	Mn	5.52 ± 0.01	5.39 ± 0.03	-25
26	Fe	7.48 ± 0.01	7.50 ± 0.07	4
27	Co	4.90 ± 0.01	4.92 ± 0.04	4
28	Ni	6.23 ± 0.01	6.25 ± 0.04	4
29	Cu	4.28 ± 0.04	4.21 ± 0.04	-15
30	Zn	4.66 ± 0.04	4.60 ± 0.08	-15
31	Ga	3.11 ± 0.02	2.88 ± (0.10)	-40
32	Ge	3.62 ± 0.04	3.41 ± 0.14	-40
33	As	2.35 ± 0.02		
34	Se	3.40 ± 0.02		
35	Br	2.61 ± 0.04		
36	Kr		3.30 ± 0.06	
37	Rb	2.40 ± 0.02	2.60 ± (0.15)	60
38	Sr	2.89 ± 0.02	2.90 ± 0.06	3

Table 3 (continued)

Element		$\log N(N(\mathrm{H}) \equiv 10^{12})$		$(N_{ph} - N_m)/N_m$
		Meteorites N_m	Photosphere N_{ph}	[%]
39	Y	2.22 ± 0.02	2.24 ± 0.03	6
40	Zr	2.60 ± 0.02	2.60 ± 0.03	1
41	Nb	1.39 ± 0.02	1.42 ± 0.06	7
42	Mo	1.96 ± 0.02	1.92 ± 0.05	−8
44	Ru	1.82 ± 0.04	1.84 ± 0.07	5
45	Rh	1.08 ± 0.08	1.12 ± 0.12	9
46	Pd	1.69 ± 0.04	1.69 ± 0.04	0.5
47	Ag	1.23 ± 0.04	(0.94 ± 0.25)	
48	Cd	1.75 ± 0.04	1.86 ± 0.15	30
49	In	0.80 ± 0.04	(1.66 ± 0.15)	
50	Sn	2.12 ± 0.04	2.0 ± (0.3)	−25
51	Sb	1.01 ± 0.04	1.0 ± (0.3)	−2
52	Te	2.22 ± 0.04		
53	I	1.50 ± 0.08		
54	Xe		2.16 ± 0.09	
55	Cs	1.12 ± 0.02		
56	Ba	2.21 ± 0.02	2.13 ± 0.05	−20
57	La	1.22 ± 0.02	1.22 ± 0.09	1
58	Ce	1.63 ± 0.02	1.55 ± 0.20	−15
59	Pr	0.80 ± 0.04	0.71 ± 0.08	−20
60	Nd	1.49 ± 0.02	1.50 ± 0.06	3
62	Sm	0.98 ± 0.02	1.00 ± 0.08	5
63	Eu	0.55 ± 0.02	0.51 ± 0.08	−9
64	Gd	1.08 ± 0.02	1.12 ± 0.04	9
65	Tb	0.34 ± 0.04	(−0.1 ± 0.3)	
66	Dy	1.16 ± 0.02	1.1 ± 0.15	−15
67	Ho	0.51 ± 0.04	(0.26 ± 0.16)	
68	Er	0.97 ± 0.02	0.93 ± 0.06	−8
69	Tm	0.15 ± 0.04	(0.00 ± 0.15)	
70	Yb	0.95 ± 0.02	1.08 ± (0.15)	35
71	Lu	0.13 ± 0.02	(0.76 ± 0.30)	
72	Hf	0.75 ± 0.02	0.88 ± (0.08)	35
73	Ta	−0.14 ± 0.04		
74	W	0.68 ± 0.03	(1.11 ± 0.15)	
75	Re	0.28 ± 0.03		
76	Os	1.38 ± 0.02	1.45 ± 0.10	20
77	Ir	1.35 ± 0.01	1.35 ± (0.10)	0.5
78	Pt	1.68 ± 0.04	1.8 ± 0.3	30
79	Au	0.86 ± 0.02	(1.01 ± 0.15)	
80	Hg	1.16 ± 0.08		
81	Tl	0.81 ± 0.04	(0.9 ± 0.2)	
82	Pb	2.06 ± 0.04	1.85 ± 0.05	−40
83	Bi	0.70 ± 0.06		
90	Th	0.079 ± 0.02	0.12 ± (0.06)	10
92	U	−0.52 ± 0.04	(< −0.47)	

3.4.9 Nuclear abundances and their decomposition in nucleosynthetic components

In Table 4a solar abundances are listed for individual isotopes of elements from H to Mn. The isotopic compositions were taken from recommendations by IUAPC [91I]. For H, N and the noble gases values by Anders and Grevesse [89A] were used, i.e., solar wind derived isotope ratios for rare gases and an estimate of a protosolar D/H-ratio. Anders and Grevesse [89A] have also provided some information on nucleosynthetic processes that have contributed to the abundance of individual isotopes.

In Table 4b we have listed isotopic compositions for elements from Fe to U and here we have given assignments to specific nucleosynthetic processes, based on calculations (see [89B, 90B1, 91B, 92B, 89K] for details). Assignments for the Fe-slope isotopes were, in part, taken from Anders and Grevesse [89A]. They are uncertain as the nucleosynthesis of these elements is complicated. S-process computations were made by applying a least-squares fit to the products of neutron capture cross section times abundance (σN) for elements exclusively produced by the s-process and then calculating the s-process fractions of the other nuclei. Results of such calculations are shown in Fig. 4. It is seen that in order to reproduce the abundance pattern from Fe to Bi a three-component s-process is required. Besides the main s-component a weak component provides significant contributions in the mass range A from 56 to 90 and a strong component in the mass range from 206 to 209. Table 4b contains results of the additional components only in these mass ranges. Uncertainties of the calculations are estimated to be between 5 and 10%, i.e., comparable to errors for meteorite-derived abundances. At branching points where a direct and strong influence of s-process neutron density, temperature and electron density is present, uncertainties of the calculations are significantly higher. Such branchings occur in the Ni-Zn, Se-Kr, Sr-Rb, Sm-Gd, Dy-Er, Lu-Hf, Os-Re and Ir-Pt regions. Normally, the abundances of short-lived unstable isotopes generated in the s-process are added to their corresponding stable isobar.

The analysis served to estimate solar-system abundances for some volatile elements whose abundances are difficult to determine otherwise. The values for Kr (56 atoms/10^6 Si-atoms, based on ^{82}Kr) and Xe (4.1 atoms/ 10^6 Si-atoms, based on ^{128}Xe) obtained in this way are used in Table 4b. The calculated value for Pb (2.49 ± 0.18 atoms/10^6 Si atoms, based on ^{204}Pb) is smaller by 24% than the meteorite-derived value. But the decomposition of the Pb and Bi isotopes is made on the basis of the Pb abundance estimate via s-process calculations. The deviation of pure s-process isotopes not located in a s-process branching reflects the scatter around the mean value represented by the curve. Note that a p-process correction has to be applied to the respective solar abundances before a comparison can be made. The deviation of calculated to observed abundances is significant, i.e., larger than the estimated uncertainty, only in the case of ^{116}Sn. From the point of view of s-process systematics a revision of the meteoritic Sn abundance would be required [91B].

If s-process abundances are subtracted from the solar abundances a kind of semiempirical distribution of abundances ascribed to the r-process nucleosynthesis is obtained. This distribution shown in Fig. 5 forms the basis for any comparison with r-process calculations. A decomposition of abundances for individual elements is shown in Fig. 6. The low-abundant neutron-poor isotopes are due to the p-process.

Fig. 7 is a plot of abundances vs. mass number. The generally higher abundances of even masses is apparent. Even and odd mass numbers, separately, plot along more or less smooth curves, with odd mass numbers forming a considerably smoother curve than even mass numbers. Historically, these so-called abundance rules, established by Suess [47S] postulating a smooth dependance of isotopic abundances on mass number A, especially of odd-A nuclei, played an important role in estimating unknown abundances and/or for detecting fractionation effects. Later this rule was modified and supplemented by two additional rules [73S] in order to make the concept also applicable to the now more accurate abundance data. However, in the last years the smoothness of odd-A nuclei abundances itself has been questioned [90B2, 89A]. For example, there are major breaks in the abundance curve of odd-A nuclei in the region of Nd, Sm and Eu (Fig. 9). It is, therefore, recommended to rely on the concepts of nucleosynthesis which provide reasonable explanations for many exceptions to the Suess' rules. The high abundance of ^{89}Y (Fig. 8), an apparent discontinuity [90B2], reflects the low neutron capture cross section of a dominantly s-process nucleus with a magic neutron number (50). But even the pure r-process distribution is not a perfectly smooth function of mass number. The structures reflect special nuclear properties of the isotopes involved (e.g., the strong odd-even staggering around mass number 80 [90K]).

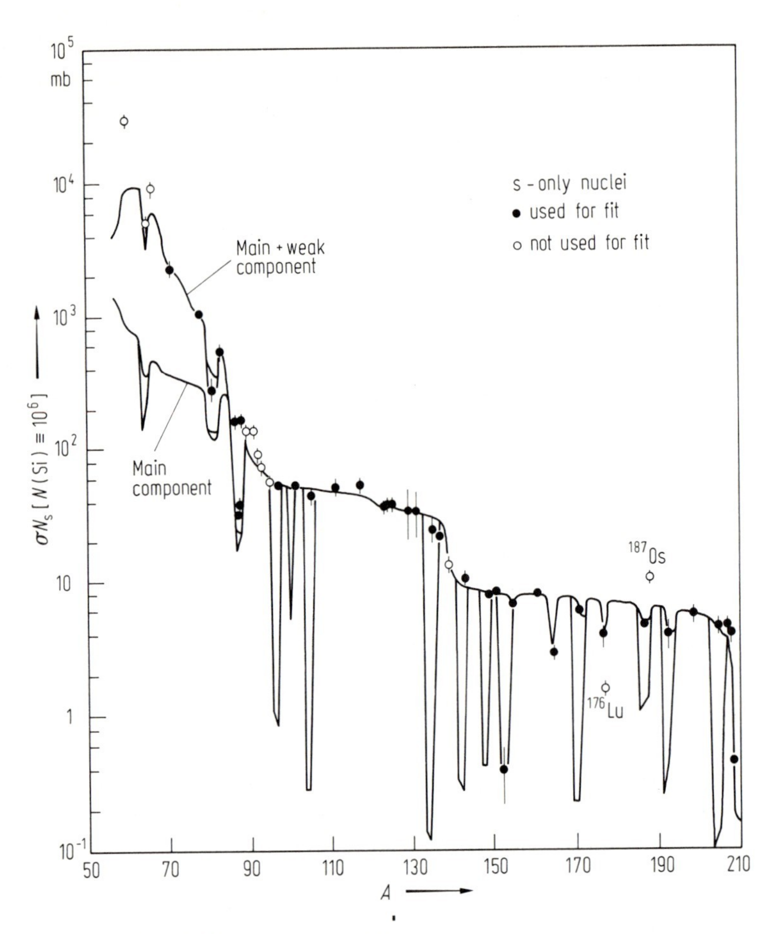

Fig. 4. Computer fit of σN (neutron cross section times abundance) versus mass number. Full symbols are s-only nuclei and are used for the fit. Open symbols are s-only nuclei which were not used for the fit. The ^{187}Os abundance has grown by decay of ^{187}Re, and ^{176}Lu has decayed since s-process synthesis, a long time before solar system formation. S-process contributions to nuclei produced by more than one nucleosynthetic processes can be interpolated. The negative spikes indicate the presence of branching points. Three different neutron exposures are required, at low, intermediate and high (^{208}Pb) masses (see text).

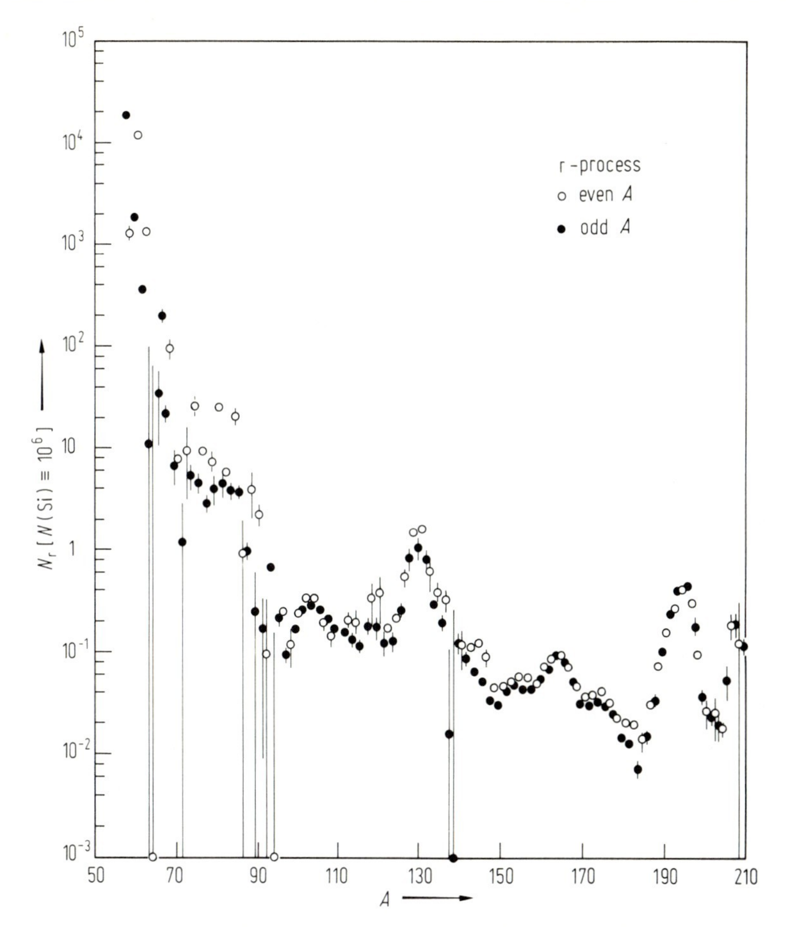

Fig. 5. Subtraction of s-process nuclei from total abundances leads to r-process abundances. Full circles denote odd mass numbers and open circles even mass numbers.

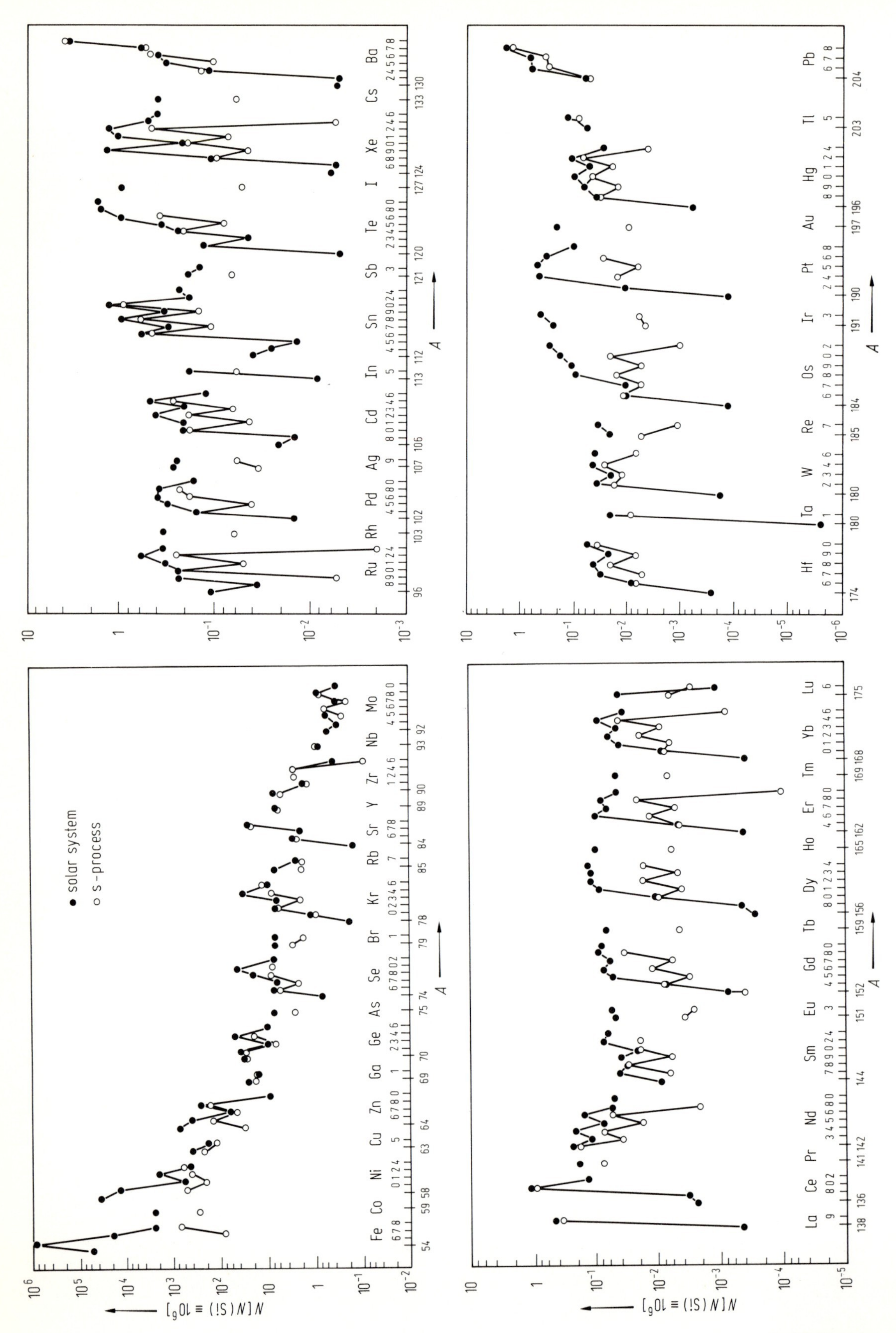
• solar system
○ s-process
N [N(Si) ≡ 10^6]
A
Fe Co Ni Cu Zn Ga Ge As Se Br Kr Rb Sr Y Zr Nb Mo
Ru Rh Pd Ag Cd In Sn Sb Te I Xe Cs Ba
La Ce Pr Nd Sm Eu Gd Tb Dy Ho Er Tm Yb Lu
Hf Ta W Re Os Ir Pt Au Hg Tl Pb

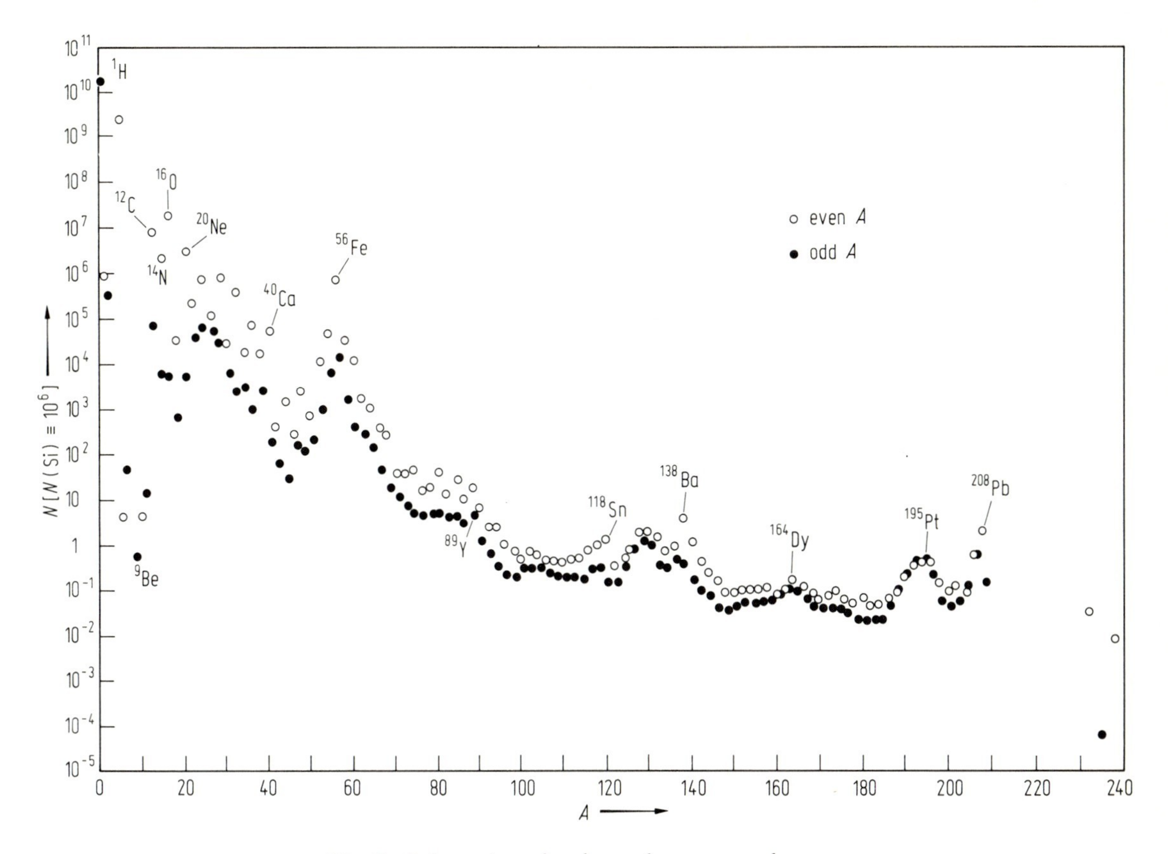

Fig. 7. Solar-system abundances by mass number.

←

Fig. 6. Solar abundances and calculated s-process contribution for elements from Fe to Pb arranged according to increasing atomic number (see text). Large differences between solar abundance and calculated s-process indicate large r-process contributions, e.g., from Sb-Xe, Eu-Yb and Re-Au. Low abundances of p-process nuclei are clearly visible.

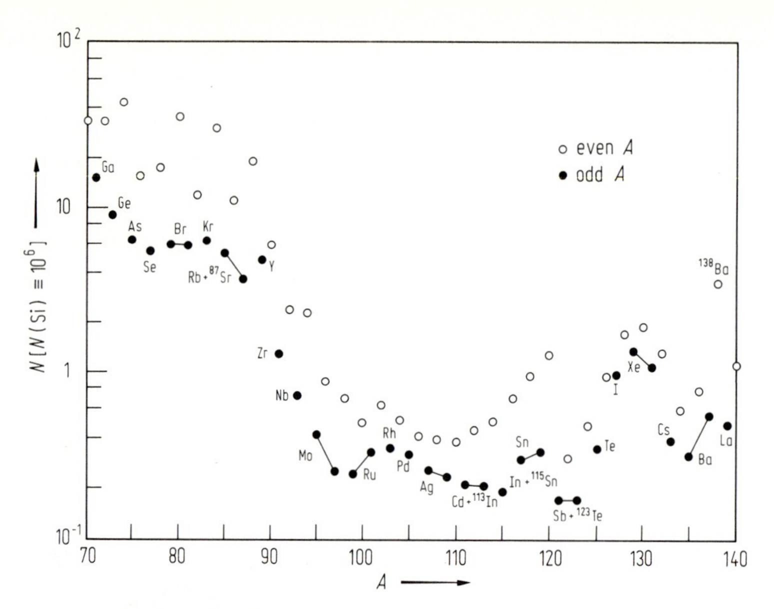

Fig. 8. Abundances from mass number 70 to 140. Odd masses form an approximately smooth curve. However, there are distinct kinks indicating limitations to the smoothness of abundance vs. mass number distribution. Chemical symbols only for odd masses.

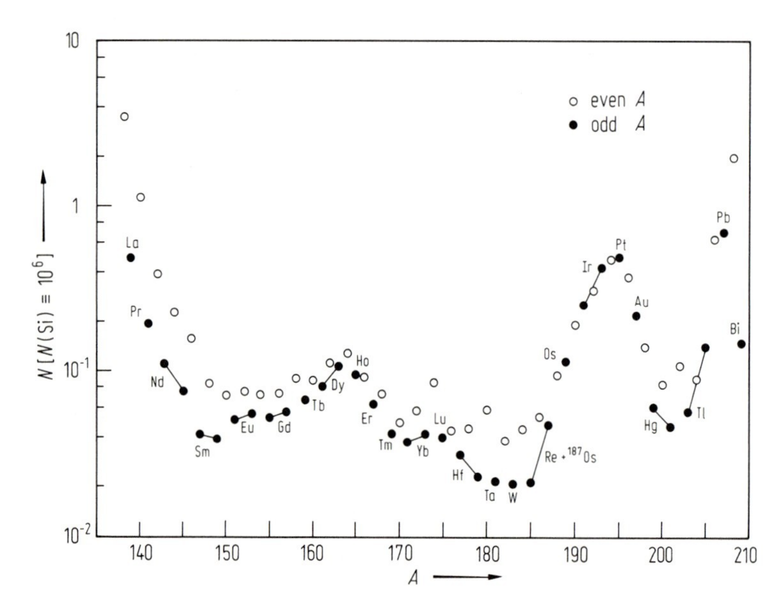

Fig. 9. Abundances from mass number 138 to 209. Analogue to Fig. 8.

Table 4a. Isotopic composition of elements and isotopic abundance N normalized to 10^6 Si-atoms for elements from H to Mn.

Element	A	Isotopic composition [%]	N [$N(\mathrm{Si}) \equiv 10^6$]
1 H	1	99.9966	$2.79 \cdot 10^{10}$
	2	0.0034	$9.49 \cdot 10^{5}$
2 He	3	0.0142	$3.86 \cdot 10^{5}$
	4	99.9858	$2.72 \cdot 10^{9}$
3 Li	6	7.5	4.23
	7	92.5	52.17
4 Be	9	100	0.73
5 B	10	19.9	4.22
	11	80.1	16.98
6 C	12	98.90	$9.99 \cdot 10^{6}$
	13	1.10	$1.11 \cdot 10^{5}$
7 N	14	99.634	$3.12 \cdot 10^{6}$
	15	0.366	$1.15 \cdot 10^{4}$
8 O	16	99.762	$2.37 \cdot 10^{7}$
	17	0.038	$9.04 \cdot 10^{3}$
	18	0.200	$4.76 \cdot 10^{4}$
9 F	19	100	$8.05 \cdot 10^{2}$
10 Ne	20	92.99	$3.20 \cdot 10^{6}$
	21	0.226	$7.77 \cdot 10^{3}$
	22	6.79	$2.34 \cdot 10^{5}$
11 Na	23	100	$5.70 \cdot 10^{4}$
12 Mg	24	78.99	$8.21 \cdot 10^{5}$
	25	10.00	$1.04 \cdot 10^{5}$
	26	11.01	$1.15 \cdot 10^{5}$
13 Al	27	100	$8.43 \cdot 10^{4}$
14 Si	28	92.23	$9.22 \cdot 10^{5}$
	29	4.67	$4.67 \cdot 10^{4}$
	30	3.10	$3.10 \cdot 10^{4}$
15 P	31	100	$9.38 \cdot 10^{3}$
16 S	32	95.02	$4.10 \cdot 10^{5}$
	33	0.75	$3.23 \cdot 10^{3}$
	34	4.21	$1.81 \cdot 10^{4}$
	36	0.02	86
17 Cl	35	75.77	$3.92 \cdot 10^{3}$
	37	24.23	$1.25 \cdot 10^{3}$
18 Ar	36	84.2	$8.50 \cdot 10^{4}$
	38	15.8	$1.60 \cdot 10^{4}$
	40	0.024	25
19 K	39	93.26	$3.41 \cdot 10^{3}$
	40	0.01	0.43
	41	6.73	$2.46 \cdot 10^{2}$
20 Ca	40	96.941	$6.05 \cdot 10^{4}$
	42	0.647	$4.04 \cdot 10^{2}$
	43	0.135	84
	44	2.086	$1.30 \cdot 10^{3}$
	46	0.004	2.50
	48	0.187	$1.17 \cdot 10^{2}$
21 Sc	45	100	34.5
22 Ti	46	8.0	$1.94 \cdot 10^{2}$
	47	7.3	$1.77 \cdot 10^{2}$
	48	73.8	$1.79 \cdot 10^{3}$
	49	5.5	$1.33 \cdot 10^{2}$
	50	5.4	$1.31 \cdot 10^{2}$
23 V	50	0.250	0.70
	51	99.750	$2.79 \cdot 10^{2}$
24 Cr	50	4.345	$5.82 \cdot 10^{2}$
	52	83.789	$1.12 \cdot 10^{4}$
	53	9.501	$1.27 \cdot 10^{3}$
	54	2.365	$3.17 \cdot 10^{2}$
25 Mn	55	100	$9.25 \cdot 10^{3}$

Abundances from Table 2; H, C, N, O, Ne, Ar from [89A]; Isotopic ratios from IUPAC [91I], except H, He, Ne, Ar [89A].

Table 4b. Isotopic abundances and assignments to specific nucleosynthetic processes.
(1) Isotopic composition from IUPAC [91I], except Kr, Xe [89A] and Pb [76T];
(2) Calculated total abundance;
(3) s-process calculations [91B, 92B];
(4) Relative contribution of s-processes (main + weak components), N_s/N, and non-s processes, $1 - N_s/N$, to total abundance.
(5) Contributing nucleosynthetic processes (see [89A] for details):
ex: explosive nucleosynthesis; e: nuclear statistical equilibrium; He: helium burning; C: carbon burning; s: s-process; r: r-process; p: p-process; (s) or (r) : possible minor contributions; r*: including transbismuth r-process abundances;

Element		A	Isotopic composition [%] (1)	N [N(Si) ≡ 10^6] (2)	N_s[N(Si) ≡ 10^6] main component (3)	weak component	N_s/N [%] (4)	$1-N_s/N$ [%]	Process (5)
26	Fe	54	5.8	49800					ex
		56	91.72	787000					ex,e
		57	2.2	18900	22	64	0.4	100	e,ex
		58	0.28	2402	62	467	22	78	He,e,C,s
27	Co	59	100	2257	19	164	8	92	e,C,s
28	Ni	58	68.077	32813					e,ex
		60	26.223	12639	24	271	2	98	e
		61	1.140	549	7	89	17	83	e,ex,C,s
		62	3.634	1752	16	208	13	87	e,ex,C,s
		64	0.926	446	18.9	374	88	12	s,ex
29	Cu	63	69.17	375	8	211	58	42	s,ex,C
		65	30.83	167	8	94	61	39	s,ex
30	Zn	64	48.6	631	4.5	30	5	95	ex,e
		66	27.9	362	11	113	34	65	e,s
		67	4.1	53	2.4	23.9	49	51	e,s
		68	18.8	244	17	120	56	44	s,e
		70	0.6	7.8					e
31	Ga	69	60.108	22	2.1	13.9	73	27	s,e,r
		71	39.892	15	2.4	12.5	102	–2	s,e,r
32	Ge	70	21.23	25	3.5	20.9	97	3	s,p,(e)
		72	27.66	33	4.9	21.3	80	20	s,e,r
		73	7.73	9	0.9	3.6	49	51	s,e,r
		74	35.94	42	4.8	15.6	48	52	s,e,r
		76	7.44	8.8					e,(r)
33	As	75	100	6.35	0.60	1.7	36	64	r,s
34	Se	74	0.89	0.6					p
		76	9.36	6.6	1.5	4.41	89	11	s,p
		77	7.63	5.4	0.60	1.53	39	61	r,s
		78	23.78	17	2.7	6.1	52	48	r,s
		80	49.61	35	2.34	5.13	21	79	r,s
		82	8.73	6.2					r

Table 4b (continued)

Element		A	Isotopic composition [%] (1)	N [N(Si) ≡ 10^6] (2)	N_s[N(Si) ≡ 10^6] (3)		N_s/N [%] (4)	1-N_s/N [%]	Process (5)
					main component	weak component			
35	Br	79	50.69	5.8	0.7	1.7	41	59	r,s
		81	49.31	5.7	0.59	1.05	29	71	r,s
36	Kr	78	0.339	0.19					p
		80	2.22	1.24	0.49	0.606	88	12	s,p
		82	11.45	6.41	2.85	3.49	99	1	s,p
		83	11.47	6.42	0.89	0.94	28	72	r,s
		84	57.11	32.0	5.6	0.87	20	80	r,s
		86	17.42	9.76	5.46	4.04	97	3	s,r
37	Rb	85	72.165	5.2	1.30	0.23	30	70	r,s
		87	27.835	2.0	1.22	0.145	69	31	s,r
38	Sr	84	0.56	0.12					p
		86	9.86	2.1	2.0	0.17	101	–1	s,p
		87	7.00	1.5	1.47	0.025	98	2	s,p, ^{87}Rb
		88	82.58	18	15.4	0.39	88	12	s,r
39	Y	89	100	4.6	4.1	0.10	91	9	s,r
40	Zr	90	51.45	5.8	3.4	0.27	64	36	s,r
		91	11.22	1.3	1.06	0.04	88	12	s,r
		92	17.15	1.9	1.8	0.08	98	2	s,r
		94	17.38	1.9	2.01	0.03	105	–5	s,r
		96	2.80	0.3	0.07		22	78	r,s
41	Nb	93	100	0.73	0.59	0.009 ^{93}Zr	82	18	s,r
42	Mo	92	14.84	0.38					p
		94	9.25	0.23					p
		95	15.92	0.40	0.18		45	55	r,s
		96	16.68	0.42	0.43		101	–1	s,(p)
		97	9.55	0.24	0.14		58	42	s,r
		98	24.13	0.61	0.49		80	20	s,r
		100	9.63	0.24					r
44	Ru	96	5.52	0.10					p
		98	1.88	0.03					p
		99	12.7	0.24	0.060	incl. ^{99}Tc	26	74	r,s
		100	12.6	0.23	0.23		98	2	s,(p)
		101	17.0	0.32	0.047		15	85	r,s
		102	31.6	0.59	0.245		42	58	r,s
		104	18.7	0.35	0.002		1	99	r,(s)
45	Rh	103	100	0.35	0.057		16	84	r,s

Table 4b (continued)

Element		A	Isotopic composition [%] (1)	N [N(Si) ≡ 10^6] (2)	N_s[N(Si) ≡ 10^6] main component (3)	weak component	N_s/N [%] (4)	1-N_s/N [%]	Process (5)
46	Pd	102	1.02	0.014					p
		104	11.14	0.15	0.155		102	−2	s,(p)
		105	22.33	0.31	0.039		13	87	r,s
		106	27.33	0.37	0.177		47	53	r,s
		108	26.46	0.36	0.22		61	39	s,r
		110	11.72	0.16					r
47	Ag	107	51.839	0.25	0.034	^{107}Pd	14	86	r,s
		109	48.161	0.23	0.057		25	75	r,s
48	Cd	106	1.25	0.020					p
		108	0.89	0.014					p
		110	12.49	0.20	0.18		91	9	s,p
		111	12.80	0.20	0.042		21	79	r,s
		112	24.13	0.38	0.18		47	53	r,s
		113	12.22	0.19	0.06		31	69	r,s
		114	28.73	0.46	0.26		57	43	r,s
		116	7.49	0.12					r
49	In	113	4.3	0.008					p,r,(s)
		115	95.7	0.17	0.056		33	67	r,s
50	Sn	112	0.97	0.036					p
		114	0.65	0.024					p
		115	0.34	0.013					p,r,s
		116	14.53	0.54	0.44		81	19	s,p
		117	7.68	0.29	0.10		35	65	r,s
		118	24.23	0.90	0.58		64	36	s,r
		119	8.59	0.32	0.14		44	56	r,s
		120	32.59	1.21	0.87		72	28	s,r
		122	4.63	0.17					r
		124	5.79	0.22					r
51	Sb	121	57.36	0.16	0.063		38	62	r,s
		123	42.64	0.12					r
52	Te	120	0.096	0.0045					p
		122	2.603	0.12	0.122		100	0	s,p
		123	0.908	0.042	0.042		99	1	s,p
		124	4.816	0.23	0.21		93	7	s,p
		125	7.139	0.33	0.076		23	77	r,s
		126	18.95	0.89	0.36		41	59	r,s
		128	31.69	1.48					r
		130	33.80	1.58					r
53	I	127	100	0.90	0.049		5	95	r,s

Table 4b (continued)

Element		A	Isotopic composition [%] (1)	N [N(Si) ≡ 10^6] (2)	N_s[N(Si) ≡ 10^6] main weak component (3)	N_s/N [%] (4)	1-N_s/N [%]	Process (5)
54	Xe	124	0.121	0.005				p
		126	0.108	0.004				p
		128	2.19	0.09	0.09	100	0	s,p
		129	27.34	1.12	0.042	4	96	r,s
		130	4.35	0.18	0.176	101	–1	s,(p)
		131	21.69	0.89	0.067	8	92	r,s
		132	26.50	1.09	0.446	40	60	r,s
		134	9.76	0.40	0.0052	1	99	r,(s)
		136	7.94	0.33				r
55	Cs	133	100	0.37	0.057	15	85	r,s
56	Ba	130	0.106	0.005				p
		132	0.101	0.005				p
		134	2.417	0.111	0.13	117	–17	s,(p)
		135	6.592	0.304	0.097	32	68	r,s
		136	7.854	0.362	0.453	125	–25	s,(p)
		137	11.23	0.518	0.49	95	5	s,r
		138	71.70	3.31	3.56	108	–8	s,r
57	La	138	0.0902	0.0004				p
		139	99.9098	0.46	0.33	69	31	s,r
58	Ce	136	0.19	0.002				p
		138	0.25	0.003				p
		140	88.48	1.06	0.92	87	13	s,r
		142	11.08	0.133				r
59	Pr	141	100	0.180	0.077	43	57	r,s
60	Nd	142	27.13	0.234	0.182	78	22	s,p
		143	12.18	0.105	0.035	33	67	r,s
		144	23.80	0.206	0.076	37	63	r,s
		145	8.30	0.072	0.017	24	76	r,s
		146	17.19	0.149	0.055	37	63	r,s
		148	5.76	0.050	0.002	4	96	r,s
		150	5.64	0.049				r
62	Sm	144	3.1	0.008				p
		147	15.0	0.040	0.006	15	85	r,s
		148	11.3	0.030	0.029	95	5	s,p
		149	13.8	0.037	0.0057	15	85	r,s
		150	7.4	0.020	0.019	95	5	s,p
		152	26.7	0.072	0.019	26	74	r,s
		154	22.7	0.061				r
63	Eu	151	47.8	0.048	0.0034	7	93	r,s
		153	52.2	0.052	0.0025	5	95	r,s

Table 4b (continued)

Element		A	Isotopic composition [%] (1)	N [N(Si) ≡ 10^6] (2)	N_s[N(Si) ≡ 10^6] main component (3)	N_s[N(Si) ≡ 10^6] weak component (3)	N_s/N [%] (4)	1-N_s/N [%]	Process (5)
64	Gd	152	0.20	0.00068	0.00038		56	44	s,p
		154	2.18	0.007	0.0072		97	3	s,p
		155	14.8	0.050	0.0029		6	94	r,s
		156	20.47	0.070	0.012		17	83	r,s
		157	15.65	0.053	0.0057		11	89	r,s
		158	24.84	0.085	0.034		40	60	r,s
		160	21.86	0.075					r
65	Tb	159	100	0.062	0.0044		7	93	r,s
66	Dy	156	0.06	0.0002					p
		158	0.10	0.0004					p
		160	2.34	0.010	0.0093		97	3	s,p
		161	18.9	0.078	0.0037		5	95	r,s
		162	25.5	0.105	0.016		15	85	r,s
		163	24.9	0.102	0.0044		4	96	r,s
		164	28.2	0.116	0.016		14	86	r,s
67	Ho	165	100	0.090	0.0057		6	94	r,s
68	Er	162	0.14	0.0004					p
		164	1.61	0.0042	0.0041		98	2	s,p
		166	33.6	0.0877	0.012		14	86	r,s
		167	22.95	0.0599	0.0048		8	92	r,s
		168	26.8	0.0699	0.020		29	71	r,s
		170	14.9	0.0389	0.0009		2	98	r,(s)
69	Tm	169	100	0.0398	0.0066		17	83	r,s
70	Yb	168	0.13	0.0003					p
		170	3.05	0.0077	0.0070		91	9	s,p
		171	14.3	0.0359	0.0058		16	84	r,s
		172	21.9	0.0550	0.017		31	69	r,s
		173	16.12	0.0405	0.0082		20	80	r,s
		174	31.8	0.0798	0.0040		5	95	r,s
		176	12.7	0.0319	0.0007		2	98	r,(s)
71	Lu	175	97.41	0.0372	0.0060		16	84	r,s
		176	2.59	0.0010	0.0025 decay of ^{176}Lu not consid.		252		s
72	Hf	174	0.162	0.0003					p
		176	5.206	0.0082	0.0063 no ^{176}Lu		77	23	s,p,^{176}Lu
		177	18.606	0.0294	0.0051		17	83	r,s
		178	27.297	0.0431	0.021		49	51	r,s
		179	13.629	0.0215	0.0069		32	68	r,s
		180	35.100	0.0555	0.036		65	35	s,r

Table 4b (continued)

Element		A	Isotopic composition [%] (1)	N [N(Si) ≡ 10^6] (2)	N_s[N(Si) ≡ 10^6] main component (3)	weak component	N_s/N [%] (4)	1-N_s/N [%]	Process (5)
73	Ta	180	0.012	0.000002					p,s,(r)
		181	99.988	0.020	0.0085		43	57	r,s
74	W	180	0.13	0.0002					p,(s)
		182	26.3	0.036	0.017		48	52	r,s
		183	14.3	0.019	0.012		62	38	s,r
		184	30.67	0.042	0.029		70	30	s,r
		186	28.6	0.039	0.007		18	82	r,s
75	Re	185	37.40	0.020	0.0055		27	73	r,s
		187	62.60	0.034	0.0012		4	96	r,s
76	Os	184	0.02	0.0001					p
		186	1.58	0.011	0.0119		112	–12	s,(p)
		187	1.6	0.011	0.0055		51	49	s,^{187}Re
		188	13.3	0.089	0.016		18	82	r,s
		189	16.1	0.108	0.0052		5	95	r,s
		190	26.4	0.177	0.021		12	88	r,s
		192	41.0	0.276	0.001				r,(s)
77	Ir	191	37.3	0.234	0.0046		2	98	r,s
		193	62.7	0.394	0.0059		1	99	r,s
78	Pt	190	0.01	0.0001					p
		192	0.79	0.011	0.010		91	9	s,p
		194	32.9	0.441	0.016		4	96	r,s
		195	33.8	0.453	0.0068		2	98	r,s
		196	25.3	0.339	0.029		9	91	r,s
		198	7.2	0.096					r
79	Au	197	100	0.203	0.0095		5	95	r,s
80	Hg	196	0.15	0.001					p
		198	9.97	0.041	0.033		81	19	s,p
		199	16.87	0.069	0.015		22	78	r,s
		200	23.1	0.095	0.047		50	50	s,r
		201	13.18	0.054	0.019		35	65	r,s
		202	29.86	0.122	0.072		59	41	s,r
		204	6.87	0.028	0.003		11	89	r,s
81	Tl	203	29.524	0.054	0.039		72	28	s,r
		205	70.476	0.130	0.082		63	37	s,r
						strong component:			
82	Pb	204	1.94	0.062	0.052		84	16	s,p
		206	19.10	0.613	0.290	0.014	50	50	r*,s,^{238}U
		207	20.59	0.661	0.317	0.023	51	49	r*,s,^{237}U
		208	58.36	1.873	0.766	0.601	72	28	s,r*,^{232}Th
83	Bi	209	100	0.146	0.018	0.019	25	75	r*,s

Table 4b (continued)

Element		A	Isotopic composition [%] (1)	N [N(Si) ≡ 10^6] (2)	N_s[N(Si) ≡ 10^6] main component (3)	N_s[N(Si) ≡ 10^6] weak component (3)	N_s/N [%] (4)	1-N_s/N [%]	Process (5)
90	Th	232	100	0.0338					r
92	U	235	0.7200	$5.98 \cdot 10^{-5}$					r
		238	99.2745		0.0082				r

3.4.10 Other sources for solar-system abundances

Emission spectroscopy of the solar corona, solar energetic particles (SEP) and the composition of the solar wind yield information on the composition of the Sun. Solar wind data were used for isotopic decomposition of rare gases. Coronal abundances are fractionated relative to photospheric abundances. Elements with high first ionization potential are depleted relative to the rest (see [89A] and literature cited there).

Data on SEP are less accurate than photospheric data but agree with them within limits of error [89A].

Recent data on the composition of the comet Halley allowed some estimate of its composition [88J]. An appropriate mixture of gas and dust [88D] yields a composition which is similar to solar abundances except for a severe depletion of H [89A].

Interplanetary dust particles (IDPs) have approximately chondritic bulk composition. Porous IDPs match the CI-composition better than non-porous (smooth) IDPs. The average C-content of porous IDPs, for example, is four times the CI-abundance and is thus approaching the solar C/Si-abundance ratio more closely than CI-condrites. Because Halley compositional data show the same trend a cometary origin has been suggested for the porous IDPs [89S]. This may offer the possibility of collecting solar system material of essentially solar composition (except H), i.e., material more primitive than CI-chondrites.

References for 3.4

47S Suess, H.E.: Z. Naturforsch. **2a** (1947) 311.
63M Mason, B.: Space Sci. Rev. **1** (1963) 621.
67R Reed, G.W., Jovanovic, S.: J. Geophys. Res. **72** (1967) 2219.
73S Suess, H.E., Zeh, H.D.: Astrophys. Space Sci. **23** (1973) 173.
76T Tatsumoto, M., Unruh, D.M., Desborough, G.A.: Geochim. Cosmochim. Acta **40** (1976) 617.
78E Evenson, N.M., Hamilton, P.J., O'Nions, R.K.: Geochim. Cosmochim. Acta **42** (1978) 1199.
79S Shima, M.: Geochim. Cosmochim. Acta **43** (1979) 353.
80B Begemann, F.: Rep. Prog. Phys. **43** (1980) 1309.
81K Kallemeyn, G.W., Wasson, J.T.: Geochim. Cosmochim. Acta **45** (1981) 1217.
82A Anders, E., Ebihara, M.: Geochim. Cosmochim. Acta **46** (1982) 2363.
82R Rammensee, W., Palme, H.: J. Radioanal. Chem. **71** (1982) 401.
84B Beer, H., Walter, G., Macklin, R.L., Patchett, P.J.: Phys. Rev. C **30** (1984) 464.
85W Wasson, J.T.: Meteorites. New York: W. H. Freeman and Company (1985) 250.
86J Jochum, K.P., Seufert, H.M., Spettel, B., Palme, H.: Geochim. Cosmochim. Acta **50** (1986) 1173.
88D Delsemme, A.H.: Philos. Trans. R. Soc. London **A325** (1988) 509.

88J Jessberger, E.K., Christoforidis, A., Kissel, J.: Nature **332** (1988) 691.
88P Palme, H., in: Review in Modern Astronomy 1, Cosmic Chemistry (G. Klare, ed.). Berlin: Springer-Verlag (1988) 28.
88T Thiemens, M., in: Meteorites and the early solar system (J.F. Kerridge, M.S. Matthews, eds.). Tucson: The University of Arizona Press (1988) 899.
88W Woolum, D., in: Meteorites and the early solar system (J.F. Kerridge, M.S. Matthews, eds.). Tucson: The University of Arizona Press (1988) 995.
88Z Zinner, E. in: Meteorites and the early solar system (J.F. Kerridge, M.S. Matthews, eds.). Tucson: The University of Arizona Press (1988) 956.
89A Anders, E., Grevesse, N.: Geochim. Cosmochim. Acta **53** (1989) 197.
89B Beer, H., Walter, G., Käppeler F.: Astron. Astrophys. **211** (1989) 245.
89K Käppeler, F., Beer, H., Wisshak, K.: Rep. Prog. Phys. **52** (1989) 945.
89S Schramm, L.S., Brownlee D.E. and Wheelock, M.M.: Meteoritics **24** (1989) 99.
90B1 Beer, H., in: Astrophysical Ages and Dating methods (E. Vangioni-Flam, M. Casse, J. Audouze, J. Tran Thanh, eds.). Gif sur Yvette, France: Editions Frontieres, (1990) 349.
90B2 Burnett, D.S., Woolum D.S.: Astron. and Astrophys. **228** (1990) 253.
90G Grevesse, N., Lambert, D.L., Sauval, A.J., van Dishoeck, E.F., Farmer, C.B., Norton, R.H.: Astron. Astrophys. **232** (1990) 225.
90J Jarosewich, E.: Meteoritics **25** (1990) 323.
90K Kratz, K.-L., Harms, V., Hillebrandt, W., Pfeiffer, B., Thielemann, F.-K., Wöhr, A.: Z. f. Phys. **A336** (1990) 357.
91B Beer, H.: Astrophys. J. **375** (1991) 823.
91G Grevesse, N., Lambert, D.L., Sauval, A.J., van Dishoeck, E.F., Farmer, C.B., Norton, R.H.: Astron. Astrophys. **242** (1991) 488.
91H Holweger, H., Bard, A., Kock, A., Kock, M.: Astron. Astrophys. **249** (1991) 545.
91I Int. Union of Appl. and Pure Chem. (IUAPC): Comission on Atomic Weights and Isotopic Abundances, Subcommittee for Isotopic Abundance Measurements: Pure and Appl. Chem. **63** (1991) 991.
92B Beer, H., Walter, G., Käppeler F.: Astrophys. J. **389** (1992) 784.
92R Richter, S., Ott, U., Begemann, F., in: Lunar and Planet. Science XXIII. Houston: Lunar and Planetary Institute, (1992) 1147.
93C Chen, J.H., Wasserburg, G.J., Papanastassiou, D.A., in: Lunar and Planet. Science XXIV. Houston: Lunar and Planetary Institute, (1993) 277.
93D Dreibus, G., Palme, H., Spettel, B., and Wänke H.: Meteoritics **28** (1993) 343.
93R Rocholl, A., Jochum, K.P.: Earth. Planet. Sci. Lett. **117** (1993) 265.
93S Spettel, B., Palme H., Wänke H.: in preparation.